FATIGUE TESTING OF WELDMENTS

A symposium
presented at
May Committee Week
AMERICAN SOCIETY FOR
TESTING AND MATERIALS
Toronto, Canada, 1-6 May 1977

ASTM SPECIAL TECHNICAL PUBLICATION 648
D. W. Hoeppner, University of Missouri,
editor

List price $28.50
04-648000-30

AMERICAN SOCIETY FOR TESTING AND MATERIALS
1916 Race Street, Philadelphia, Pa. 19103

Library of Congress Catalog Card Number: 78-51630

NOTE
The Society is not responsible, as a body,
for the statements and opinions
advanced in this publication.

Printed in Mechanicsburg, Pa.
July 1978

Foreword

The symposium on Fatigue Testing of Weldments was presented at the May Committee Week of the American Society for Testing and Materials held in Toronto, Canada, 1-6 May 1978. ASTM Committee E-9 on Fatigue sponsored the symposium. D. W. Hoeppner, University of Missouri, presided as symposium chairman and served as editor of this publication. C. Hartbower, U. S. Department of Transportation, H. Reemsnyder, Bethlehem Steel Corporation, and D. Mauney, Alcoa Laboratories, served as session chairmen.

Related ASTM Publications

Achievement of High Fatigue Resistance in Metals and Alloys, STP 467 (1970), $28.75, 04-467000-30

Handbook of Fatigue Testing, STP 566 (1974), $17.25, 04-566000-30

Manual on Statistical Planning and Analysis for Fatigue Experiments, STP 588 (1975), $15.00, 04-588000-30

Fatigue Crack Growth Under Spectrum Loads, STP 595 (1976), $34.50, 04-595000-30

A Note of Appreciation to Reviewers

This publication is made possible by the authors and, also, the unheralded efforts of the reviewers. This body of technical experts whose dedication, sacrifice of time and effort, and collective wisdom in reviewing the papers must be acknowledged. The quality level of ASTM publications is a direct function of their respected opinions. On behalf of ASTM we acknowledge their contribution with appreciation.

ASTM Committee on Publications

Editorial Staff

Contents

Introduction

Testing of materials to determine their fatigue properties is an extremely challenging aspect of engineering design and the development of materials. The development of ASTM standards by which fatigue tests can be conducted has been the broad goal of ASTM Committee E-9 on Fatigue. In the last few years standards for unnotched fatigue testing in the "long life" region and strain cycling fatigue testing have emerged. In addition, ASTM Committee E-24 on Fracture Testing of Metals is developing a recommended practice for fatigue-crack growth testing utilizing a precracked specimen. At the time that this symposium on fatigue testing of weldments was planned, the aforementioned standards and recommended practice were well along in their development.

In engineering fatigue design, however, we frequently are faced with joining one or more objects together. One of the more common methods of joining is by welding. It is commonly used in ground transportation equipment, bridges, aircraft, space vehicles, pressure vessels, piping, etc. All too frequently engineers are forced to utilize welds in fatigue design situations with an inadequate amount of information on the fatigue properties of the welds. Consequently, ASTM Committee E-9 planned this symposium to focus attention on the many facets of welding that would impact the fatigue properties of weldments. In addition, it was believed desirable to focus on methods by which welds are fatigue tested to evaluate their properties.

As you read the papers contained herein you will undoubtedly agree that the broad goals of the symposium were met. A discussion of the numerous factors that influence the fatigue behavior of weldments is provided. In addition, fatigue testing of simple elements is covered with emphasis on unnotched, notched, and precracked specimens. Fracture mechanics concepts as related to the fatigue-crack growth behavior of weldments also are presented. A clear recognition of the need for testing welded structural components and full-scale welded structure is presented. Thus, this volume will serve as a guide to those persons who are required to perform fatigue tests on weldments. The review papers contained herein present excellent background and a brief state-of-the-art review on this subject. An adequate number of references are cited to provide excellent background on this timely subject.

The papers must be studied carefully to obtain their full meaning. A clear need for integration of welding technology, inspection, materials analysis, experimental mechanics, fatigue design, and joint failure criteria emerge from reading the papers. It also is obvious that welds must be designed as a complex system and no simple (and low cost) method of design, manufacturing, fatigue analysis, and inspection has emerged. This symposium was intended to juxtapose the numerous engineers and technologists that are charged with fatigue testing of weldments. It accomplished that goal—the papers contained here also accomplished that goal.

As with so many of these endeavors, we hope to provide a summary of the current state of the art in order to develop insight into exactly where efforts must be placed in the future to provide more reliable and durable structures. This volume also will serve that function. Finally, it becomes clear from this effort that the fatigue testing of weldments and fatigue design of welded structures are extremely complex subjects, deserving of much more attention in future years.

D. W. Hoeppner
Professor of Engineering
University of Missouri
Columbia, Mo. 65201,
editor

H. S. Reemsnyder[1]

Development and Application of Fatigue Data for Structural Steel Weldments

REFERENCE: Reemsnyder, H. S., "**Development and Application of Fatigue Data for Structural Steel Weldments,**" *Fatigue Testing of Weldments, ASTM STP 648*, D. W. Hoeppner, Ed., American Society for Testing and Materials, 1978, pp. 3–21.

ABSTRACT: Traditionally, designs of and design specifications for welded joints have been based on test data. This paper describes axially loaded specimens containing butt and fillet welds, welded beams with typical beam details, and tubular welded-joint specimens, as well as the characteristics of the fatigue testing systems. Parameters affecting the fatigue resistance of structural steel weldments, for example, notches, residual stresses, and tensile strength, are reviewed in the light of test results. Fatigue design criteria for weldments are also discussed. Recent applications of linear elastic fracture mechanics and the concepts of strain-cycled fatigue to quantify the initiation and propagation of fatigue cracks in weldments are presented. Continued and refined application of these analytical techniques, quantification of environmental effects and of acceptance levels of internal discontinuities, and the application of load spectra to the design of weldments are recognized as research goals for the future.

KEY WORDS: fatigue tests, weldments, fatigue of metals, residual stress, stress concentration, crack initiation, crack propagation, fracture mechanics, environments

Nomenclature

N	Fatigue life, cycles
R	Stress ratio, S_{min}/S_{max}
S_{max}	Maximum nominal stress per cycle
S_{min}	Minimum nominal stress per cycle
S_m	Nominal mean stress
S_r	Nominal stress range
S_u	Tensile strength
ΔK	Range of stress intensity factor

Traditionally, designs of and design specifications for welded joints have

[1] Supervisor, Fatigue and Fracture, Research Department, Bethlehem Steel Corporation, Bethlehem, Pa. 18016.

been based on fatigue test data. In general, these fatigue tests have been conducted on relatively large specimens, the size of which has been limited only by the capacity of the test system. Parameters affecting the fatigue resistance of structural steel weldments, for example, notches, residual stresses, and tensile strength, have been studied in tests on: (*a*) axially loaded and plate cantilever beam specimens containing welds, (*b*) model and full-scale welded beams and pressure vessels, and (*c*) tubular welded joints.

Extensive compilations, discussion, and reviews of weldment fatigue test results have been published [*1-7*].[2] A bibliography on the fatigue strength of welded joints for the years 1950 to 1971 [*8*] has been expanded and extended to 1976 [*9*].

Weldment Fatigue Specimens and Tests

Specimens

Axially loaded fatigue specimens containing either butt or fillet welds have been the most widely used test configurations [*4*]. These specimens are generally fabricated with plate thicknesses and weld sizes and processes identical to those of full-scale structures. However, the specimen width is usually limited by the load capacity of the test system to the order of 25 to 150 mm (1 to 6 in.). Butt welds may be either transverse or longitudinal (respectively, Fig. 1(*a*) and (*b*)). Transverse and longitudinal fillet welds are

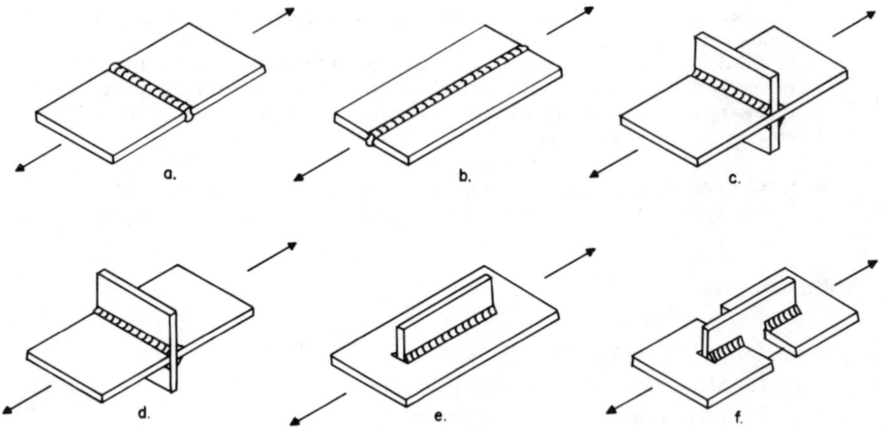

FIG. 1—*Axially loaded weldment specimens.*

[2] The italic numbers in brackets refer to the list of references appended to this paper.

either nonload carrying (Fig. 1(*c*) and (*e*)) or load carrying (Fig. 1(*d*) and (*f*)). The longitudinal nonload carrying fillet weldment of Fig. 1(*e*) simulates only those cases where the fillet weld is terminated or intermittent in a region of high stress. A continuous longitudinal fillet-welded tee specimen has been developed [*10*] in which the weld termination—a severe notch—is carried beyond the highly stressed or test section (Fig. 2). This specimen simulates the flange-web weld of built-up beams, axially loaded box members, or attachments. The tee specimen has been used in the investigations of carbon steel [*11*], constructional alloy steel [*12*], and the effects of internal weld discontinuities [*13*].

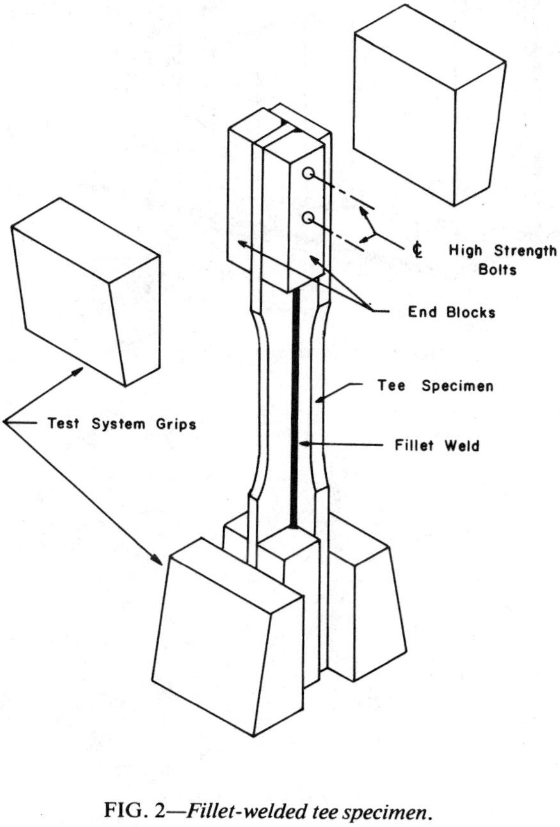

FIG. 2—*Fillet-welded tee specimen.*

Fatigue tests of pressure vessel steels and weldments have been performed on cantilever beam specimens [*14*] and on biaxially bent plates [*15*].

Small axially loaded cylindrical specimens have been machined from weldments so that the welded zone is located in either a straight or hourglass shaped test section. Such specimens have been used to investigate both the high-cycle [*16*] and low-cycle [*13,17*] fatigue resistance of weldments.

Welded beam details (Fig. 3), for example, stiffeners, flange splices, cover plates, and flange-web welds, have been studied in fatigue tests of both small-scale beams [*4,12,18*] and full-size welded girders [*19-22*]. Full-size fatigue tests have been conducted on pressure vessels [*23*] and on tubular K-joints (Fig. 4) typical of offshore structures [*24-26*].

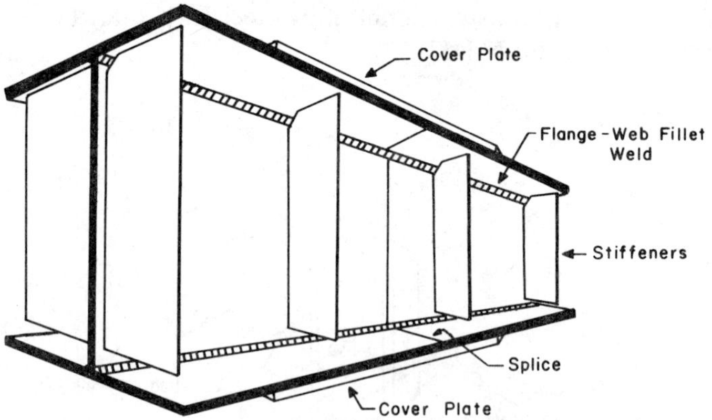

FIG. 3—*Welded built-up beam.*

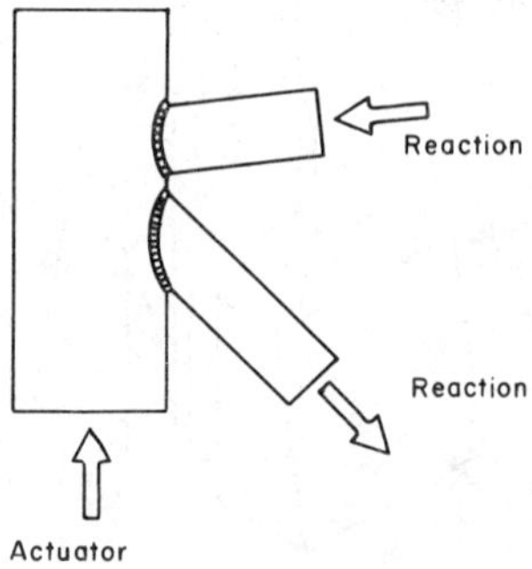

FIG. 4—*Tubular K-joint specimen.*

Tests

Most of the axial-load, beam, and girder fatigue tests have been performed on constant-displacement-amplitude test systems, either the eccentric crank-and-lever [*4*] or the hydraulic pulsator [*11,12,18-21*]. In the case of the crank-and-lever systems, both the mean load and load range will vary with the change in stiffness of the specimen concomitant with crack growth.

On the other hand, hydraulic pulsator systems automatically maintain either the maximum or minimum load at the desired level, and only the load range varies with specimen stiffness. With the advent of closed-loop servo-hydraulic test systems, both the mean load or mean displacement and the load range or displacement range at desired levels are maintained throughout the fatigue test. Such systems have been used to test axially loaded weldments [*13*] and welded girders [*22*]. Servohydraulic systems are used almost exclusively to perform the crack propagation and strain-cycled fatigue tests of weldments described as follows.

The majority of the weldment fatigue tests to date have been conducted in one of two ways:

1. By maintaining a constant stress ratio, R, and varying the maximum stress, S_{max}, from test to test [*4,5*]

2. Maintaining a constant minimum stress, S_{min}, and varying the stress range, S_r, from test to test [*18*]

The former approach is favored by vehicle and machine designers, whereas the latter is favored by structural engineers. The data are presented then graphically as stress-life (*S-N*) curves or constant life diagrams (Fig. 5) derived from such curves. The fatigue life, N, is defined either as complete fracture (for example, axially loaded specimens) or as inability to continue to carry the prescribed load (for example, beams).

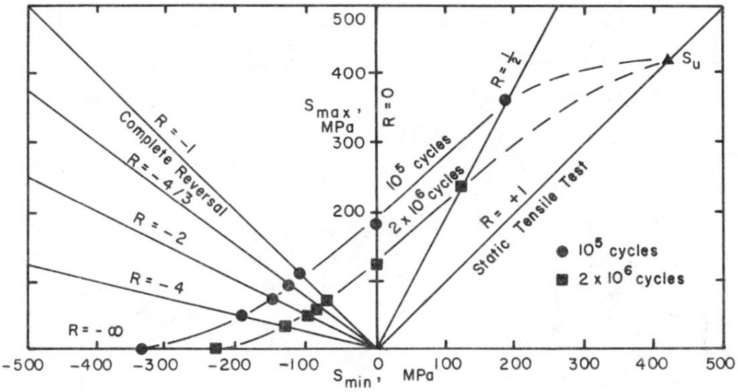

FIG. 5—*Constant-life diagram, carbon steel butt welds.*

Tabular data presentation also has been common for comparison of weld details, defect severity, steel grades, etc. Fatigue strengths for lives of 100 000 cyles or 2 000 000 cycles, or both, are determined from *S-N* curves and listed [*4,5,7*]. However, both the variability and nonlinearity of the fatigue data can be masked by this procedure. During the rewriting of the British design rules for welded joints—Steel Girder Bridges, BS153—the format of the tabular presentation was improved [*27*]. The regression coefficients, standard deviation, correlation coefficients, and number of observations for each type of joint[3] were presented. Mean values and lower 95 percent confidence limits for the fatigue strengths at 100 000 and 2 000 000 cycles of the various joint details were also listed.

Recognition of the importance of separation of fatigue life into the macroscopic crack initiation and propagation phases has introduced state-of-the-art fatigue concepts to weldment investigations. Crack initiation is displayed as local strain amplitude versus cycles or reversals, and crack propagation is presented as range of stress intensity factor versus crack growth rate. These concepts are discussed next.

Fatigue Characteristics of a Weldment

Notches and Crack Initiation

In general, notches have a greater effect on fatigue resistance than any other parameter, and a weldment generally contains notches (Fig. 6). Notches include: (*a*) changes in section due to reinforcement or weld geometry, (*b*) surface ripples, (*c*) undercuts, and (*d*) lack of penetration. In addition, welds may be subject to internal heterogeneities such as shrinkage cracks, lack of fusion, porosity, and inclusions.

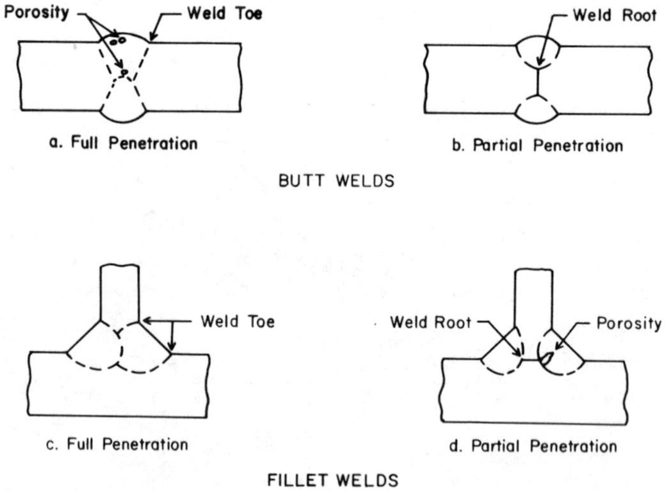

FIG. 6—*Typical weldment notches.*

Shrinkage cracks are caused by excessive restraint exerted by adjacent material on shrinkage of the weld zone upon cooling. The presence of slag or scale on the surfaces to be welded may lead to lack of fusion. Porosity is

[3] Logarithm of life expressed as a linear function of the logarithm of stress range.

caused by gas entrapped during solidification of the weld metal, excessive moisture in the electrode coating, or disturbance of the arc shield by drafts. Slag inclusions from the electrode coating are one of the most common weld discontinuities encountered. One cause of such inclusions is imperfect cleaning of the weld between successive passes.

In full-penetration transverse butt welds with reinforcement intact, fatigue cracks are initiated at the weld toe where a geometric stress raiser exists (Fig. 6(*a*)). However, in the case of partial-penetration transverse welds, cracks are initiated at the weld root (Fig. 6(*b*)). With the reinforcement removed, cracks are initiated either in the base metal or in the weld at inclusions or gas pockets (porosity). Cracks are initiated in longitudinal butt welds at weld-bead surface imperfections such as ripples or at a point of change of electrode in manual welding. Internal discontinuities, such as lack of penetration, are not critical when aligned with the principal stress.

The reduction in life of a continuous longitudinal fillet weld is generally due to a notch such as porosity (Fig. 6(*d*)) or a crater due to change of electrode. On the other hand, transverse or intermittent longitudinal fillet welds combine the detrimental effects of both external and internal notches. The fatigue strengths for a given life of transverse and intermittent longitudinal fillet welds are about one half of those for continuous longitudinal fillet welds. In transverse fillet welds cracks are initiated at either the weld toe or root (Fig. 6(*c*) and (*d*)). Cracks are initiated at the toe of intermittent longitudinal fillet welds where there is a severe mechanical notch.

Fatigue cracks in welded built-up beams are generally initiated in the flange-web weld at craters due to change of electrode or at weld discontinuities such as porosity (Fig. 6(*d*)). Weld roots of partial-penetration fillet welds are not critical as long as the discontinuity is parallel to the principal stress. The severe geometric notches of both the transverse fillet welds and the intermittent longitudinal fillet welds that join cover plates to beam flanges significantly reduce the fatigue resistance of welded beams. On the other hand, well executed beam splices have little or no effect on the fatigue resistance of welded beams. In the case of stiffeners welded to the web, fatigue cracks are usually initiated at the termination of the web-stiffener fillet weld. The crack propagates up into the web in the panel toward the load point and along the flange-web fillet weld in the panel away from the load point. When the stiffeners are welded to the tension flange, cracks are initiated at the toe of the fillet weld on the flange.

The fatigue strengths of butt and fillet weldments and of the various beam details are compared in Ref *28*.

External Notches

The importance of external notches on the fatigue resistance of weldments is illustrated by the fact that the removal of reinforcement raises the fatigue resistance of transverse butt welds to that of the base metal (Fig. 7).

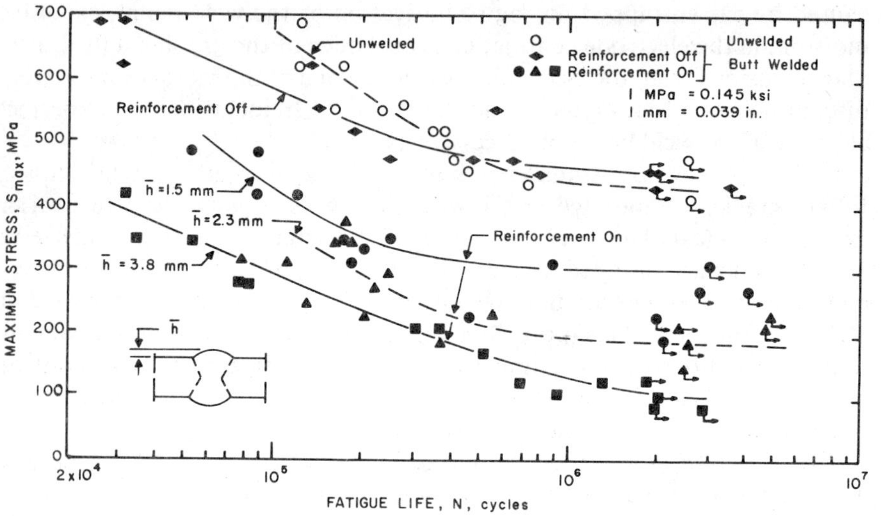

FIG. 7—*Transverse butt welds, quenched and tempered carbon steels,* R = 0.

This phenomenon has been demonstrated for hot-rolled carbon steels [*4,5,11*], quenched and tempered carbon steels [*29,30*], and quenched and tempered constructional alloy steels [*31,32*]. The stress concentrations due to transverse butt weld geometry has been the subject of much study [*5,33-36*], and it has been shown that fatigue life is significantly affected by the weld toe radius, reinforcement shape, and reinforcement height. The effect of reinforcement height is shown in Fig. 7 for a quenched and tempered carbon steel, 19.1 mm (0.75 in.) thick, with a tensile strength of 780 MPa (114 ksi). Doubling the height of reinforcement reduced the fatigue strength at 2 000 000 cycles by approximately 67 percent.

The geometric notch severity of the weld reinforcement accounts for the frequently observed phenomenon that the fatigue strength of transverse butt welds with reinforcement intact (or of transverse fillet welds) is insensitive to tensile strength. This effect is demonstrated in Fig. 8, where test results are plotted for hot-rolled carbon steels [*6,36*], high strength low alloy structural steel [36], quenched and tempered carbon steels [*29*], and constructional alloy steels [*29,35,37-39*]. Although the tensile strengths of the steels shown in Fig. 8 varied from 400 to 1020 MPa (58 to 148 ksi), there were no significant differences among the fatigue strengths of the various steels. In all likelihood, this sensitivity to reinforcement geometry contributed to the significant variability seen from one investigator to another (Fig. 8).

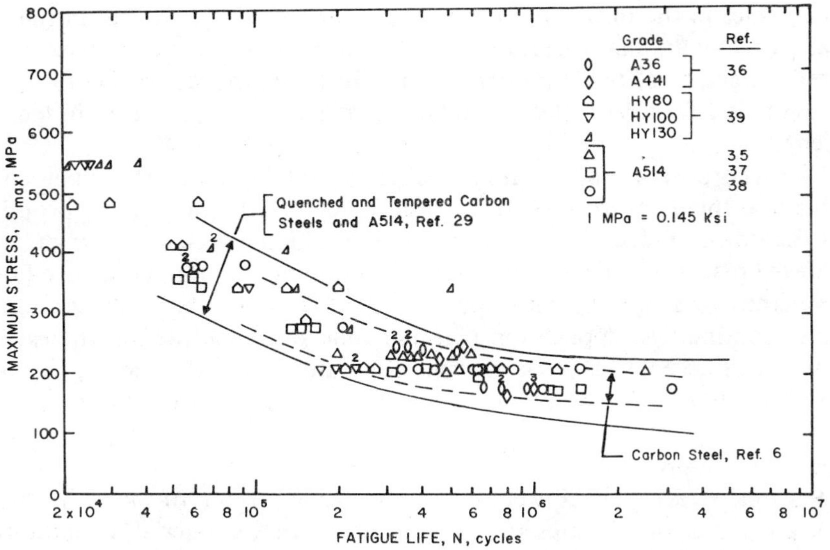

FIG. 8—*Effect of grade, transverse butt welds, reinforcement intact,* R = 0.

Internal Notches

Experimental techniques and a lack of uniform criteria for judging the severity of internal discontinuities have made it difficult in the past to discuss in quantitative terms the effects of these discontinuities on fatigue strength. It is possible, however, to draw some general conclusions from reviews of existing tests [*5*]. Lack of penetration lowers the fatigue strength of transverse welds significantly but has relatively little effect on longitudinal welds. Porosity and slag inclusions decrease fatigue resistance in proportion to the decrease in effective weld area of transverse welds. Microstructural changes due to severe quenching concomitant with the sudden extinguishing of the welding arc may initiate a fatigue crack at the point of change of electrode. Severe quenching also results from stray flashes and weld spatter. Such stress raisers may be reduced or eliminated by control of the welding procedure.

The presence of internal weld discontinuities can contribute to the variability observed in the fatigue testing of weldments. Wide variations in fatigue life have been observed when cracks were initiated at internal discontinuities. This variability decreases markedly for specimens in which cracking is initiated at the toe of the weld [*39*].

Through-thickness fatigue properties of a normalized steel (American Bureau of Shipping Grade EH32) were investigated [*40*]. These properties control the fatigue resistance of a welded detail that transfers the load to a plate perpendicular to that detail. Fatigue cracks were initiated at bands of inclusions running parallel to the rolling direction. The fatigue strength at

1000 cycles in the through-thickness direction was less than that in the in-plane or longitudinal direction by an amount equal to the difference in tensile strengths of the two orientations. However, the reduction in fatigue strength at 1 000 000 cycles was much greater than the reduction in tensile strength.

The incorporation of fracture mechanics and strain-cycled fatigue concepts into the quantification of internal notch effects shows great promise for the future. Indeed, these concepts are used in the present British efforts to develop acceptance levels of weld discontinuities for fatigue service [*41*]. Given the expected fatigue life and severity of service of the joint containing the discontinuities, application of these concepts would permit the establishment of acceptance levels of weld discontinuities, for example, permissible percent porosity by volume and maximum slag inclusion length.

Residual Stresses

Residual stresses due to welding are formed as a result of the differential in heating and cooling rates at various locations in the material. In addition, due to these thermal gradients, some of the regions will be elastic while other regions will be plastic. The interaction between these regions results in residual stresses after cooling. These stresses may be quite large and will be tensile in the vicinity of the weld where their magnitude is approximately equal to the yield strength of the weld metal.

Thermal stress relief has little or no effect on the weldment fatigue resistance of hot-rolled carbon steel [*11*] or constructional alloy steel [*35*] for the case of pulsating tension. However, thermal stress relief can improve the fatigue strength of weldments in the case of pulsating compression and in the case of alternating stress where the tensile stress component is small compared to the compressive component. Such improvement has been shown for fillet welded carbon steel specimens similar to that shown in Fig. 1(*e*) [*42*]. The increases in fatigue strength at 2 000 000 cycles due to thermal stress relief for stress ratios of 0, −1, and −4 were, respectively, 9, 73, and 140 percent.

The introduction of compressive residual surface stresses at stress raisers can increase the fatigue resistance of weldments. For example, shot peening of nonload-carrying fillet-welded carbon steel [*5*] and butt-welded constructional alloy steel [*43*] has increased the fatigue strengths at 2 000 000 cycles by 20 to 40 percent. Grit blasting under controlled conditions was observed to raise the fatigue strength of carbon steel transverse fillet weldments (Fig. 1(*c*)) to that of unwelded carbon steel [*44*]. The efficacy of shot peening for fatigue resistance is strongly influenced by shot size, arc height, and percent coverage [*45*]. For example, the 20 to 40 percent improvement in the case of quenched and tempered constructional alloy steel was achieved by peening to an arc height of 0.010 to 0.012 C[4] [*43*]. On the other hand, an improve-

[4] Almen C strip.

ment of only 7 percent was observed for a steel of similar tensile strength peened to an arc height of 0.005 to 0.007 C [*30*]. Therefore, for an improvement in fatigue resistance to be significant and repeatable, shot peening must be closely controlled.

Hammer peening has been observed to improve the 2 000 000-cycle fatigue strength of carbon steel butt welds by 15 to 25 percent and that of fillet welds by 20 to 50 percent [*5*]. In contrast, hammer peening of a quenched and tempered carbon steel was observed to reduce the long-life fatigue strength by 9 percent [*30*]. In general, hammer peening should not be considered the equivalent of a carefully controlled shot-peening program in the fabrication of cyclically loaded elements.

Spot heating to introduce compressive residual stresses of fillet weld toes [*5*] and of beams with welded attachments [*46*] has been observed to double the fatigue strength at 2 000 000 cycles. Proof loading also increases the fatigue resistance of weldments at lives greater than 1 000 000 cycles [*47*]. This increase is probably the result of favorable alteration of the residual stress distribution.

It should be noted, however, that periodic compressive loads in a variable-amplitude spectrum could eliminate the aforementioned beneficial effects of compressive residual stresses.

Weld Process

As compared with manual welding, semiautomatic (gas metal arc) and automatic (submerged arc and electroslag) welding processes generally provide greater fatigue resistance because they produce welds with fewer internal discontinuities and with a smoother surface [*7*]. Electroslag welding has been observed to produce transverse butt welds with up to 90 percent of the base metal fatigue strength, but fatigue strength was sensitive to procedural parameters [*48*].

Gas tungsten arc welding (TIG) of the toes and roots of welds laid by other processes increases the fatigue resistance of the weld. This increase is probably due to both an alteration of residual stresses and a modification of the stress raiser at the root or toe radius. Increases in fatigue strength of 100 percent have been observed [*49,50*].

Mean Stress and Stress Range

Although it is widely accepted that the fatigue resistance of weldments is primarily a function of stress range and is insensitive to mean stress, such a conclusion is not applicable to all weld details and stress ratios. For weldments containing severe stress raisers, and subjected to pulsating tension, the fatigue resistance is insensitive to mean stress. However, weldment fatigue strength can be sensitive to mean stress as well as to stress range, especially at stress ratios less than zero or in the absence of severe notch effects, or both [*11,28,51*]. Indeed, it is recognized that mean-stress independence is a conservative design approach, and further research is required to define the fatigue behavior of welded joints at stress ratios less than -1 (complete reversal) [*52*]. Obviously, if mean stresses were unimportant, the aforementioned effects of residual stresses would be meaningless.

Environment

A corrosive environment can have a deleterious effect on weldment fatigue strength. Increasing concern for fatigue in high-performance ships and offshore structures has led to the fatigue testing of welded specimens in either a salt solution or seawater. Crack initiation in carbon steel weldments tested at 0.1 Hz was not affected significantly by seawater, although the total life was reduced by a factor of 3 [*53*]. The presence of seawater reduced by about 20 percent the 2 000 000-cycle fatigue strength of a high-strength low-alloy (HSLA) steel tested at 1/8 Hz [*54*]. Presence of saltwater accelerates fatigue crack growth both in low-cycle [*55*] and high-cycle fatigue [*53*] at 0.1 Hz. The saltwater crack growth rate of a quenched and tempered pressure-vessel steel was similar to the crack growth rate in air at 24°C (75°F) but was lower than that in air at -1°C (30°F) [*56*].

The cathodic protection of weldments in a sea- or saltwater environment restored the fatigue strength of both carbon steel [*53*] and a quenched and tempered carbon steel [*57*] to that of air. However, the overprotection by cathodic means can in some cases lead to hydrogen embrittlement and the loss of low-cycle fatigue resistance [*58*].

Corrosion fatigue resistance is sensitive to even small changes in test environment factors such as pH [*58*] and test frequency [*55*]. Very little long-life data have been developed to date, and the existing long-life data have been obtained at high test frequencies.

Weldment Fatigue Design Criteria

Current weldment fatigue design criteria consider the effects of maximum and minimum stress or stress range, material tensile strength, and the stress concentration effect of various details on the service life of a structural element. A few criteria, principally those for vehicle and machine design, show constant-life diagrams for each type of weld joint and material. On such a diagram, for example, Fig. 9, the allowable stress envelope is defined by the allowable static tensile stress (Line DF), the allowable compressive stress (Line AC), the allowable cyclic stress (Line CD or BE), and the ray, $R = +1$. The allowable cyclic stress is a function of life and generally represents the cases: $N \leqslant 100\,000$, $100\,000 < N \leqslant 500\,000$, $500\,000 < N \leqslant 2\,000\,000$, and $N > 2\,000\,000$. The allowable cyclic-stress lines are offset by safety factors from the constant-life lines derived from test data *S-N* curves [*28*].

The current criteria governing land and marine structures in the United States and Great Britain assume weldment fatigue resistance to be inde-

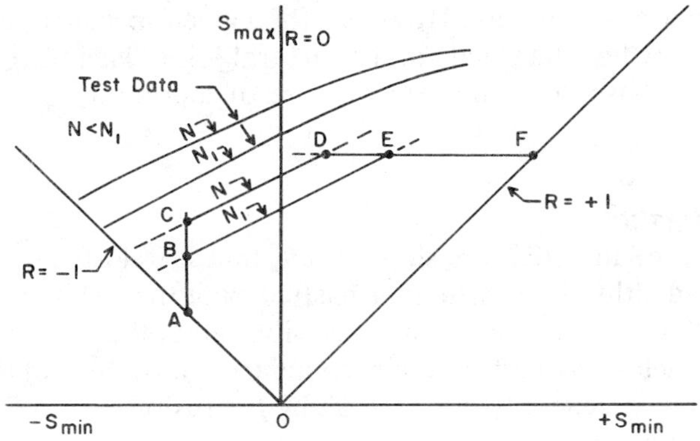

FIG. 9—*Design criteria.*

pendent of mean stress and cite only stress range. In such criteria, the Lines CD and BE would be parallel to the ray $R = +1$ in Fig. 9. However, these criteria do not show constant-life diagrams for each detail. Instead, they show sketches of the various weld-joint configurations from which the designer selects the stress category closest to his detail. The designer then enters a tabular array with this stress category and the desired service life, for example, 100 000 to 500 000 cycles, and selects the allowable stress range for his particular case.

All allowable stresses cited in the various criteria are nominal net section normal or shear stresses. Many of the structural design criteria furnish guidelines for handling spectrum loading via the Miner-Palmgren linear cumulative damage hypothesis. None of the structural criteria consider the effects of environment.

Mechanics of Crack Initiation and Propagation in Weldments

The present fatigue design criteria for welded joints are based on laboratory tests in which the failure criterion is "cycles to separation into two pieces." Although the welded specimens are large as compared with typical laboratory fatigue specimens, they are small compared to welded structural joints.

The fatigue life of a test specimen, structural joint, or machine component consists of three phases:

1. Initiation of a macroscopic crack
2. Propagation of the crack to a critical size
3. Exceeding of the residual strength of the cracked element resulting in complete fracture

The bulk of the fatigue life consists of the first two phases. For a given weldment detail, crack-initiation life in a laboratory specimen may be similar to that of a structure. However, for a given specimen, Phase 2 could be considerably less than that for a structural joint. Therefore, the reliability of criteria based on complete fracture of laboratory specimens is uncertain at best.

Crack Propagation

One group of investigators showed [*60*] that a weld-fusion zone, where the metal had either been melted or pasty during the welding process, contained a high concentration of slag inclusions and other nonmetallics, both as isolated inclusions and as grain-boundary films. In addition, a slight undercutting was generally observed along the fusion boundary. Following the conclusion of these investigators, that is, sharp notches exist in the weld fusion zone, much effort has been and continues to be expended on the application of linear elastic fracture mechanics (LEFM) to weldment fatigue. In this approach, it is assumed that all weldments initially contain crack-like flaws and that the significant portion of life is crack growth, that is, Phase 2 just mentioned.

Stress-intensity factors have been developed (Fig. 10) for a longitudinal butt weld in a residual stress field [*61*], a transverse butt weld [*62*], a partial penetration fillet weld [*63*], a "three-corner" crack in a beam web-flange junction [*64*], and transverse fillet welds, both nonload-carrying [*65*] and load-carrying [*66*].

Crack propagation has been described by LEFM in carbon steel base

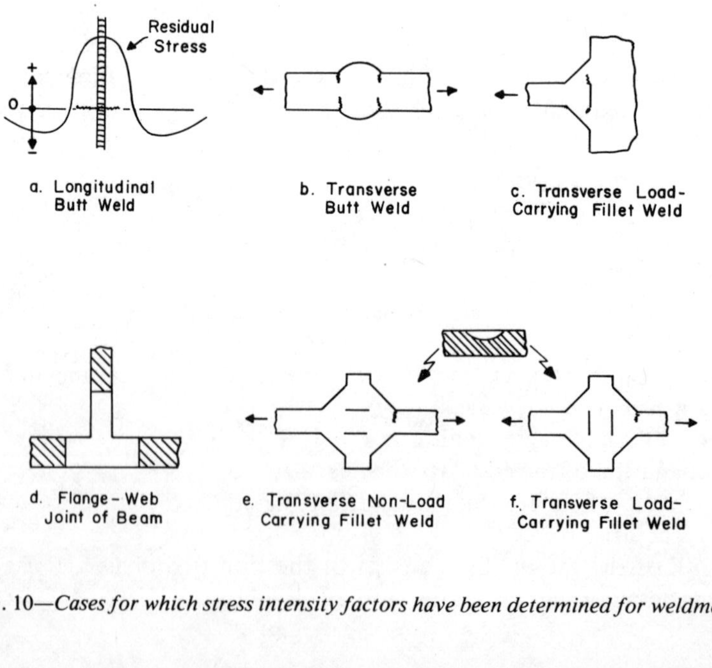

FIG. 10—*Cases for which stress intensity factors have been determined for weldments.*

metal, weld metal and HAZ [*67,68*], low-alloy weld metal [*69*], quenched and tempered constructional alloy steel base metal and weldment [*70*], quenched and tempered pressure vessel steel base metal and HAZ [*56*], and carbon and HSLA steels with electroslag welds [*71*]. The effect of mean stress was included in the study of crack growth in carbon and HSLA butt and transverse fillet welds [*72*]. LEFM has been used to describe crack growth from lack of penetration in transverse butt welds [*73,74*] and fillet welds [*66*] and from various discontinuities in butt welds, for example, inclusions, lack of fusion, porosity [*75*].

Cumulative-damage predictions for weldments have been made using LEFM and the Miner-Palmgren hypothesis for the block loading of nonload-carrying fillet welds [*76*] and for the narrow band random loading of partial-penetration fillet welds [*77*].

The fatigue behavior of welded beams has been evaluated with LEFM for cracks originating at gas pockets in the web-flange weld [*78*] and at stiffeners welded to the web of a rolled beam [*79*]. The stress-intensity factor for a surface crack in the fillet weld of a tubular K-joint has been experimentally determined and used to predict fatigue test results of similar specimens [*80*].

The complete process of fatigue life estimation for weldments is illustrated for butt and fillet welds in, respectively, Refs *62* and *66*.

Crack Initiation

Fracture mechanics concepts may be used to estimate crack initiation in weldments [*28*]. For small notch root radii, the stress field ahead of the notch is approximately described by the stress-intensity factor. For a given range of stress-intensity factor ΔK, the cycles to crack initiation decrease with notch-root radius until the root radius equals 0.25 mm (0.01 in.). At root radii less than 0.25 mm (0.01 in.), where crack initiation is a function only of ΔK and is independent of root radius, hot-rolled carbon steels exhibit slightly shorter initiation lives than quenched and tempered alloy steels. Typical weld toe radii are equal to or less than 0.25 mm (0.01 in.), and therefore a plot of ΔK versus cycles to initiation may be used to predict crack initiation in weldments at lives where the notch behavior is essentially elastic [*28*].

According to one investigator [*62*], LEFM models of crack growth in full-penetration butt welds adequately describe the fatigue lives of carbon and HSLA steels but underestimate the lives of quenched and tempered steels. This study concluded that the initiation time is relatively short for carbon and HSLA steels but not for quenched and tempered steels. Another study found that the crack initiation time for partial-penetration butt welds in carbon steels was a significant portion of the total life [*74*]. Strain-cycled fatigue concepts are now being applied to predictions of crack initiation at the toes of full-penetration butt welds and at the roots of partial-penetration butt welds [*81,82*]. Such an approach not only models observed behavior adequately but also quantifies both the cyclic relaxation of mean or residual stresses and the effects of weld shape and internal discontinuities.

Strain-cycled fatigue techniques also have been used successfully to predict crack initiation at porosity in longitudinal fillet welds [*13*], transverse nonload-carrying fillet welds [*72*], and tubular K-joints for offshore structures [*83*].

Assessment for the Future

Weldment fatigue testing and data interpretation are now separating fatigue life into three component phases: crack initiation, propagation, and fracture. This recognition has led to the increasing application of fracture mechanics and strain-cycled fatigue concepts to quantifying the effects of both the weldment shape and internal discontinuities upon life. However, recognition of the problem is only the first step toward its solution. Continued and refined application of analytical techniques corroborated by experimentation is required to define quantitatively the interaction of geometry, environmental effects (especially at long lives), residual stresses, and variable amplitude service loading in their effects on the fatigue resistance of weldments. Such efforts are required by the increasing demand for land and marine vehicles and structures optimally designed to serve in hostile environments.

The effects of internal weld discontinuities on the fatigue resistance of weldments must be quantitatively defined. Acceptance levels of discontinuities must be based on "fitness for purpose," that is, acceptance level is a function of the severity of the total service environment.

Sophisticated analytical methods and testing systems to evaluate weldments are available. Let us use these tools to establish rational, realistic design criteria that will result in safe, economical weldments.

References

[*1*] Spraragen, W. and Claussen, G. E., *Welding Journal,* Vol. 16, No. 1, Jan. 1937, pp. 1-44.

[*2*] Spraragen, W. and Rosenthal, D., *Welding Journal,* Vol. 21, No. 7, July 1942, pp. 297-s-348-s.

[*3*] Reemsnyder, H. S., "The Fatigue Behavior of Structural Steel Weldments—A Literature Survey," Fritz Engineering Laboratory Report No. 284.1, Lehigh University, Bethlehem, Pa., Nov. 1961.

[*4*] Munse, W. H. and Grover, L., *Fatigue of Welded Structures,* Welding Research Council, New York, N.Y., 1964.

[*5*] Gurney, T. R., *Fatigue of Welded Structures,* Cambridge University Press, London, England, and New York, N.Y., 1968.

[*6*] Reemsnyder, H. S., "Fatigue Data for Plain and Welded Structural Steels," Bethlehem Steel Corp., Bethlehem, Pa., 26 April 1968.

[*7*] Pollard, B. and Cover, R. J., *Welding Journal,* Vol. 51, No. 11, Nov. 1972, pp. 544-s-554-s.

[*8*] Larson, C. F., *Welding Journal,* Vol. 51, No. 9, Sept. 1972, pp. 457-s-460-s.

[9] Reemsnyder, H. S., "Bibliography on Fatigue of Weldments," Bethlehem Steel Corp., Bethlehem, Pa., 1977.

[10] Reemsnyder, H. S., *Proceedings,* American Society for Testing and Materials, Vol. 65, 1965, pp. 729-735.

[11] Reemsnyder, H. S., *Welding Journal,* Vol. 48, No. 5, May 1969, pp. 213-s-220-s.

[12] Reemsnyder, H. S. *Welding Journal,* Vol. 44, No. 10, Oct. 1965, pp. 458-s-465-s.

[13] van der Zanden, A. M., Robins, D. B., and Topper, T. H. in *Testing for Prediction of Material Performance in Structures and Components, ASTM STP 515,* American Society for Testing and Materials, 1972, pp. 268-284.

[14] Gross, J. H., Tsang, S., and Stout, R. D., *Welding Journal,* Vol. 32, No. 1, Jan. 1953, pp. 23-s-30-s.

[15] Blaser, R. U., Tucker, J. T., Jr., and Kooistra, L. F., *Welding Journal,* Vol. 31, No. 3, March 1952, pp. 161-s-168-s.

[16] Buchanan, R. A. and Young, D. M., *Welding Journal,* Vol. 54, No. 9, Sept. 1975, pp. 296-s-301-s.

[17] Miller, W. R. and Amin, H. S., *Experimental Mechanics,* Vol. 15, No. 6, June 1975, pp. 230-233.

[18] Hirt, M. A., Yen, B. T., and Fisher, J. W., *Proceedings,* American Society of Civil Engineers, Vol. 97, No. ST7, July 1971, pp. 1897-1911.

[19] Yen, B. T. and Mueller, J. A., "Fatigue Tests of Large-Size Welded Plate Girders," Bulletin No. 118, Welding Research Council, New York, N.Y., Nov. 1966.

[20] Mueller, J. A. and Yen, B. T., "Girder Web Boundary Stresses and Fatigue," *Bulletin No. 127,* Welding Research Council, New York, N.Y., Jan. 1968.

[21] Toprac, A. A., *Welding Journal,* Vol. 34, No. 5, May 1969, pp. 195-s-202-s.

[22] Reemsnyder, H. S. and Demo, D. A. "Fatigue Cracking in Welded Crane Runway Girders: Causes and Repair Procedures." submitted to *Iron and Steel Engineer.*

[23] Pickett, A. G. and Grigory, S. C., "Cyclic Pressure Tests of Full-Size Pressure Vessels," Bulletin No. 135, Welding Research Council, New York, N.Y., Nov. 1968.

[24] Becker, J. M., Gerberich, W. W., and Bouwkamp, J. G., *Proceedings,* American Society of Civil Engineers, Vol. 98, No. ST1, Jan. 1972, pp. 37-60.

[25] Kurobane, Y. and Konomi, M., "Fatigue Strength of Tubular K-Joints," Document XV-340-73, International Institute of Welding, Paris, 1973.

[26] Bouwkamp, J. G., "Cyclic Loading of Full-Size Tubular Joints," Paper No. OTC 2605, Eighth Annual Offshore Technology Conference, Houston, Tex., May 1976.

[27] Gurney, T. R. and Maddox, S. J., *Welding Reserach International,* Vol. 3, No. 4, April 1973, pp. 1-54.

[28] Reemsnyder, H. S. in *Structural Steel Design,* 2nd. ed., L. Tall, Ed., Ronald Press, New York, 1974, Chapter 16, pp. 519-551.

[29] Reemsnyder, H. S., "Fatigue Properties of Welded RQ Steels," Bethlehem Steel Corp., Bethlehem, Pa., 13 Nov. 1972.

[30] Reemsnyder, H. S., "Fatigue Properties of Welded RQ Steels—Second Phase," Bethlehem Steel Corp., Bethlehem, Pa., 11 Jan. 1974.

[31] Hartmann, A. J. and Munse, W. H., "Fatigue Behavior of Welded Joints and Weldments in HY-80 Steel Subjected to Axial Loading," Structural Research Series No. 250, University of Illinois, Urbana, Ill., July 1962.

[32] Sahgal, R. K., Stallmeyer, J. E., and Munse, W. H., "Effect of Welding on the Axial Fatigue Properties of High Strength Structural Steels," Structural Research Series No. 172, University of Illinois, Urbana, Ill., March 1963.

[33] Green, G. J. and Marlin, D. H., *Welding Journal,* Vol. 30, No. 7, July 1951, pp. 607-617.

[34] Sanders, W. W., Jr., Derecho, A. T., and Munse, W. H., *Welding Journal,* Vol. 44, No. 2. Feb. 1965. pp. 49-s-55-s.

[35] Selby, K. A., Stallmeyer, J. E., and Munse, W. H., "Influence of Geometry and Residual Stress on Fatigue of Welded Joints," Structural Research Series No. 297, University of Illinois, Urbana, Ill., June 1965.

[36] Williams, H. E., Ottsen, H., Lawrence, F. V., Jr., and Munse, W. H., "The Effects of Weld Geometry on the Fatigue Behavior of Welded Connections," Structural Research Series No. 366, University of Illinois, Urbana, Ill., Aug. 1970.

[37] Rone, J. W., Stallmeyer, J. E., and Munse, W. H., "Fatigue Behavior of Plain Plate

and Butt Welded Joints in T-1 Steel," University of Illinois, Urbana, Ill., March 1963.
[38] Munse, W. H., Stallmeyer, W. H., and Bruckner, W. H., "Fatigue Behavior of Plain Plates and Butt Welded Joints in T-1 and T-1A Steel," University of Illinois, Jan. 1966.
[39] Radziminski, J. B. and Lawrence, F. V., Jr., *Welding Journal,* Vol. 49, No. 8, Aug. 1970, pp. 365-s-375-s.
[40] Adams, C. J. and Popov, E. P., "Through Thickness Fatigue Properties of Steel Plate," Bulletin No. 217, Welding Research Council, New York, N.Y., July 1976.
[41] Boulton, C. F., *Welding Journal,* Vol. 56, No. 1, Jan. 1977, pp. 13-s-22-s.
[42] Gurney, T. R., *British Welding Journal,* Vol. 15, No. 6, June 1968, pp. 276-282.
[43] Doty, W. D., *Welding Journal,* Vol. 34, No. 9, Sept. 1955, pp. 425-s-441-s.
[44] Salter, G. R. and Gurney, T. R., *Research Bulletin,* Vol. 17, No. 12, Dec. 1976, pp. 319-323.
[45] Campbell, J. E., "Shot Peening for Improved Fatigue Properties and Stress Corrosion Resistance," Metals and Ceramics Information Center, Battelle-Columbus, Columbus, Ohio, Dec. 1971.
[46] Puchner, O., *Welding Research Abroad,* Vol. 6, No. 6, June-July 1960, pp. 32-44.
[47] Welter, G. and Choquet, J. A., *Welding Journal,* Vol. 33, No. 1, Jan. 1967, pp. 39-s-48-s.
[48] Harrison, J. D., *Metal Construction and British Welding Journal,* Vol. 1, No. 8, Aug. 1969, pp. 366-370.
[49] Harrison, J. D., Watkinson, F., and Bodger, P. H. in *Fatigue of Welded Structures,* The Welding Institute, Cambridge, England, July 1970, pp. 97-113.
[50] Millington, D., *Metal Construction and British Welding Journal,* Vol. 5, No. 4, April 1973, pp. 134-139.
[51] Trufiakov, V. I., *Fatigue of Welded Joints,* Science Council, Kiev, USSR, 1973.
[52] Gurney, T. R. and Maddox, S. J., *Metal Construction and British Welding Journal,* Vol. 4, No. 12, Nov. 1972, pp. 418-422.
[53] Kochera, J. W., Tralmer, J. P., and Marshall, P. W., "Fatigue of Structural Steel for Offshore Platforms," OTC Paper No. 2604, Eighth Annual Offshore Technology Conference, Houston, Tex., May 1976.
[54] Berge, S., "Corrosion Fatigue Testing of Welded Joints at Low Frequencies," presented at the Symposium on Corrosion Fatigue, American Society for Testing and Materials, Denver, Colo., Nov. 1976.
[55] Crooker, T. W. and Lang, E. A. in *Fatigue Crack Propagation, ASTM STP 415,* American Society for Testing and Materials, 1967, pp. 94-127.
[56] Socie, D. F. and Antolovich, S. D., *Welding Journal,* Vol. 53, No. 6, June 1974, pp. 267-s-271-s.
[57] Havens, F. E. and Bench, D. M., "Fatigue Strength of Quenched and Tempered Carbon Steel Plates and Welded Joints in Sea Water," OTC Paper No. 1046, First Annual Offshore Technology Conference, Houston, Tex., May 1969.
[58] Marshall, P. W. in *Welding in Offshore Constructions,* The Welding Institute, Cambridge, England, 1975, pp. 10-34.
[59] Marshall, P. W., "Basic Considerations for Tubular Joint Design in Offshore Construction," Bulletin No. 193, Welding Research Council, New York, N.Y., April 1974.
[60] Signes, E. G., Baker, R. G., Harrison, J. D., and Burdekin, F. M., *British Welding Journal,* Vol. 14, No. 3, March 1967, pp. 108-116.
[61] Terada, H., *Engineering Fracture Mechanics,* Vol. 8, 1976, pp. 441-444.
[62] Lawrence, F. V., *Welding Journal,* Vol. 52, No. 5, May 1973, pp. 212-s-220-s.
[63] Frank, K. H., "The Fatigue Strength of Fillet Welded Connections," Ph.D. thesis, Lehigh University, Bethlehem, Pa., 1971.
[64] Marek, P. J., Perlman, M., Pense, A. W., and Tall, L., *Welding Journal,* Vol. 49, No. 6, June 1970, pp. 245-s-253-s.
[65] Maddox, S. J., *International Journal of Fracture,* Vol. 11, No. 2, April 1975, pp. 221-243.
[66] Maddox, S. J., *Welding Journal,* Vol. 53, No. 9, Sept. 1974, pp. 401-s-409-s.
[67] Maddox, S. J., *Metal Construction and British Welding Journal,* Vol. 2, No. 7, July 1970, pp. 285-289.
[68] Maddox, S. J., *Welding Research International,* Vol. 4, No. 1, Jan. 1974, pp. 36-60.
[69] Griffiths, J. R., Mogford, I. L., and Richards, C. E., *Metal Science Journal,* Vol. 5,

1971, pp. 150-154.

[*70*] Parry, M., Nordberg, H., and Hertzberg, R. W., *Welding Journal,* Vol. 51, No. 10, Oct. 1972, pp. 485-s-490-s.

[*71*] Kapadia, B. M. and Imhof, E. J., Jr. in *Flaw Growth and Fracture, ASTM STP 631,* American Society for Testing and Materials, 1977, p. 159.

[*72*] Smith, K. N., Haddad, M. El., and Martin, J. F., *Journal of Testing and Evaluation,* Vol. 5, No. 4, 1977, p. 327.

[*73*] Boulton, C. F., *International Journal of Pressure Vessels and Piping,* Vol. 4, No. 3, July 1976, pp. 171-194.

[*74*] Lawrence, F. V. and Munse, W. H., *Welding Journal,* Vol. 52, No. 5, May 1973, pp. 221-s-225-s, 232-s.

[*75*] Lawrence, F. V., Jr. and Radziminski, J. B., *Welding Journal,* Vol. 49, No. 10, Oct. 1970, pp. 445-s-452-s.

[*76*] Maddox, S. J., *Welding Research International,* Vol. 4, No. 2, Feb. 1974, pp. 1-30.

[*77*] Pook, L. P., *Welding Research International,* Vol. 4, No. 3, March 1974, pp. 1-24.

[*78*] Hirt, M. A. and Fisher, J. W., *Engineering Fracture Mechanics,* Vol. 5, 1973, pp. 415-429.

[*79*] Tanaka, K. and Matsuoka, S., *Engineering Fracture Mechanics,* Vol. 7, 1975, pp. 79-99.

[*80*] Pan, R. B. and Plummer, F. B., "A Fracture Mechanics Approach to Nonoverlapping Tubular K-Joint Fatigue Life Prediction," OTC Paper No. 2645, Eighth Annual Offshore Technology Conference, Houston, Tex., May 1976.

[*81*] Mattos, R. J. and Lawrence, F. V., "Estimation of the Fatigue Crack Initiation Life in Welds Using Low Cycle Fatigue Concepts," FCP Report No. 19, Fracture Control Program, University of Illinois, Urbana, Ill., Oct. 1975.

[*82*] Higashida, Y. and Lawrence, F. V., "Strain Controlled Fatigue Behavior of Weld Metal and Heat-Affected Base Metal in A36 and A514 Steel Welds," FCP Report No. 22, Fracture Control Program, University of Illinois, Urbana, Ill., Aug. 1976.

[*83*] Atsuta, T., Toma, S., Kurobane, Y., and Mitsui, Y., "Fatigue Design of an Offshore Structure," OTC Paper No. 2607, Eighth Annual Offshore Technology Conference, Houston, Tex., May 1976.

W. W. Sanders, Jr.[1] *and F. V. Lawrence, Jr.*[2]

Fatigue Behavior of Aluminum Alloy Weldments

REFERENCE: Sanders, W. W., Jr. and Lawrence, F. V., Jr., "**Fatigue Behavior of Aluminum Alloy Weldments,**" *Fatigue Testing of Weldments, ASTM STP 648,* D. W. Hoeppner, Ed., American Society for Testing and Materials, 1978, pp. 22–34.

ABSTRACT: During the last eight years, a number of research studies have been conducted on the fatigue behavior of aluminum alloy weldments under the guidance of the Aluminum Alloys Committee of the Welding Research Council. The initial study was a comprehensive review of the current (1970) state of knowledge and the development of research needs. The additional studies have included investigations into the effect of weld orientation (longitudinal versus transverse), thickened plates, and weld defects on fatigue behavior. This paper provides a summary of these five studies.

KEY WORDS: fatigue tests, fatigue (materials), aluminum alloys, weldments, defects, thickness, mechanical properties, evaluation

About ten years ago, the Welding Research Council formed the Aluminum Alloys Committee to coordinate research on the behavior of aluminum alloy weldments. As one of its first efforts, the committee had state-of-the-art reports [*1-3*][3] prepared on three of the major areas of research interest. These three reports summarized the current state of knowledge and recommended needed research. The reports [*1,2*] recommended a number of areas of research related to the fatigue behavior of aluminum weldments. These areas included studies of the effects of porosity and other severe defects on fatigue behavior, effects of the nature of reinforcement shape on fatigue resistance, fatigue resistance of groove welded joints in thicker weldments, and evaluation of postweld treatments and environments on fatigue resistance.

As a result of the state-of-the-art report on fatigue behavior [*2*], two subsequent investigations [*4,5*] on the fatigue behavior of sound, thick alumi-

[1] Professor, Department of Civil Engineering, Iowa State University, Ames, Iowa 50011.

[2] Professor, Departments of Civil Engineering and Metallurgy and Mining Engineering, University of Illinois at Urbana-Champaign, Urbana, Ill. 61801.

[3] The italic numbers in brackets refer to the list of references appended to this paper.

num alloy weldments were undertaken. The state-of-the-art report on weld defects [1] led to a study [6] of the effect of weld defects on static behavior of weldments and, with the support of the basic fatigue report [2], to two studies [7,8] of their effects on fatigue behavior of aluminum alloy weldments.

Thus, to date, the Aluminum Alloys Committee has coordinated five separate investigations [2,4,5,7,8] on the fatigue behavior of aluminum alloy weldments. This paper will summarize these investigations and their findings. For the details of the studies, the reader is referred to the individual project reports.

Literature Survey

The literature survey [1] included reviews of published papers and reports, as well as numerous unpublished reports furnished by a number of aluminum companies. Of the nearly 300 papers and reports reviewed, only about 80 were found to contain information pertinent to the investigation. About 40 reports had quantitative data, of which many were unpublished company reports. Over 400 test series and about 5000 individual tests were reported.

Current research to update the survey indicates that less than 75 test series have been conducted since the original report [2]. The bulk of these, except for those referred to in this paper, are on aluminum alloys for marine use.

The majority of the data available in the original survey [2] (and in the current research) was from studies of 5000 series alloys (or foreign equivalents to this series) with significant information of 6000 and 7000 series alloys. Only very limited data were available on 2000 and 3000 series alloys. Most of the data collected were on butt-welded joint tests, although significant information was obtained on fillet-welded joints.

Using computer analyses of all data, statistical analyses were made to determine the factors affecting fatigue behavior. A detailed bibliography was developed and an appendix with a detailed summary of the reviewed data was prepared and included in the survey report [2].

The average diagrams relating stress and number of cycles for failure (*S-N* diagrams) for these data on as-welded 5000 series butt welds are shown in Fig. 1. These diagrams summarize the bulk of the available data. The majority of the tests was conducted on thin plates or sheets (9.5 mm (3/8 in.) or less in thickness). The most common type of joint was the single-V groove butt weld. Approximately 1760 tests were available on transverse single-V groove welds, with about 500 tests on transverse square groove welds, 460 tests on transverse double-V groove welds, and 590 tests on all three types of longitudinal welds.

The results do not indicate that the specific joint detail (square groove, single-V groove, or double-V groove) is a major factor. However, the orientation of the weld is a significant factor as the longitudinal welds have con-

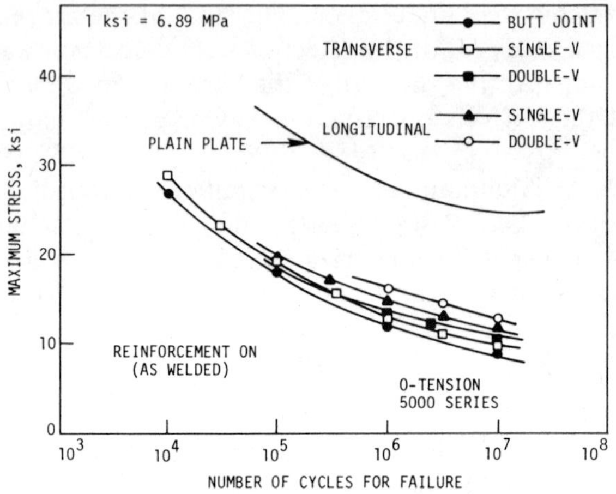

FIG. 1—*Average* S-N *diagrams for butt-welded joints with reinforcement on.*

sistently better fatigue behavior than transverse welds. In both cases, the double-V groove welds are consistently slightly superior.

Similar behavior is noted for the tests of welded joints with the reinforcement removed (Fig. 2). The curves are based on the average of available data. The shape of the reinforcement is probably the single most significant factor affecting fatigue behavior. Combining the results in a summary diagram (Fig. 3), the marked rise in fatigue strength from reinforcement removal and plain plate (no reinforcement) is evident. These results are indicative only for good quality welds. For poor quality welds, the removal of reinforcement may simply shift the point of fracture initiation from the external notch to an internal notch (defect) without the significant increase shown in fatigue strength for sound welds.

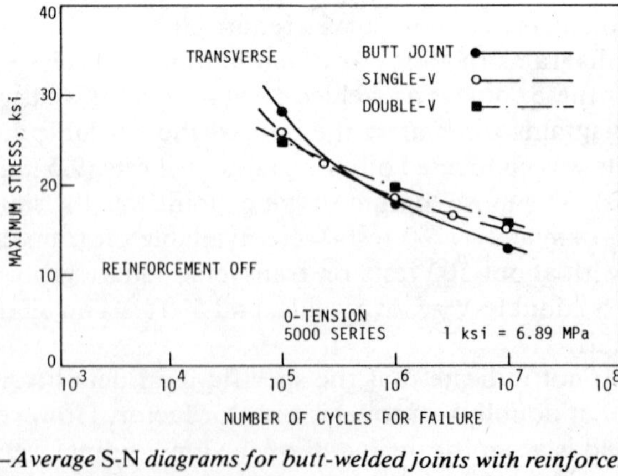

FIG. 2—*Average* S-N *diagrams for butt-welded joints with reinforcement off.*

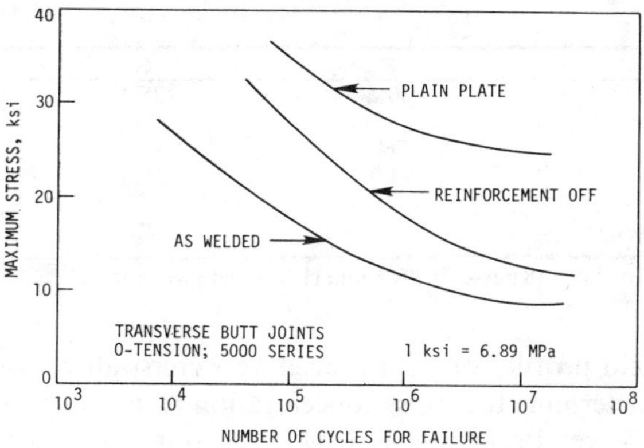

FIG. 3—*Effect of reinforcement removal on fatigue strength.*

The project report [2] also discusses the data available on effect of other variables, such as welding procedure, test environment, alignment, and defects.

Behavior of Thick Weldments in Air

The purpose of this study [4] was to determine the fatigue behavior of sound weldments in aluminum alloy 5083-0 plate with thicknesses approaching the upper limits of most usage. As noted earlier, most of the available data was on weldments in plate thicknesses 9.5 mm (3/8 in.) or less. In this study, particular emphasis was placed on the effect of weld geometry and orientation.

The basic program consisted of 54 fatigue tests in axial tension, including 14 longitudinally welded plates in as-welded condition, 4 longitudinally welded plates with reinforcement removed, 20 transversely welded plates in as-welded condition, 6 transversely welded plates with reinforcement removed, and 10 plain plates. All tests were conducted in a servocontrolled fatigue testing system. Of particular interest here were supplemental programs of plastic casts of all welds to determine the angle at the toe of the weld at points of fracture and a study of photoelastic models of typical transverse welds (from the three fabricators) to determine the weld concentration at the weld toe.

The specimens were all of 5083-0 aluminum alloy and prepared from 356 by 1524 mm (14 by 60 in.) rectangular plates. The specimens were typical reduced-section shape with a width at the critical section of 125 mm (5 in.). Three different welding procedures (Table 1) were used (designated F, G, and H). The basic differences were in the welding speed and number of passes used to make the 60-deg double-V groove welds, but the primary result of the different procedures was a variation in the shape of the reinforcement.

TABLE 1—*Welding procedures.*

Procedure	F	G	H
Plate thickness, mm (in.)	30.5 (1.2)	25.4 (1.0)	25.4 (1.0)
Welding speed, mm/min (in./min)	760 (30)	460 (18)	460 to 860 (18 to 34)
Number of passes	28[a]	12	8

[a] Equivalent to about 18 passes in 25.4-mm (1-in.) plate weldment.

Typical weld profiles of each procedure were studied using photoelastic methods to determine the stress concentration factor at the toe of the weld. The factors were: Procedure F–1.69, Procedure G–1.84, and Procedure 4–1.34.

The results of the fatigue tests of the transverse butt-welded plates are shown in Fig. 4. It can be seen that there is considerable scatter with the welds, with Procedure H performing consistently better than the other procedures. This results from the improved external geometry (lower profile) from this procedure. This effect is demonstrated in Fig. 5 where the maximum nominal stresses have been adjusted by the stress concentration factor, thus, providing a measure of the actual stress at point of fracture initiation (weld toe). The plot of "actual stress" results in an *S-N* curve with minimum scatter.

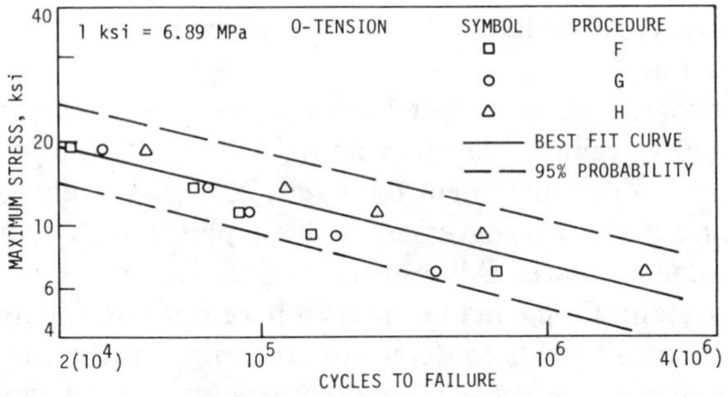

FIG. 4—S-N *diagram for transverse welded joints, as welded.*

The results of the longitudinal welded joints are plotted in Fig. 6. Since the weld is oriented parallel to the stress, the effect of weld shape is minimized and the scatter is small. The effect of joint configuration is shown in Fig. 7 and compares favorably to that found in the literature survey (Fig. 3).

From this study, the following conclusions were reached.

1. Orientation of the weld in the direction of stress applications is beneficial to fatigue strength of a joint.

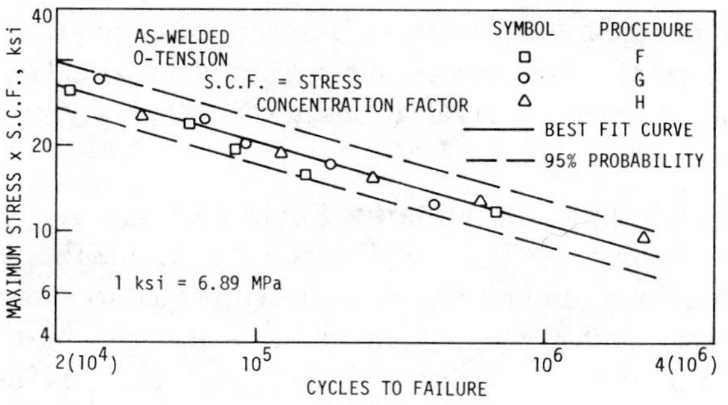

FIG. 5—S-N *diagram for transverse welded joints considering stress concentration factor.*

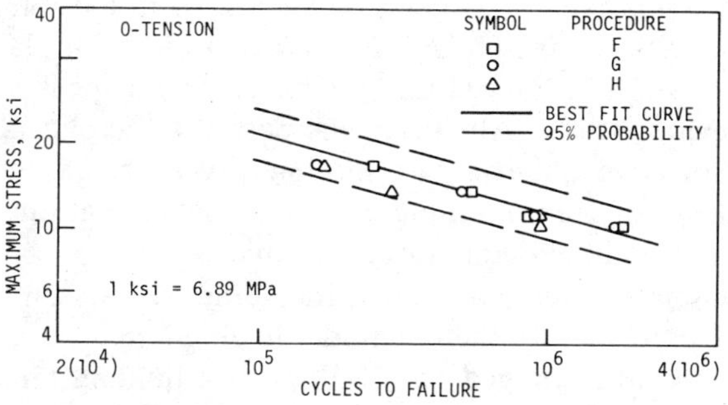

FIG. 6—S-N *diagram for longitudinal welded joints, as welded.*

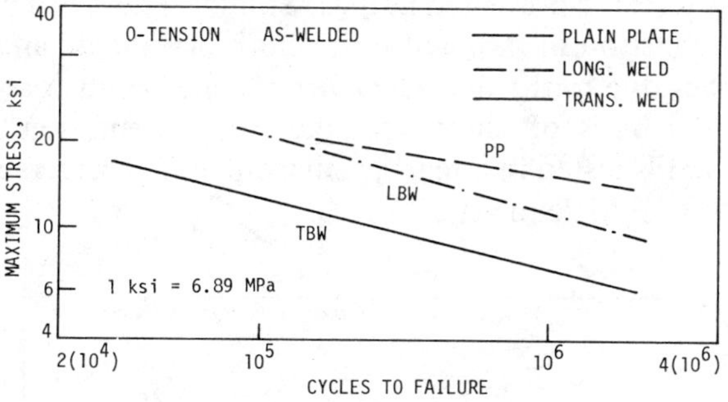

FIG. 7—*Comparison of* S-N *diagrams for different joint configurations.*

2. The angle at the toe of the weld reinforcement is the most critical factor in the determination of the fatigue life of an as-welded transverse joint: the larger the weld angle, the greater the stress concentration factor, and the lower the resulting fatigue life.

3. The actual maximum stress at the weld toe, determined from the nom-

inal stress and the stress concentration factor, appears to give a much better indication of expected fatigue life. This method provides correlation between different specimens with variations in weld geometry.

Behavior of Thick Weldments in Marine Environments

The purpose of this investigation [5] was to determine the fatigue behavior of 5000 series aluminum alloy weldments subjected to a marine environment. Tests were conducted on plain plate, transverse butt-welded, and longitudinal butt-welded specimens of 5086–H116, 5456–H116, and 5456–H117 aluminum alloys. These alloys are generally used in marine environments. All welds were made using normal shop fabrication practices. The specimens were full-thickness plates of 19.1 and 25.4 mm thickness (3/4 and 1 in. thickness), axially fatigued under a zero-to-tension stress cycle. The test section was machined to a 102 mm (4 in.) width for plain plates and a 127 mm (5 in.) width for all welded specimens. The shape is similar to that used in the previous study. All specimens passed all applicable codes and specifications for weld quality. The specimens were representative of the physical conditions of aluminum plates used in field situations for the construction of ships and liquefied natural gas tanks.

Sixty full-scale tests were conducted, including 39 tests on specimens in the marine environment with the remainder in an air environment. The marine specimens were submerged for 30 days in a holding tank containing substitute seawater (ASTM Specification for Substitute Ocean Water (D 1141–75)). These specimens were also enclosed in a tank during the testing, allowing the specimen test section to be submerged completely.

Results of key tests are presented in Figs. 8 and 9. The *S-N* diagrams show the significant reduction in fatigue life for both plain plate and weldments at all stress ranges due to the marine environment. Results are further differentiated on the basis of alloy type and weld orientation. Failures of transverse butt welds tested in a marine environment occurred at maximum stresses as low as 41.3 MPa (6 ksi).

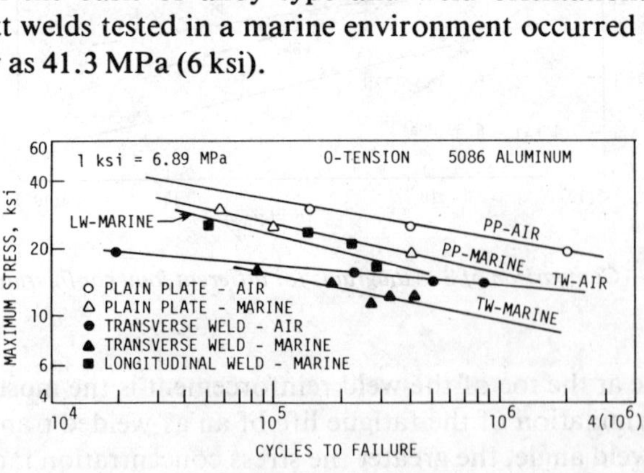

FIG. 8—S-N *curves for 5086 aluminum alloy plain plate and transverse and longitudinal weld specimens exposed to air or salt water.*

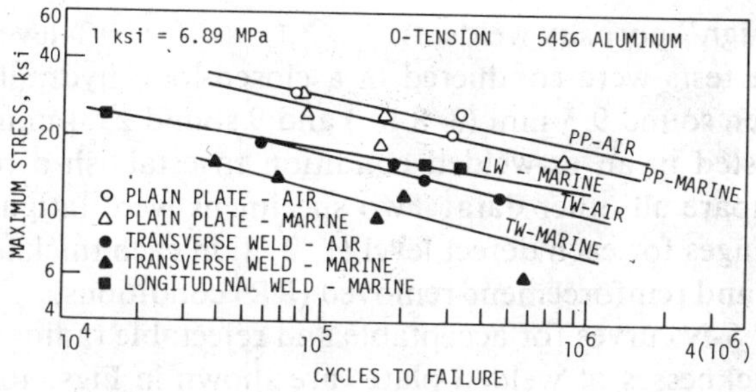

FIG. 9—S-N *curves for 5456 aluminum alloy plain plate and transverse and longitudinal weld specimens exposed to air or salt water.*

The main fatigue test program was supplemented by additional investigations. Six plain plate specimens of ABS Class C steel were tested in axial fatigue. Three of these specimens were tested in air with the other three tested in the marine environment. The results show a reduction in fatigue behavior due to exposure to salt water similar to that just indicated for aluminum.

Two major conclusions were drawn from the study.

1. Saltwater corrosion significantly reduces the fatigue strength of 5456 and 5086 weldments for all stress ranges. However, the reduction in fatigue life is not constant but generally increases with decreasing applied stress range.

2. Orientation of the weld in the direction of stress application is beneficial to the fatigue strength of a joint exposed to either air or salt water.

Effects of Porosity

The objective of this investigation was to study the effect of distributed porosity on the fatigue resistance of 5083-0 double-V groove butt weldments subjected to a constant amplitude, 0-tension stress cycle. Porosity levels were recorded by normal incidence radiography prior to testing and measured directly on the fatigue fracture surfaces.

The test program included 92 specimens. The specimens were reduced sections with a width of 50.8 mm (2 in.) and thicknesses of 9.5 or 25.4 mm (3/8 in. or 1 in.). All welded specimens were fabricated in the flat position. The sound welds (0-level) were fabricated using conventional procedures. The other welds (1-, 2-, and 3-levels) were fabricated using procedures intentionally designed to induce high levels of porosity in excess of that normally encountered in acceptable welds. The four levels of porosity are identified by

"sound" welds:	0-level
"low" porosity welds:	1-level
"intermediate" porosity welds:	2-level

"high" porosity welds: 3-level

The fatigue tests were conducted in a closed-loop hydraulic testing apparatus. Eleven sound 9.5-mm (3/8-in.) and 9 sound 25.4-mm (1-in.) specimens were tested in an as-welded condition to establish a reference with which to compare all other data. Two specimens were fatigued at each of three stress ranges for each defect level (1, 2, 3) in each thickness for the as-welded (AW) and reinforcement-removed (RR) conditions.

The best fit *S-N* curves for acceptable and rejectable radiographic ratings in the two thicknesses of welded plates are shown in Figs. 10 and 11. Test results for 9.5-mm (3/8-in.) reinforcement-removed tests exceed the results for 9.5-mm as-welded tests by a small amount at 130 MPa (19 ksi) and substantially at 83 MPa (12 ksi). The difference between 25.4-mm (1-in.) reinforcement-removed and 25.4-mm (1-in.) as-welded specimens is less pronounced, however. The fatigue lives of the 25.4-mm (1-in.) as-welded and reinforcement-removed tests were less than those of the 9.5-mm (3/8-in.) as-welded and reinforcement-removed tests, respectively.

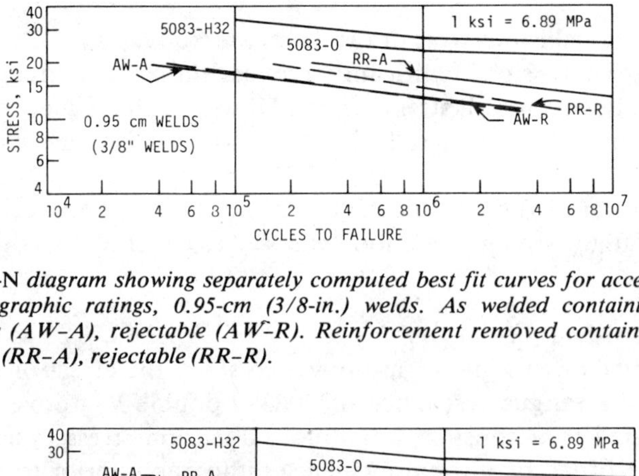

FIG. 10—S-N *diagram showing separately computed best fit curves for acceptable and rejectable radiographic ratings, 0.95-cm (3/8-in.) welds. As welded containing acceptable porosity levels (AW–A), rejectable (AW–R). Reinforcement removed containing acceptable porosity levels (RR–A), rejectable (RR–R).*

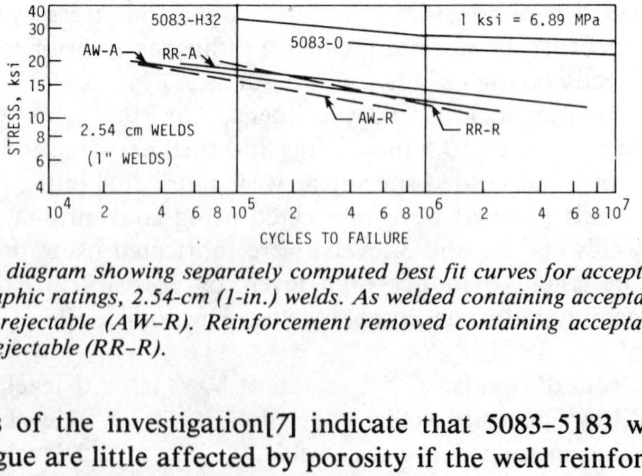

FIG. 11—S-N *diagram showing separately computed best fit curves for acceptable and rejectable radiographic ratings, 2.54-cm (1-in.) welds. As welded containing acceptable porosity levels (AW–A), rejectable (AW–R). Reinforcement removed containing acceptable porosity levels (RR–A), rejectable (RR–R).*

The results of the investigation[7] indicate that 5083–5183 welds subjected to fatigue are little affected by porosity if the weld reinforcement is left in place. The weld reinforcement itself is the critical and fatigue limiting notch. Most welds tested with their reinforcement removed gave longer fatigue lives than as-welded tests regardless of porosity level. Porosity most influenced the fatigue lives of the reinforcement removed tests at the lowest stress levels. The radiographic standards currently in use by the U.S. Navy were found to be effective in ensuring superior results with reinforcement-removed welds. Conversely, few reinforcement-removed welds which failed these standards gave shorter fatigue lives than porosity-free, as-welded welds.

The key conclusions were as follows.

1. Welds with their reinforcement intact were little affected by porosity unless the weld reinforcement was shallow and the porosity level was very high.

2. Welds with their reinforcement removed gave longer lives than sound as-welded welds except in cases of very high porosity. Porosity influenced the fatigue lives of reinforcement removed tests most significantly at the lowest stress level, 0 to 83 MPa (0 to 12 ksi).

Effects of Lack of Penetration and Lack of Fusion

In this study [8], 112 zero-to-tension fatigue tests were performed on double-V groove butt welds of 5083-0 aluminum alloy which contained full-length lack-of-penetration (LOP) defects and "natural," less-than-full-length lack-of-fusion (LOF) defects. The LOP and LOF defects were incorporated in the welds using "improper" welding methods.

The specimen shape and testing procedure were similar to those used in the previous study on the effect of porosity. The testing program used is summarized in Table 2.

TABLE 2—*Test program: LOP and LOF defects.*

Reinforcement[a]	Thickness, mm (in.)	LOP Defect Level 1 (sound)	2	3	4	LOF
RI	9.5 (3/8)	12	6	6	6	...
	25.4 (1)	6	6	6	6	...
RR	9.5 (3/8)	...	6	6	6	12
	25.4 (1)	...	6	6	6	10

[a] RI = reinforcement intact, and RR = reinforcement removed.

The defect levels indicated for LOP defects are similar to those indicated previously for porosity levels.

The results of the fatigue tests of the LOP and LOF specimens are shown in Figs. 12 and 13. The results show that full length LOP defects have a profound effect on fatigue life. Larger width defects are substantially more serious than smaller width defects. Both the reinforcement intact (RI) and reinforcement-removed specimens gave about the same lives which were usually considerably less than those found for sound welds.

The figures also indicate the effect of LOF. The effect of an LOF defect is essentially similar to the effect of a full length LOP of similar width if the projected width of the LOF is considered.

The fracture surfaces of the specimens showed that large amounts of porosity were associated with the LOP. However, examination of the fracture surfaces revealed that fatigue crack initiation occurred predominately at the LOP defect and not from the associated porosity. Unlike the LOP defects, porosity was not associated with the LOF defects.

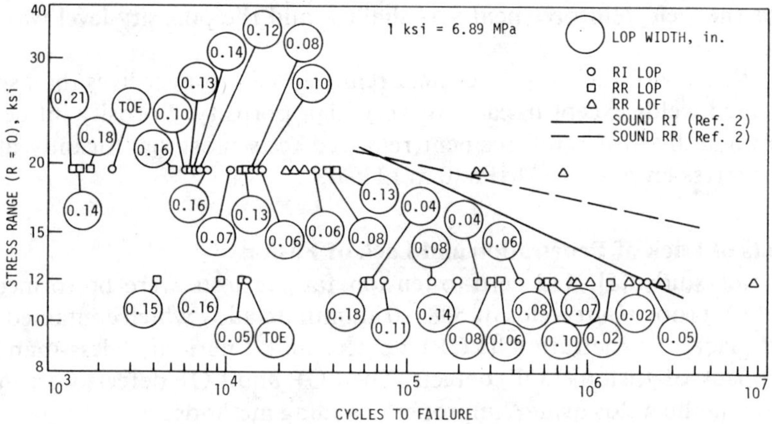

FIG. 12—S-N *diagram for 0.95-cm (3/8-in.) RI and RR specimens with LOP and LOF defects. The LOP defect length ($2c_o$) is noted in the figure. Results for sound RI and RR welds* [2] *are also plotted.*

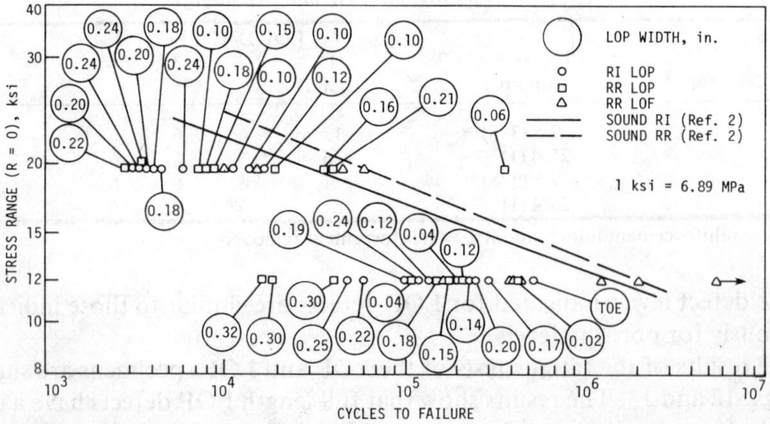

FIG. 13—S-N *diagram for 2.54-cm (1-in.) RI and RR specimens with LOP and LOF defects. The LOP defect length ($2c_o$) is noted in the figure. Results for sound RI and RR welds* [2] *are also plotted.*

From these findings the following conclusions were drawn.

1. Lack-of-penetration (LOP) defects can seriously reduce the fatigue life of both reinforcement intact and reinforcement-removed welds. The magnitude of this reduction is determined by the through-thickness dimension of the LOP. The porosity associated with these defects seemed to exert no influence.

2. Less than full length, inclined LOF defects were generally less serious than LOP defects, but if compared with LOP defects on the basis of width projected normal to the tensile stress, they were found to follow the same trends as equivalent width LOP defects. The length of the LOF along the axis of the weld seemed to have little influence on their effect.

Summary

This paper has presented a summary of five investigations [*2,4,5,7,8*] into the fatigue behavior of aluminum alloy weldments.

The first of these studies [*2*] was a review of the literature to determine the current state of the art and recommend further research. As a result of that study, four subsequent investigations [*4,5,7,8*] were undertaken in which more than 300 specimens were tested. These investigations included studies of thick weldments in air and marine environments, and of the effects of porosity, LOP and LOF defects in several thickness weldments.

The general outline of each study is given along with the key findings and conclusions.

Acknowledgments

The research summarized in this paper was coordinated by the Subcommittees on Fatigue and on Weld Defects of the Welding Research Council's Aluminum Alloys Committee. Funding was provided by the Welding Research Council, National Science Foundation, and Naval Sea Systems Command with materials and weld fabrication provided by several aluminum companies and industrial firms. The support and assistance of the committee, sponsors, and contributors are gratefully acknowledged.

The authors also wish to acknowledge the assistance of a number of university associates who worked with them on the various studies. These associates are listed in the individual references.

The literature survey study and the investigations of the behavior of thick weldments in air and marine environments were conducted by the Engineering Research Institute of Iowa State University. The studies of the effects of porosity and the effects of lack of penetration and lack of fusion were conducted by the Department of Civil Engineering of the University of Illinois at Urbana-Champaign.

References

[*1*] Pense, A. W. and Stout, R. D., "Influences of Weld Defects on the Mechanical Properties of Aluminum Alloy Weldments," *WRC Bulletin 152,* Welding Research Council, New York, July 1970.

[*2*] Sanders, W. W., Jr., "Fatigue Behavior of Aluminum Alloy Weldments," *WRC Bulletin 171,* Welding Research Council, New York, April 1971.

[*3*] Masukuchi, K., "Residual Stresses and Distortion in Welded Aluminum Structures and Their Effects on Service Performance," *WRC Bulletin 174,* Welding Research Council, New York, July 1972.

[*4*] Sanders, W. W., Jr. and Gannon, S. M., "Fatigue Behavior of Aluminum Alloy 5083 Butt Welds," *WRC Bulletin 199,* Welding Research Council, New York, Oct. 1974.

[*5*] Sanders, W. W., Jr. and McDowell, K. A., "Fatigue Behavior of 5000 Series Aluminum Alloy Weldments in Marine Environment," Report to Welding Research Council and National Science Foundation, ISU-ERI-Ames-78055, Engineering Research Institute, Iowa State University, Sept. 1977.

[*6*] Lawrence, F. V., Jr. and Munse, W. H., "Effects of Porosity on the Tensile Properties of 5083 and 6061 Aluminum Alloy Weldments," *WRC Bulletin 181,* Welding Research Council, New York, Feb. 1973.

[*7*] Lawrence, F. V., Jr., Munse, W. H., and Burk, J. D., "Effects of Porosity on the Fatigue Properties of 5083 Aluminum Alloy Weldments," *WRC Bulletin 206,* Welding Research Council, New York, June 1975.

[*8*] Burk, J. D. and Lawrence, F. V., Jr., "Effects of Lack of Penetration and Lack-of-Fusion on the Fatigue Properties of 5083 Aluminum Alloy Welds," Final Report to Naval Sea Systems Command, Dept. of Civil Engineering, University of Illinois at Urbana-Champaign, March 1977.

K. J. Pascoe[1] *and P. R. Christopher*[2]

Investigations of the Short Transverse Monotonic and Fatigue Strengths of Various Ship-Quality Steels

REFERENCE: Pascoe, K. J. and Christopher, P. R., "**Investigations of the Short Transverse Monotonic and Fatigue Strengths of Various Ship-Quality Steels,** *Fatigue Testing of Weldments, ASTM STP 648,* D. W. Hoeppner, Ed., American Society for Testing and Materials, 1978, pp. 35–56.

ABSTRACT: This paper summarizes a cooperative research program undertaken by the Naval Construction Research Establishment and the Department of Engineering of Cambridge University. It gives a tentative generalized picture of the monotonic and zero-to-tension low cycle fatigue (up to about 5000 cycles) behavior of cruciform welded ship quality plates in range 25 to 76 mm thick. The test results given were obtained over the past few years in a special rig capable of applying loads of 2 MN repeatedly.

The rig was designed to accommodate short transverse specimens welded to loading ends so as to apply tension, combined tension and shear, or shear loading, the object being to develop short transverse failures (that is, failures associated with inclusions rolled into the plate).

The results indicate the dependence of the short transverse properties on the method of loading, the strength of the steel, and the cleanness of the material.

Although there were several forms of failure ranging from deep short transverse tears to failure in attachment welds, which occurred in some of the cleaner steels, it has been possible to rationalize the plots of stress range-logarithm of life to failure in the form of straight lines. This was particularly the case for the tension tests where there was less scatter than for the other two types of test.

KEY WORDS: fatigue tests, weldments, steels, low cycle fatigue, welded joints, mechanical properties, short transverse strength, inclusions, fractures (materials), carbon-manganese steels, low alloy steels

The work described in this paper was carried out at Cambridge University Engineering Laboratory in conjunction with the Naval Construction Research Establishment. It started as an investigation of the influence of inclusions on the low-cycle fatigue behavior of QT35 steel plate for tensile and

[1] University lecturer, Department of Engineering, Cambridge University, Cambridge, England.

[2] Principal scientific officer, Naval Construction Research Establishment, Dunfermline, Fife, Scotland.

shear loading applied to the plate thickness.

As a preliminary to the detailed planning, simple tension and shear loads were applied to specimens. Later, a specialized repeated loading machine was built to carry out the fatigue tests [*1*].[3]

The program was continued then with similar tests being carried out on a number of other ship-quality steels.

This paper presents the results obtained and attempts to rationalize the monotonic and zero-tension low cycle fatigue results.

Procedure

Initial Simple Tension and Shear Tests (0 and 90 deg)

For the initial tests, 76-mm-wide coupons cut from the plate under investigation (a 38-mm-thick QT35 steel plate) were welded to loading arms cut from plate of a similar material. Standard welding procedures were used. The coupons and adjacent portions of the loading arms then were machined to the shapes shown in Figs. 1 and 2 so that the final dimensions of the test

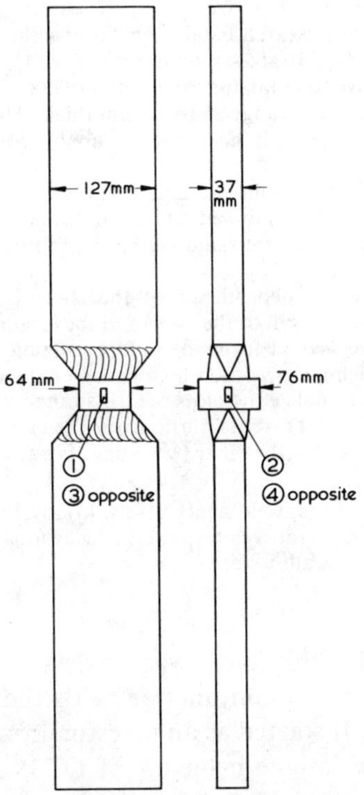

FIG. 1—*Tension specimen. Numbers in circles indicate strain gage positions.*

[3] The italic numbers in brackets refer to the list of references appended to this paper.

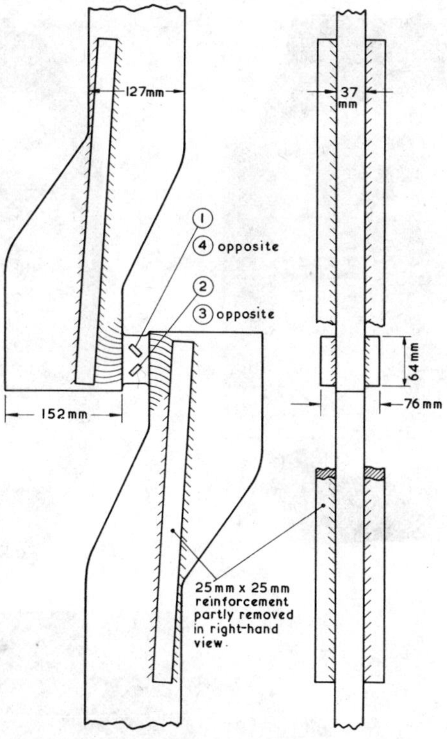

FIG. 2—*Shear specimen. Numbers in circles indicate strain gage positions.*

coupons were 76 mm wide by 64 mm long in each case. For the shear tests, the loading arms were reinforced with extra side strips to avoid bending of the loading arms.

The specimens were loaded by gripping the outer ends of the loading arms in the jaws of a 2-MN testing machine.

The fractures in each case occurred in the coupon, in the heat-affected zone (HAZ), just below the weld in a series of subsurface layers, see Fig. 3. It will be seen in these cases that for the tension specimen, the penetration of the weld was not complete, and for the shear specimen, failure was in the weld on one side and within the specimen on the other.

Further Developments of Monotonic Testing

Following the satisfactory start of the program, further tests were decided upon and carried out.

1. An additional set of tests was made with the specimen at an angle other than 0 or 90 deg to the direction of loading; 45 deg was chosen, the consequence being that there were then equal tensile and shear loads on the specimen. Figure 4 shows a complete specimen.

FIG. 3—*Fracture surfaces:* (top) *tension specimen,* (bottom) *shear specimen.*

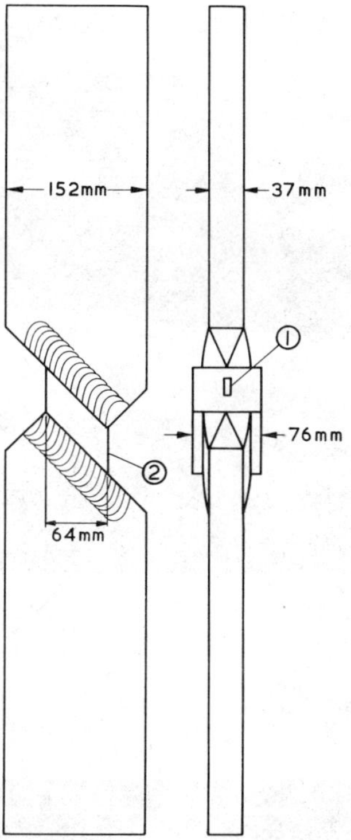

FIG. 4—*45-deg specimen. Numbers in circles indicate strain gage positions.*

2. Separate tests were carried out for each material with the plate rolling direction parallel and perpendicular to the plane of the loading arms.

3. A welding jig was constructed by which the test coupon could be located with respect to the loading arms and held in position until all welding was completed. Views of the jig are shown in Fig. 5.

In all tests, electric resistance strain gages were affixed to two sides of the specimen. In the 0 and 45-deg tests, they were placed longitudinally in the plane of the loading arms (see Figs. 1 and 4). In the 90-deg tests, they were placed on the sides of the specimen at 45 deg to the loading axis (Fig. 2) thereby recording tensile and compressive strains. In all cases, measurements of strains were made during loading.

The load at failure was divided by the area of contact of the weld with the coupon to give an average value of stress at that plane section. Although the stress obviously would not be constant over that section, and the detailed geometry of individual specimens differed, the average stress was taken to be as good a value for comparison purposes as any other.

Some of the early 0-deg tests were made with strain gages on all four sides of the coupon. These showed, as expected, that the strain was not uniform,

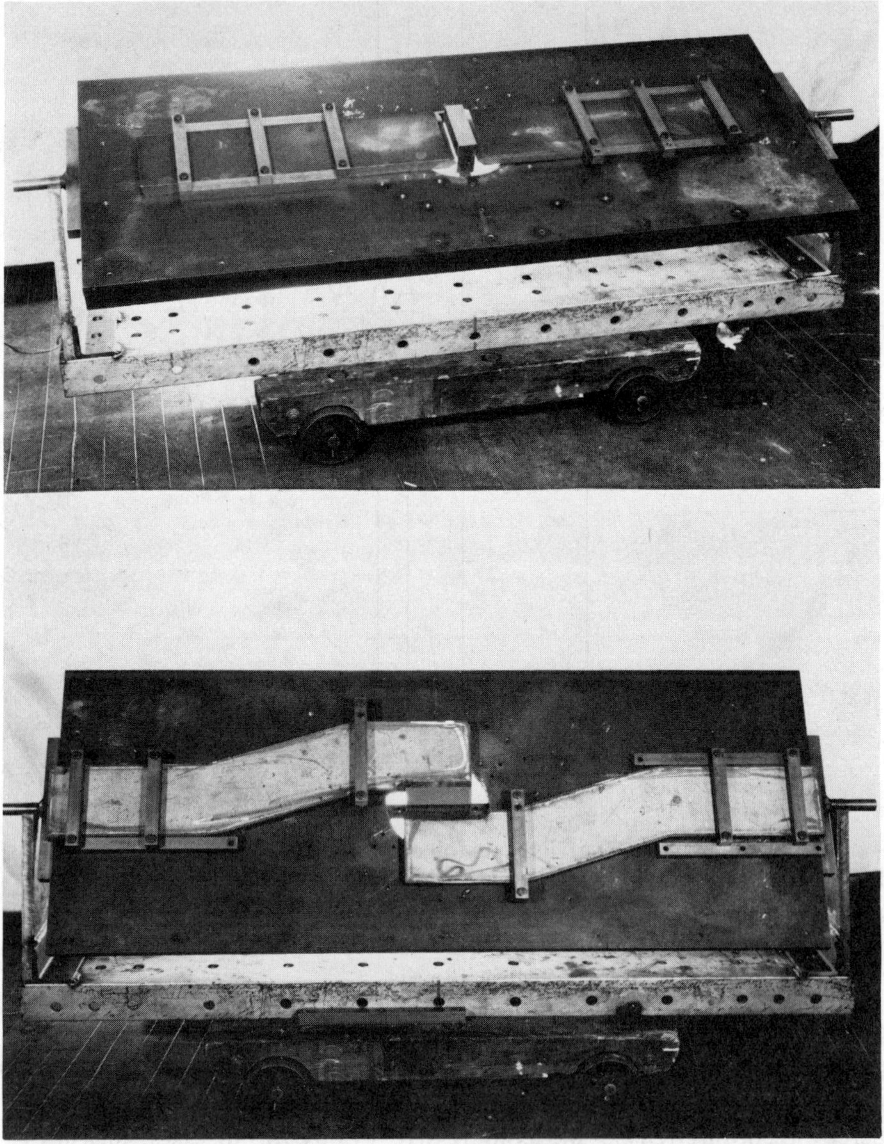

FIG. 5—*Welding jig:* (top) *with 0-deg specimen in position,* (bottom) *with 90-deg specimen in position.*

being very small on the sides furthest from the plane of the loading arms. Tests were made to see whether the width of the coupon, that is, the dimension perpendicular to the plane of the loading arms, was critical. Four 0-deg specimens were prepared with coupons 152 mm wide and 127 mm long. In the width direction, one specimen was left at the full 152 mm, and the others were machined equally from both sides to 76, 51, and 25 mm, respectively. In the first two, the fracture ran from the weld-coupon junction and through the HAZ, the mean stresses differing by less than 1 percent. In the narrower specimens, the extremities of the welds had been removed by the machining and the fractures were entirely in the coupon material. The mean stresses at fracture were greater by 20 and 37 percent, respectively. This indicated that a test coupon with a width of 76 mm might be considered to be sufficiently representative of greater widths for all further tests to be made with 76-mm-wide coupons.

Fatigue Tests

No existing machine was available to carry out automatic repeated loading tests on specimens of the types and sizes described heretofore. Three preliminary tests were made in a conventional tension testing machine by manual control, each test involving several hours for lives of only 300 cycles.

A specialized machine was designed and constructed for this purpose at the Naval Construction Research Establishment, Rosyth, and put into commission at Cambridge [*1*]. It consists of two short, deep beams. The upper movable one rests on ball seats at one end and is loaded by a 1.5-MN capacity hydraulic jack at the other end, the ball seatings and the hydraulic jack being supported on the lower beam which also acts as a base. The specimen is secured between two spherically seated loading lugs, one at the midlength of each beam, see Fig. 6. A load cell in series with the hydraulic jack supplies a signal to the control unit which causes the jack to operate between predetermined load limits.

The welding jig was modified to take the loading arms which are hinged at the loading lugs as shown in Fig. 7, the load being transmitted through a shear pin. This diagram shows a 0-deg specimen. The loading arms for 45 and 90-deg specimens are shown in Figs. 8 and 9, respectively. It can be seen that this arrangement allowed the same loading lugs to be used for all specimens.

When a specimen is set up in the machine and the desired stress chosen, the required load is calculated and the control unit set accordingly. Fracture of the specimen causes the machine to shut down. The number of loadings is recorded on a counter.

Materials

The steels on which tests were made include various types now in use for naval construction purposes and civil application. For some, plates of different degrees of cleanness and different thicknesses were tested. The compositions are given in Table 1.

FIG. 6—*Views of testing rig and auxiliary equipment.*

TABLE 1—*Compositions of steels (all values are percent by weight) and electrodes used.*

Steel	A	A	B	B	Clean B	BS 968	QT 35		HY 80		HY 100	Q1	Q1	Q2
Plate	3901	5522	3875	5759	5823	5380	4982	5292	5573	5675	5651	5689	5785	6470
C	0.16	0.125	0.18	0.18	0.20	0.20	0.10	0.125	0.13	0.13	0.16	0.16	0.16	0.16
Mn	1.55	1.39	0.51	1.33	1.37	1.35	0.96	0.90	0.26	0.23	0.26	0.33	0.32	0.37
Si	0.18	0.285	0.17	0.36	0.33	0.04	0.24	0.16	0.29	0.24	0.19	0.29	0.28	0.28
P	0.023	0.016	0.02	0.023	0.015	0.025	0.013	0.023	0.004	0.011	0.013	0.004	0.005	0.011
S	0.034	0.024	0.043	0.030	0.016	0.028	0.025	0.019	0.009	0.011	0.015	0.011	0.007	0.005
Ni	0.08		0.09	0.05	0.15	0.05	1.07	1.10	2.46	2.56	3.04	2.52	2.51	3.38
Cr	0.04		0.04	<0.05	0.04	0.08	0.79	0.83	1.29	1.42	1.61	1.30	1.32	1.56
Mo	0.01		0.01	<0.05	<0.05	<0.005	0.40	0.41	0.32	0.44	0.49	0.36	0.33	0.42
V	<0.05		<0.01	<0.05	<0.01	0.005	0.11	0.04	<0.03	<0.03	0.02	<0.03	<0.01	0.07
W						<0.05								
Ti									<0.02			<0.02		
Cu				0.05	0.21	0.045				0.04	0.04			0.10
Al	<0.005	0.038	<0.053	0.056		0.007	0.006		0.026	0.012	0.011	0.023	0.028	0.015
N	0.0053		0.0046			0.005	0.009			0.007	0.008	0.012	0.0097	0.0126
O	0.0054		0.0030							0.0035	0.002	0.001	0.0029	0.0031
Mg														
Sb									<0.005	<0.005				
Pb														
Nb						0.025								
Electrode used	Fortrex 35						Fortrex B				Tensitrode 12018	Tensitrode 11018		Tensitrode 12018

The test coupons for the monotonic and fatigue tests were welded to the loading arms by multirun manual metal arc. The type of electrode used for each steel also is given in Table 1.

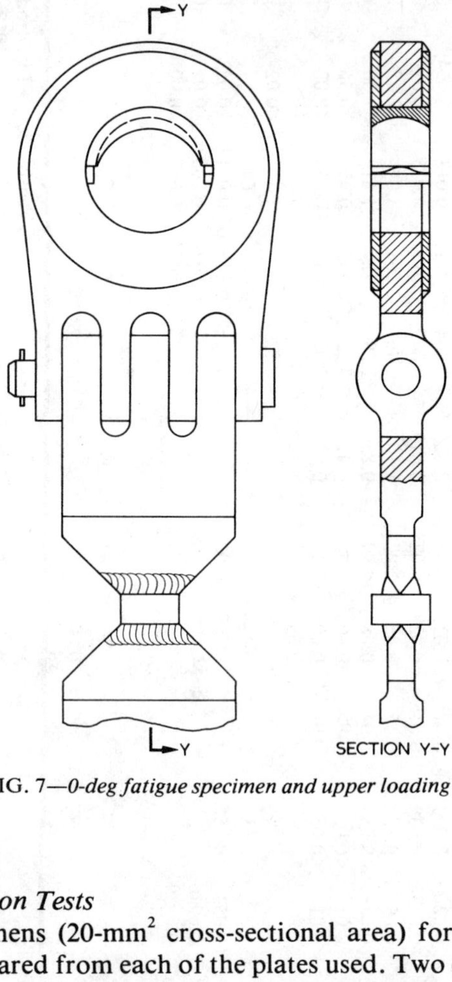

FIG. 7—*0-deg fatigue specimen and upper loading lug.*

Results

Directional Tension Tests

Tension specimens (20-mm^2 cross-sectional area) for testing in a tensiometer were prepared from each of the plates used. Two each were prepared with their axes: (*a*) longitudinal (L), that is, parallel to the direction of rolling, (*b*) transverse (T), that is, perpendicular to the direction of rolling, and (*c*) short transverse (X), that is, perpendicular to the plate surfaces.

The averaged results of the tests are given in Table 2.

There was little difference in strength and ductility between the in-plane longitudinal and transverse specimens. The short transverse specimens, however, showed lower ductilities in all cases and lower strengths in most cases. The most significant reductions in ductilities were found in Plates 3875, 4982, 5292, and 5759, these being steels with high inclusion counts

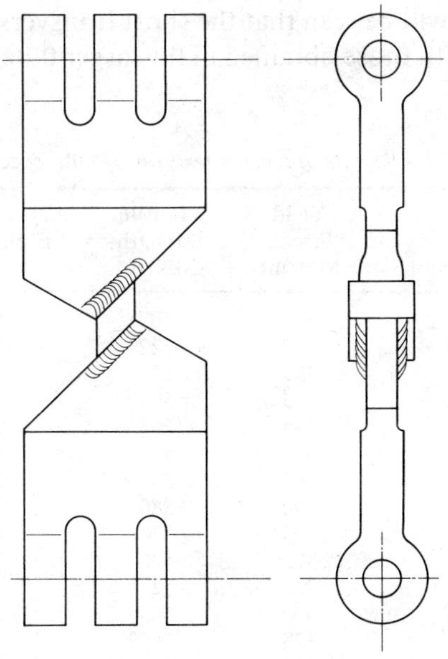

FIG. 8—*45-deg fatigue specimen.*

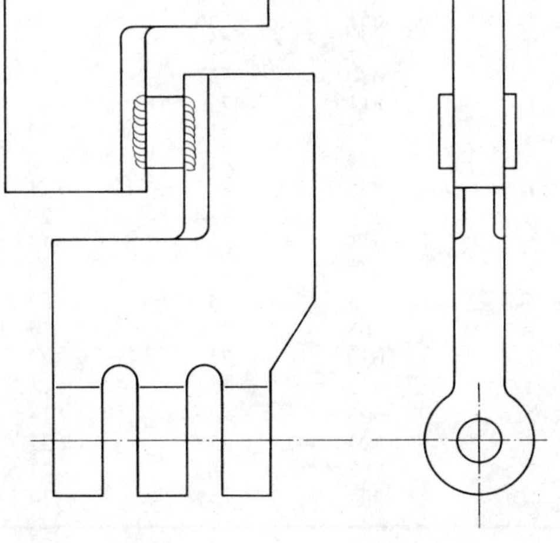

FIG. 9—*90-deg fatigue specimen.*

(see next section). It will be seen that the short transverse tension test results compare very well with those obtained in the larger 0-deg tests.

TABLE 2—*Results of tension tests on 20-mm² specimens.*

Plate	Direction	Yield Stress, MN/m²	Tensile Strength, MN/m²	Elongation, %	Reduction of Area, %
A	L	313	507	56	66
	T	337	522	54	72
3901	X	399	526	27	33
A	L	365	505	60	72
	T	348	508	58	68
5522	X	357	500	43	48
A	L	200	393	37.5	67
(51 mm	T	190	380	35	60
thick)	X	195	378	19	32
B	L	373	520	34	63
	T	325	523	35.5	70
3875	X	360	463	10	7.5
B	L	378	545	33	69
	T	393	545	33	67
5759	X	365	425	7	9
Clean B	L	425	575	35	72
	T	400	568	30	66
5823	X	395	553	21.5	30
BS 968	L	363	500	37.5	73
	T	360	500	34	70
5380	X	355	490	16	23
QT 35	L	583	663	22	68
	T	595	660	20.5	52
4982	X	533	563	5	5
QT 35	L	670	749	39	70
	T	663	740	32	61
5292	X	645	663	6	5
HY 80	L	651	756	40	68
	T	651	775	42	70
5573	X	620	732	26	40
HY 80	L	611	750	45	68
	T	614	750	42	70
5675	X	589	712	45	60
HY 100	L	755	823	21	73
	T	753	838	20	68
5651	X	715	820	15	43
Q1	L	665	770	23	77
	T	653	745	22.5	77
5689	X	668	770	17	57
Q1	L	630	753	23	77
	T	615	733	21.5	75
5785	X	605	743	17	63
Q2	L	793	858	21	72
	T	788	863	21.5	73
6470	X	803	845	9/14.5	– /40

Inclusion Counts

The results of inclusion counts for each of the plates used for fatigue tests are given in Table 3; some steels were "dirty."

These inclusion counts were obtained on sections at right angles to the surface in the transverse plane. All inclusions above 8 μm in length were counted on a section 85 by 115 mm at a magnification of x250. The major inclusion types, elongated sulfides, sulfides, and oxides were identified. Low counts were noted for the steels, HY80, HY100, Q1, and Q2.

TABLE 3—*Inclusion data.*

Steel	Plate	Thickness, mm	Inclusions, per mm^2	Major Inclusion Types[a]
A	3901	46	17.3	S
	5522	25	38.1	S
		51	33.3	E
B	3875	37	38.1	E
	5759	50	40.2	E
Clean				
B	5823	38	18.1	S
BS 968	5380	38	71.4	S
QT 35	4982	63	64.5	E
	5292	38	62	E
HY 80	5573	38	5.2	O
	5675	38	5.2	O
HY 100	5651	38	1.9	E
Q1	5689	38	3.3	O
	5785	63	1.4	S
Q2	6470	76	3.3	O

[a] E = elongated sulfides,
S = sulfides, and
O = oxides.

Fatigue Test Results

In the case of the stronger steels Q1 and Q2, fatigue failure occurred in the weld metal. Failures did not start from internal defects in the welds.

In almost all cases, both parallel and perpendicular, 0, 45, and 90-deg monotonic tests were done. Fatigue tests were done, in some cases, for 0, 45, and 90-deg specimens, though in several of these steels they were confined to perpendicular coupons only, in order to reduce the number of tests. Only a limited program was used for steels where failure always occurred in the welds.

The test results are summarized in Table 4 and rationalized in Table 5. Table 4 gives the average stress under the weld, the life and mode of failure for each case. It will be seen that for the fatigue tests, when failure occurs in the coupons, there is an increase of life with decreasing stress. Only a few tests were made on each steel, but, despite the scatter which might be ex-

TABLE 4—*Results of monotonic and fatigue tests.*

Steel, Plate, Thickness, mm	0 deg						45 deg						90 deg					
	Parallel			Perpendicular			Parallel			Perpendicular			Parallel			Perpendicular		
	σ	N	F	σ	N	F	σ	N	F	σ	N	F	σ	N	F	σ	N	F
QT 35, 5292, 38	415	1	SM	556	1	SM	286	1	SM	221	1	SM	323	1	SC	355	1	SC
A Quality, 5522, 25	412	1	DC	442	1	DC	511	1	DC	550	1	DC	256	1	SC	255	1	SC
A Quality, 3901, 46	205	1	SC	338	1	SC	252	1	SC	264	1	SC	168	1	SC	178	1	SC
A Quality, 51	352	1	S	360	1	S	449	1	D	457	1	D	188	1	S	190	1	S
B Quality, 3875, 37	477	1	D	500	1	D	405	1	D	391	1	D	238	1	S	227	1	S
				426	1	D				432	40	D				259	1	S
				419	8	D				402	818	S				278	5	S
				388	281	S				420	930	W				216	129	S
				366	1604	D				386	1363	S				193	439	S
				358	1568	D				354	1501	S				178	333	S
																162	3738	S
B Quality, 5759, 50	411	1	DM	442	1	DM	547	1	D	554	1	D				281	1	S
	341	1	DM	394	16	DM				508	34	D				208	2	S
				371	1279	DM				486	394	D				178	120	S
				355	418	DM				463	1839	D				158	543	S
				309	3103	DM										139	3334	SW
Clean B, 5823, 38	551	1	S	607	1	DM	656	1	D	661	1	D	321	1	S	405	1	S
	500	78	D	525	1	DM	602	13	D	612	10	D	263	153	S	280	9	D
	496	137	DM	486	79	DM	556	296	D	571	12	D	227	754	S	264	105	D
	485	123	DM	477	497	DM	510	864	S	550	484	D	201	959	D	232	257	S
	462	265	D	480	813	D				479	2232	S						
	457	858	D	465	1457	D												
BS 968, 5380, 38	442	1	SC	453	1	SC	510	1	S	537	1	S	230	1	S	215	1	S
	409	8	S	411	2	S	494	10	DS	502	36	DS	255	1	S	242	1	S
	394	111	D	417	7	D	392	743	DS	456	172	DS	216	77	S	216	34	S
	371	365	D	432	21	S	371	2481	S	426	213	S	185	210	S	185	637	S

				394	61	D							154	1066	S	170	1732	S
				371	998	S												
QT 35,	593	1	S	558	1	DC				710	1	SC				266	1	SC
4982, 63				573	17	DC				618	25	SC				186	11	SC
				556	53	DC				571	75	SC				124	224	SC
				541	71	DC				494	1235	SC				92	5028	SC
				517	286	DC												
				502	18	SC												
HY 80,	522	1	L	571	1	L	709	1	W	675	1	L	304	1	W	309	1	W
5573, 38									L									
HY 80,	646	1	LS	744	1	LS	670	1	W	689	1	W	283	1	W	287	1	W
5675, 38									S									
Q1,	544	1	W	666	1	W				727	1	W	383	1	W	359	1	W
5689, 38	695	1	W	593	1	W				709	4	W				309	618	L
	673	1	W	587	88	W				695	126	W				309	1564	W
				609	156	W				672	139	W						
				556	666	W												
				541	951	W												
Q1,	735	1	W	746	1	W												
5785, 63				746	1	W												
HY 100,	712	1	W	808	1	W												
5651, 38																		
Q2,	894	1	W				925	1	W									
6470, 76	772	13	W				802	445	W									
	618	330	W				693	1187	W									
	695	336	W															
	541	2358	W															

NOTE—Explanation of symbols

0, 45, and 90 deg = angle of plane of coupon to plane perpendicular to loading direction.
parallel, perpendicular = roll direction parallel or perpendicular to plane of loading arms.
σ = stress = load/area of weld at plate surface in MN/m^2 (for 45-deg specimens, value quoted is equal to both the tensile and the shear components of stress).
N = number of loadings to failure.
F = mode of failure.
S = just below plate surface,
SC = just below plate surface with coarse laminations,
SM = just below plate surface with medium laminations,
D = deeper in plate,
DC = deeper in plate with coarse laminations,
DM = deeper in plate with medium laminations,
DS = partly just below plate surface, partly deeper,
L = fracture in loading arm,
LS = fracture in loading arm and crack in specimen, and
W = fracture in weld.

TABLE 5—*Mean stress under weld to give various lives (MN/m²).*

Specimen Type	Life, Cycles	Steel and Plate B 3875	B 5759	Clean B 5823	BS 968 5380	QT 35 4982	Q1 5689	Q2 6470
0 deg	1	469	429	564	439	600	658	892
	10	432	399	526	415	571	623	793
	50	409	380	503	398	547	599	723
	100	399	372	493	391	539	588	693
	500	376	354	471	375	517	564	623
	1000	366	346	462	368	508	554	593
45 deg	1	398(?)	551	653	536	710	724	931
	10	412(?)	525	618	491	641	704	866
	50	434	507	592	460	592	692	820
	100	425	499	581	446	571	686	800
	500	402	480	554	415	522	672	755
	1000	391	473	544	402	502	666	735
90 deg	1	304	248	355	238	249		
	10	263	216	306	217	202		
	50	235	194	272	201	170		
	100	223	184	258	195	157		
	500	195	161	224	180	124		
	1000	183	151	185	176	110		

pected, there were reasonably linear relationships between stress and log (cycles to failure) as may be seen in Figs. 10, 11, and 12. The lines shown are the best fit derived by a least-squares analysis.

Discussion

The general results of fatigue tests indicate that, for in-plane specimens, under constant amplitude zero-tension loading, the applied cyclic stress below yield is related exponentially to the cyclic life [2]: namely

$$\sigma N^m = C$$

or

$$\log \sigma = \log C - m \log N$$

Above yield, the failure line, in terms of a plot of log σ versus log N, is of gradually decreasing slope up to the point of intersection, at a life of one quarter cycle, with the nominal ultimate stress.

This behavior may be modified in the case of cracks which follow lamellar planes under loads normal to the plate surface or under loads which apply shear forces along the lamellar planes. This is because, at initiation and in the early stages of fatigue crack growth from a defect or inclusion, at the edge of a specimen or inside the specimen, the stress intensity will be influenced by the size, sharpness, type, and population of the inclusions [3]. Also, the internal as-rolled structure may offer preferred paths for propagation, for example, along planes containing inclusions or planes with a weaker metallurgical structure.

Thus, the actual behavior of fatigue cracks, in the short transverse sense, may differ from in-plane behavior. The stages of linear and nonlinear behavior in the log σ − log N plot may be dependent on material and clean-

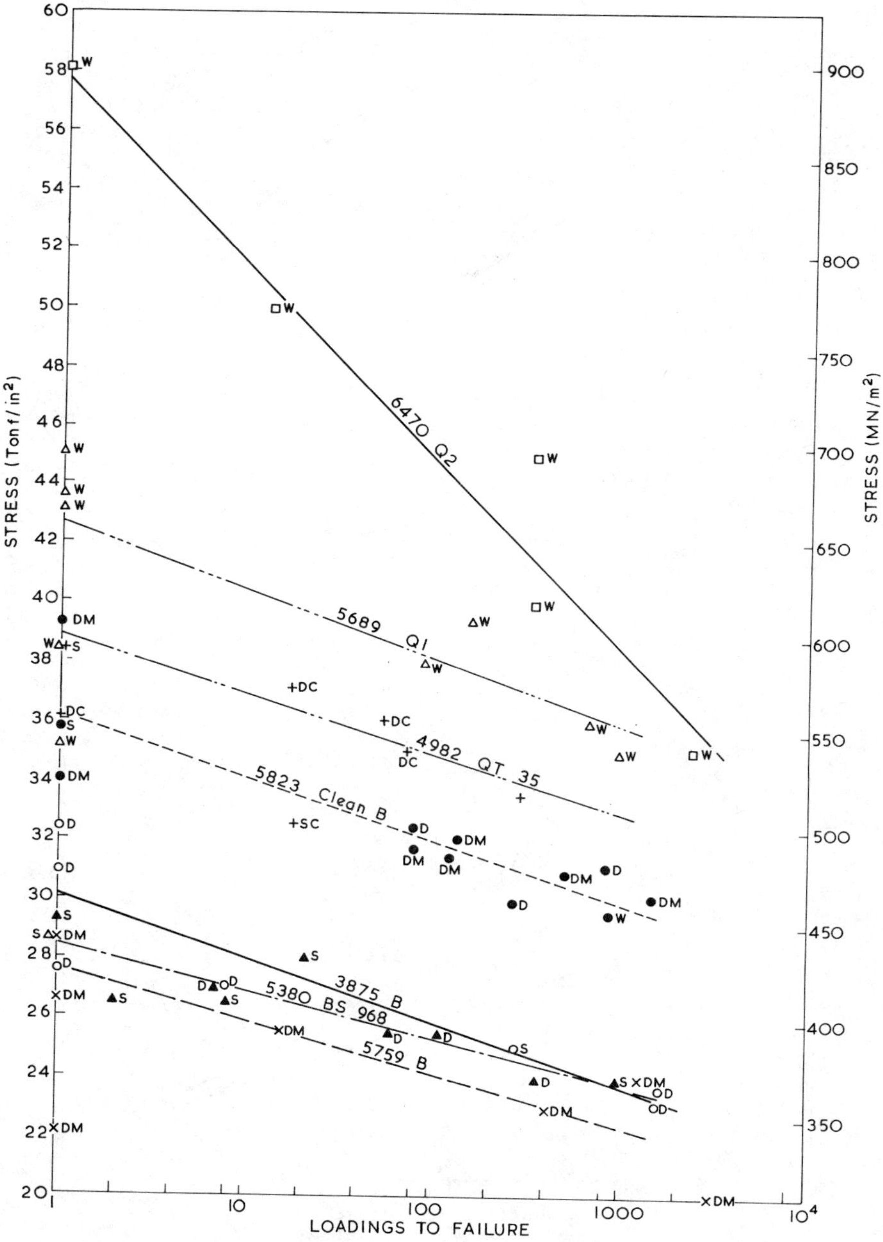

FIG. 10—*Variation of life with stress for 0-deg tests (see Table 4 for explanation of symbols).*

ness [4,5]. It was anticipated that the investigation might provide information relating to this for a number of steels tested under three conditions of loading, 0, 45, and 90 deg.

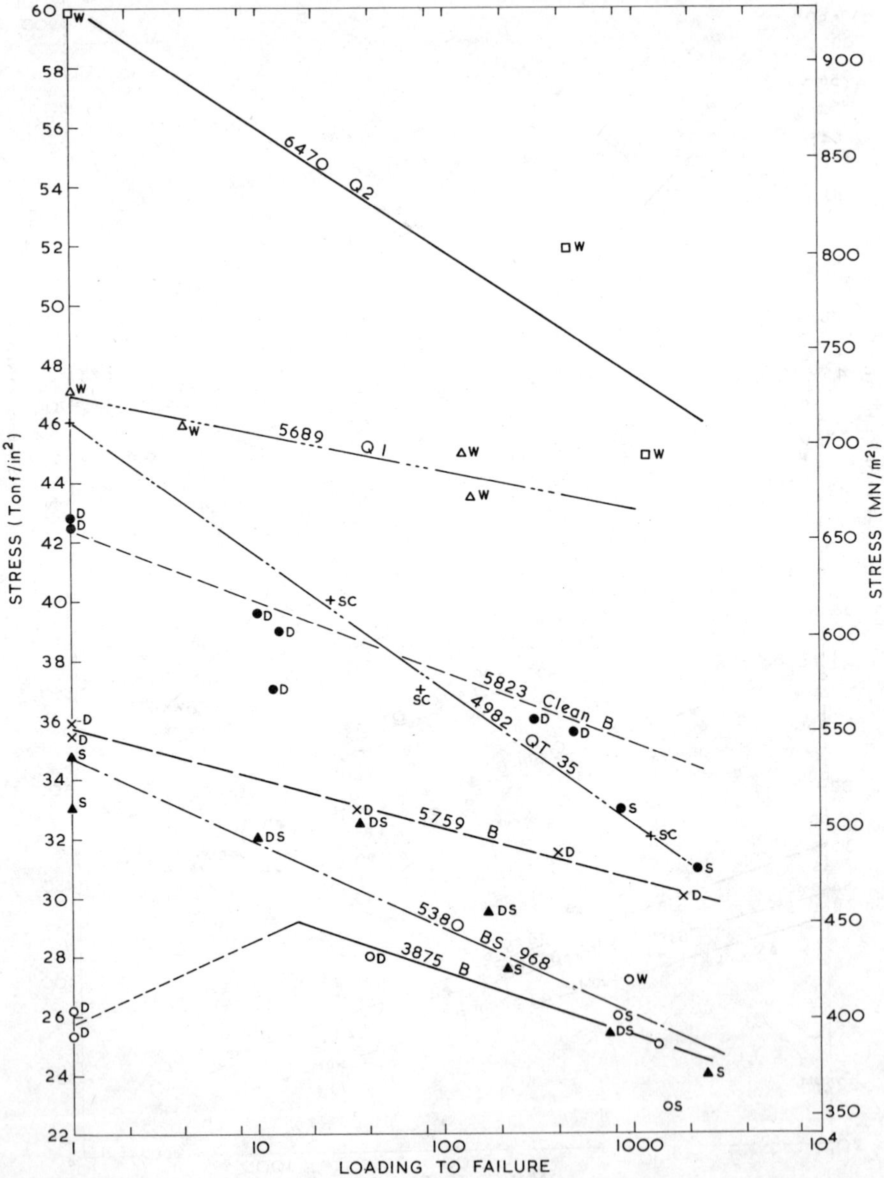

FIG. 11—*Variation of life with stress for 45-deg tests (see Table 4 for explanation of symbols).*

Before considering the actual results given in Figs. 10 through 12 and Tables 4 and 5, the following assumptions might be made in forecasting the expected behavior.

(*a*) The static strength and fatigue strength at a given life (low cycle) are functions of the in-plane tensile strength and cleanness for a given thickness.

(*b*) The mean slope of the graphs is dependent upon material strength, cleanness, method of loading, and direction of loading relative to the weld.

(*c*) Thicker steels of much the same cleanness will fail at a lower strength or have a shorter fatigue life due to larger welds.

(*d*) The 45-deg tests will give higher strength values than the 0-deg tests which will, in turn, give higher values than the 90-deg tests.

(*e*) Scatter would be greater in dirty steels than in clean steel, and 90-deg tests would, for a given steel, give more scatter than 45 or 0-deg tests because the stress concentration inherent at one end of the weld in the 90-deg specimen design would be variable.

(*f*) The strength of a specimen failing in the weld would be dependent on weld inclusions and defects.

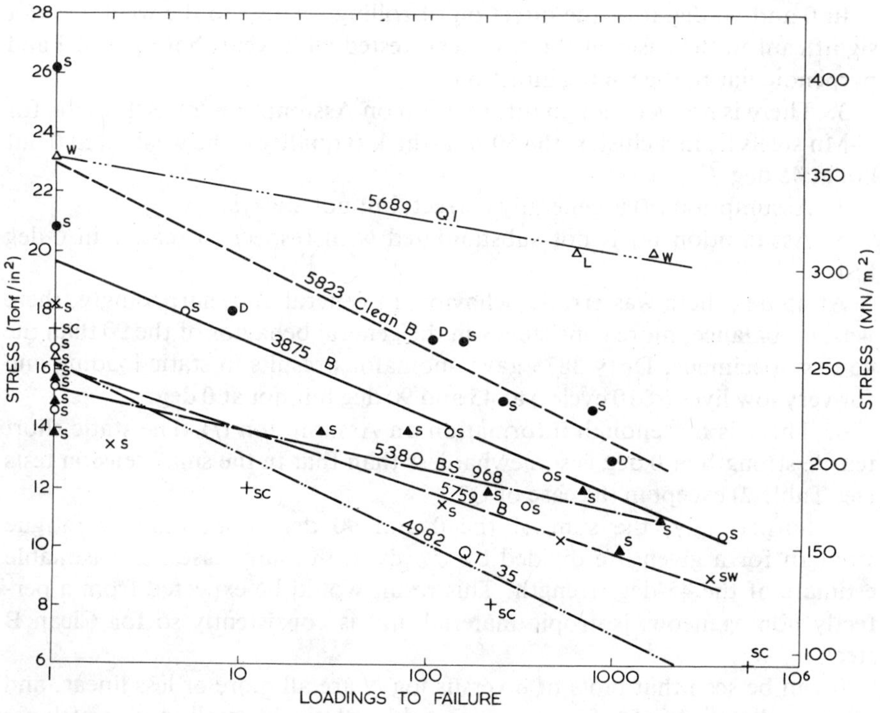

FIG. 12—*Variation of life with stress for 90-deg tests (see Table 4 for explanation of symbols).*

The results in relation to these assumptions can be compared as follows.

1. Assumption (*a*) is reasonable; Clean B steel stands out in a cluster of normalized carbon-manganese (C-Mn) steels. QT35, a dirty steel, still gives reasonable short transverse strength at 0 and 45 deg (but is lowest at 90 deg).

2. Assumption (*b*) is not valid for 0-deg tests, all lines except one (Q2) having fairly similar slopes. If σ_T is the short transverse tensile strength (as given in Table 2), then each line, except two, lies close to $\sigma = \sigma_T - 26 \log_{10} N$ where N is the fatigue life at stress σ(MN/m^2). Clearly this relationship is subject to scatter and will depend upon the population of results. The exceptions were the results for Q1, and Q2, where all fractures were in weld metal. The equations are $\sigma = 0.85\ \sigma_T - 26 \log_{10} N$ for Q1 and $\sigma = 1.04\ \sigma_T - 100 \log_{10} N$ for Q2.

In the 45-deg tests, the scatter of results precludes any firm view, but with the exception of the Q1 steel, which gave a flatter slope and where all the failures were in the weld, the slopes of the lines were in the range 26 to 46.

The 90-deg tests also show a range of slopes. (No Q2 results are yet available). The order of strengths is different from that for the 0-deg tests, the thick QT35 steel now being the weakest. Further tests are planned to discover whether there is a contribution to the failure stress due to bending stresses which are larger with thicker plates.

In 0 and 90-deg tests the direction of rolling relative to the weld was not significant in the cases of the two steels tested with welds both parallel and perpendicular to the rolling direction.

3. There is not yet enough information on Assumption (*c*). All results for C-Mn steels lie in a cluster; the 50-mm-thick B quality is the weakest at 0 but not at 45 deg.

4. Assumption (*d*) is generally correct, but not always.

5. Assumption (*e*) is not substantiated with respect to scatter in 0-deg tests.

At 45 deg there was erratic behavior, in general and, surprisingly, there was, on balance, more consistency in the general behavior of the 90 than the 45-deg specimens. Dirty 3875 gave anomalous results in static loading and for very low lives (< 10 cycles) at 45 and 90 deg but not at 0 deg.

6. There is not enough information on Assumption (*f*). The static short tensile strength at 0 deg is somewhat less than that in the small tension tests (see Table 2) except in the case of Q2.

7. Surprisingly, the sum of the 0 and 90 deg monotonic or fatigue strength for a given life divided by $\sqrt{2}$ gives, in many cases, a reasonable estimate of the 45-deg strength. This result would be expected from a perfectly homogeneous isotropic material and is consistently so for Clean B steel.

It can be seen that plots of σ versus log N are all more or less linear, and because the range of σ for any one steel is relatively small, a plot of log σ versus log N would also be approximately linear. The scatter in results, especially under static loading precludes any real distinction being made between linear and gradually decreasing slopes, supposing the latter to exist.

Conclusions

The static and fatigue results obtained indicate that the 0-deg tension tests were the most consistent and for all except one steel (Q2) might be represented by a series of parallel lines on a plot of stress against logarithm of life. The value of the static short transverse strength is close to that obtained in a normal tension test, and, in the case of B quality steel, increases with increasing cleanness.

The results for 45-deg tension and shear tests, it is suspected, would conform to a similar but steeper trend to those obtained for 0-deg tests except that there is much more scatter. The 45-deg results in many cases appear to be compatible with those at 0 and 90 deg.

The results for 90-deg shear tests likewise are subject to scatter. The order of strengths shows a similar trend as for the 0-deg tests except that one steel (thick QT35) which was almost the strongest at 0 deg was the weakest at 90 deg.

The dirty C-Mn steels gave poorer results than the clean variety, the difference being less marked in the 90-deg tests.

Except in the 90-deg tests, the dirty higher strength steel (QT 35) gave better results than the less dirty C-Mn steels; in fact, there was little difference between the strength of Clean B quality and Dirty QT 35.

These conclusions have given a tentative picture of the generalized static and low cycle fatigue behavior of welded plates subjected to short transverse tensile, combined tension and shear, and shear stress loadings even though the number of test results is limited. They have indicated the relative importance of cleanness of material in relation to failure. Steels of the C-Mn normalized type with inclusion counts in the range 38.1 to 71.4 inclusion/mm^2 were grouped together. Results for Q2 suggest that this picture may change for such higher strength steels.

Acknowledgments

The authors wish to thank their colleagues for their help in this work, in particular A. D. E. Thomson, formerly of NCRE, for his continued support and G. Wilson, NCRE, who helped to design the testing machine, built at NCRE, and procured the specimens. The specimens were welded by K. Brett and tested by R. J. Brand at Cambridge University Department of Engineering.

References

[1] Christopher, P. R. and Pascoe, K. J., *Engineering*, Vol. 210, 1970, pp. 124-125.

[2] Basquin, O. H., *Proceedings*, American Society for Testing and Materials, Vol. 10,

1910, pp. 625-630.
[3] Thomson, A. D. E., Christopher, P. R., and Bird, J., *Transactions*, American Society of Mechanical Engineers, *Journal of Engineering for Industry,* Vol. 90, 1968, pp. 627-655.
[4] Heisser, F. A. and Hortzberg, R. W., *Transactions*, American Society of Mechanical Engineers, *Journal of Basic Engineering*, Vol. 93, 1971, pp. 211-217.
[5] Christopher, P. R. and Crabbe, D. R., *Ocean Engineering*, Vol. 1, 1969, 497-520.

D. T. Raske[1]

Low-Cycle Fatigue and Cyclic Deformation Behavior of Type 16–8–2 Weld Metal at Elevated Temperature

REFERENCE: Raske, D. T., "**Low-Cycle Fatigue and Cyclic Deformation Behavior of Type 16–8–2 Weld Metal at Elevated Temperature,**" *Fatigue Testing of Weldments, ASTM STP 648*, D. W. Hoeppner, Ed., American Society for Testing and Materials, 1978, pp. 57–72.

ABSTRACT: The low-cycle fatigue behavior of Type 16–8–2 stainless steel weld metal deposited by the automatic submerged-arc process at 593°C was investigated, and the results are compared with existing data for Type 316 stainless steel base metal. Tests were conducted under axial strain control and at a constant axial strain rate of $4 \times 10^{-3}\ s^{-1}$ for continuous cyclic loadings as well as hold times at peak tensile strain. Uniform-gage specimens were machined longitudinally from the surface and root areas of a 25.4-mm-thick welded plate and tested in the as-welded condition. Results indicate that the low-cycle fatigue resistance of this weld metal is somewhat better than that of the base metal for continuous-cycling conditions and significantly better for tension hold-time tests. This is attributed to the fine duplex delta ferrite-austenite microstructure in the weld metal. The initial monotonic tensile properties and the cyclic stress-strain behavior of this material were also determined. Because the cyclic changes in mechanical properties are strain-history dependent, a unique cyclic stress-strain curve does not exist for this material.

KEY WORDS: fatigue tests, weldments, fatigue (materials), austenitic stainless steels, weld metal, cyclic deformation, elevated-temperature testing, test equipment, cyclic straining, strains, stresses

Type 16–8–2 (16Cr–8Ni–2Mo) stainless steel weld metal has been used in the fabrication of elevated-temperature structural components since the 1950s [*1*].[2] More recently, this material is also undergoing tests to determine its suitability as a filler metal for liquid metal fast breeder reactor (LMFBR) components fabricated from Type 316 stainless steel. As a result, Type 16–8–2 weld metal is well characterized in terms of its microstructural,

[1] Mechanical engineer, Materials Science Division, Argonne National Laboratory, Argonne, Ill. 60439.

[2] The italic numbers in brackets refer to the list of references appended to this paper.

TABLE 1—*Chemical composition of Type 16-8-2 stainless steel ASA weld metal (weight percent).*[a]

	C	Mn	P	S	Si	Cu	Ni	Cr	Mo	Ti	Nb	Ta	Co	N	
Weld center	0.047	1.48	0.026	0.012	0.89	0.09	10.03	14.85	1.76	0.010	0.030	0.03	0.09	0.028	<0.001	0.04
Final pass	0.047	1.39	0.029	0.014	0.87	0.04	9.27	14.43	1.80	0.013	0.003	0.03	0.09	0.028	<0.001	0.04

[a] From Ref 6.

creep-rupture, and monotonic tensile properties [*2–8*]. However, LMFBR components are operated under conditions where creep and fatigue damage can occur simultaneously, and, therefore, the creep-fatigue properties, as well as the cyclic stress-strain properties, are of interest to designers.

The principal objective of the present study was to determine the creep-fatigue and cyclic stress-strain properties of Type 16–8–2 stainless steel weld metal deposited by the automatic submerged-arc (ASA) process and compare the results with existing data on Type 316 stainless steel tested at 593°C. Another goal was to continue development of the equipment and techniques necessary to perform creep-fatigue tests on weld metals.

This paper presents the results of continuous-cycling and tension hold-time (360, 1.8×10^3, and 1.8×10^4 s) fatigue tests at 593°C on as-welded Type 16–8–2 stainless steel specimens that were machined longitudinally from the surface and root areas of the weld. The initial monotonic tensile data and the cyclic stress-strain curves are also included.

Experimental Program

Type 316 Stainless Steel Base-Line Data

Because variations between different heats and heat treatments can affect the continuous-cycling and creep-fatigue behavior of austenitic stainless steels [*9–11*], the comparison base-metal data was confined to that from one heat and heat treatment. The data used as the basis to compare the fatigue behavior of the weld metal were from Heat 65808 of Type 316 stainless steel in the solution-annealed condition [*12,13*] and are shown in Fig. 1.

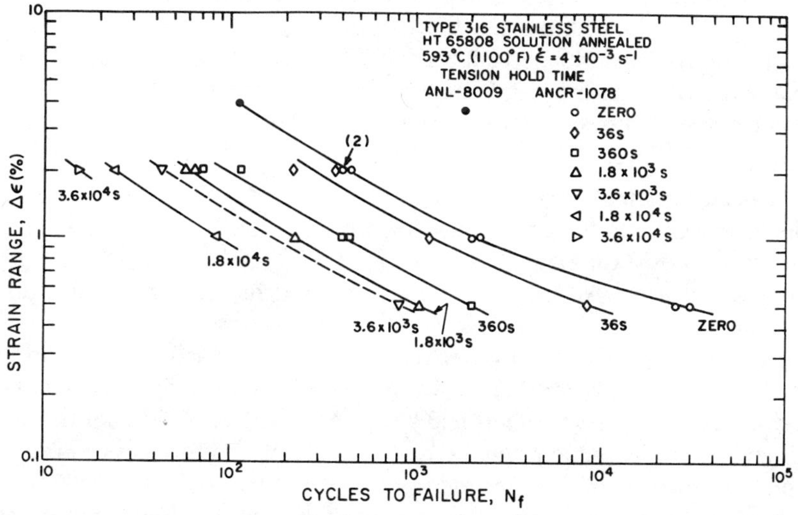

FIG. 1—*Total strain range versus cycles to failure showing effect of tension hold time on fatigue life for Type 316 stainless steel base-line data (open symbols from Ref* 12, *solid symbol from Ref* 13).

Materials and Specimens

Type 316 stainless steel base-metal plates 25.4 mm thick were welded together using the ASA process. The weld joint was a single V-groove with a 19-mm root opening and a 20-deg included angle. Approximately 20 passes were necessary to complete the weld. An analysis of the chemical composition of the Type 16–8–2 stainless steel weld metal is listed in Table 1. The microstructure of the weld metal consisted of ~0.8 to 3.0 percent ferrite distributed in an austenite matrix [*6*]. Photomicrographs of both the weld and base metal (Heat 65808) are shown in Fig. 2.

Axially loaded uniform-gage specimens in the as-welded condition were machined from the weld as shown in Fig. 3(*a*). Specimens are designated longitudinal surface (LS) and longitudinal root (LR). Dimensions of the final specimen design are shown in Fig. 3(*b*). This specimen design was used in lieu of the hourglass specimen traditionally employed in elevated-temperature tests because the weld metal is highly anisotropic and, consequently, a single diametral extensometer and strain computer cannot uniformly control the axial strain. Initially, these specimens were designed without a taper over the 16.76-mm gage section and were mechanically polished. However, room-temperature tests with strain gages in the gage section indicated the strain at the center of the straight gage section was ~15 percent lower than the strain controlled by the axial extensometer. Subsequent tests indicated that the taper shown in Fig. 3(*b*) would most nearly result in a uniform strain distribution over the gage section. Mechanical polishing of this material to obtain the surface finish necessary to avoid premature failures proved to be excessively time consuming. Mechanically polished surfaces with a roughness of $\gtrsim$0.2 μm resulted in failures that appeared to initiate at fine circumferential tool marks. Electropolishing with a solution of 60 percent phosphoric and 40 percent sulfuric acid for ~120 s after an initial mechanical polish with 600-grit paper eliminated these problems, although the minimum roughness obtainable was only 0.4 μm.

Apparatus and Test Procedure

The tests were conducted in a closed-loop hydraulic test machine in air at 593°C with a constant axial strain rate, $\dot{\varepsilon}$, of 4×10^{-3} s^{-1} using a triangular loading waveform.[3] All but one test was started in the compressive direction, usually at a strain rate of 4×10^{-4} s^{-1}. This low strain rate was used because it resulted in an *x–y* plotter record that made calculation of the elastic modulus easier. The strain rate was increased after the first quarter cycle. This procedure was later abandoned because it was believed to contribute to specimen buckling, which will be discussed later. Failure was defined as complete separation of the specimen.

The majority of the specimens were tested in a loading fixture that had a

[3] These were the same test parameters used for the comparison base-metal tests.

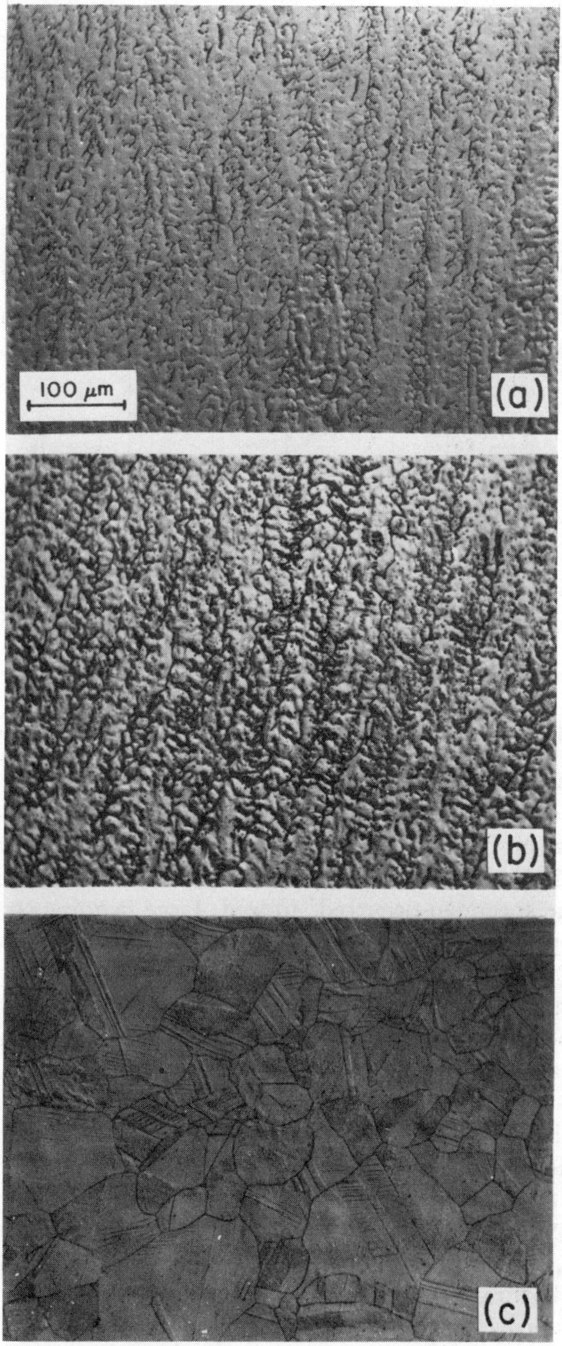

FIG. 2—*Microstructures of Type 16–8–2 stainless steel ASA weld metal and Type 316 stainless steel base metal. Oxalic acid.* (a) *Root weld metal*, (b) *surface weld metal, and* (c) *base metal.*

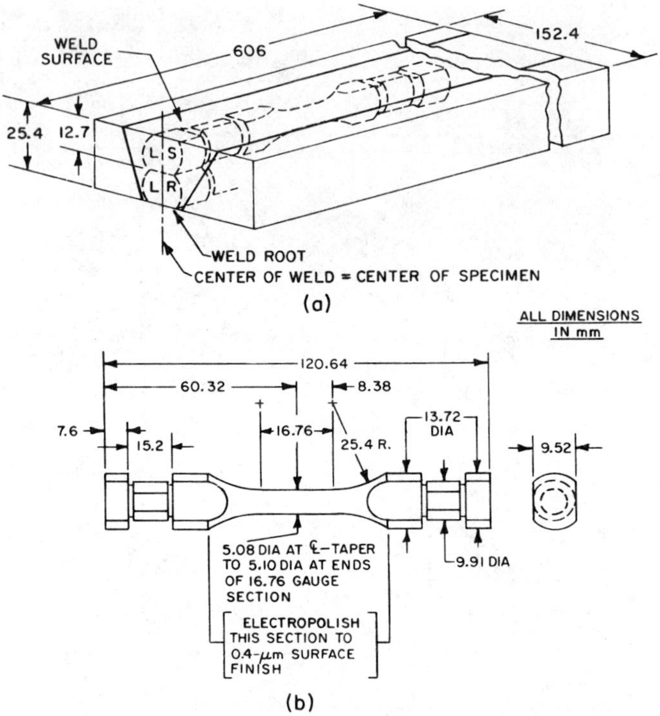

FIG. 3—*Type 16–8–2 stainless steel ASA weld-metal specimens.*

three-post die set to provide accurate alignment during tests [*14*]. A number of experimental problems were encountered with this gripping arrangement. Unless the entire load train and specimen were in exact alignment, delayed buckling of the specimen occurred. Indeed, large bending strains could be induced in a specimen if the upper clamping plate bolts were tightened improperly. For these tests, two strain gages 90 deg apart were used on each specimen in an area outside of the gage section to facilitate mounting with a minimum of bending prior to testing. This procedure was found to be effective for elimination of the buckling problems experienced in earlier tests. At present, a self-aligning liquid metal grip, shown in Fig. 4, is used for testing. Using a precision-machined alignment specimen with strain gages attached, the bending strains caused by this gripping arrangement were found to be ~2 percent of the total strain range.

Specimen heating was accomplished by an induction coil operated at 455 kHz. The coil, shown in Fig. 4, was constructed to provide a uniform temperature distribution (±4°C) over the gage section of the specimen. Power to the coil was provided by a high-frequency induction heater and was controlled by a thermocouple welded to the specimen. One additional thermocouple was placed on each side of the gage section to monitor the tempera-

FIG. 4—*Photograph of liquid metal grip showing a specimen in place surrounded by induction heating coil and the axial extensometer attached.*

ture during tests. All thermocouples were located in the transition region 11.4 mm from the center of the gage section to avoid initiation of cracks at the thermocouple welds. The temperature profile over the gage section and the control thermocouple temperature were determined from 16 equally spaced thermocouples welded to a facsimile Type 316 stainless steel specimen. However, a later temperature profile on a Type 16–8–2 stainless steel specimen indicated that the control thermocouple on specimens of this material must be operated at a higher temperature to obtain the desired 593°C over the gage section. Consequently, several tests were conducted at a gage- section temperature of 584°C. This, however, did not appear to affect the results, as will be shown in the next section.

An axial extensometer with a gage length of 11.68 mm was used to control longitudinal strain in the specimen. This extensometer, shown in Fig. 4, is supported by counterweights, and the extensometer tips are held in place on the specimen by friction. To date, no slippage has been observed nor have any failures initiated under the extensometer tips. Moreover, the locations of the crack-initiation sites around the circumference and along the gage section of the specimens are random.

Results and Discussion

Continuous-Cycling and Tension Hold-Time Fatigue Tests

The results of this portion of the investigation for ten surface weld and eight root weld metal specimens are listed in Table 2. A comparison of these results and the strain-life curves from the Type 316 stainless steel base-line data is shown in Fig. 5. As observed in this figure, the resistance of the weld metal to the loadings is greater than that of the base metal. One significant exception to this observation is a continuous-cycling test at a strain range of 1 percent (Specimen LS-2). This specimen was mechanically polished, and, although it is difficult to determine, the failure is believed to have initiated at a fine circumferential tool mark in the gage section. As indicated in Table 2, the fatigue lives for the specimens tested at 584 and 593°C are not signif-

TABLE 2—*Results[a] of fatigue tests on Type 16-8-2 stainless steel ASA weld metal as-welded longitudinal specimens at 593°C and* $\dot{\varepsilon} = 4 \times 10^{-3}\,s^{-1}$.

Specimen	Location in Weld	Axial Strain Range, % $\Delta\varepsilon_t$	Axial Strain Range, % $\Delta\varepsilon_p$	Tensile Hold Time, s	Stress at $N_f/2$, MPa $\Delta\sigma$[b]	Stress at $N_f/2$, MPa $\Delta\sigma_r$	Fatigue Life N_f	Fatigue Life t_f, ks	Remarks
LR-12	root	2.03	1.43	0.0	302.6/299.7	...	367	3.70	...
LS-7	surface	1.95	1.32	0.0	306.2/303.3	...	636	6.41	*c*
LS-13	surface	1.50	0.95	0.0	262.5/267.6	...	1 407	11.04	*d,e*
LR-4	root	1.50	1.00	0.0	255.6/250.7	...	1 567	11.88	...
LS-2	surface	0.99	0.42	0.0	245.3/243.1	...	1 335	6.78	*d, e* mechanically polished
LS-4	surface	1.00	0.52	0.0	246.3/255.5	...	2 836	14.22	*d,e*
LR-5	root	1.00	0.53	0.0	233.3/238.8	...	3 634	18.06	*e*
LS-12	surface	0.51	0.15	0.0	187.4/191.0	...	27 906	69.78	*d,e*
LR-3	root	0.50	0.12	0.0	185.7/170.9	...	31 155	77.58	*e*
LS-15	surface	0.51	0.15	0.0	172.3/172.3	...	67 021[f]	168.84	*d,e*
LR-13	root	1.00	0.61	360	198.2/218.3	397.1	1 747	637.32	...
LS-14	surface	1.00	0.64	360	185.8/219.6	383.8	2 759[g]	1 006.38	*d,e*
LR-7	root	0.51	0.16	360	176.6/203.3	374.9	5 874	2 128.98	...
LS-6	surface	0.50	0.17	360	166.1/154.9	312.8	9 947	3 605.80	...
LS-10	surface	1.50	0.74	1.8×10^3	368.7/366.5	682.3	329	592.86	...
LR-11	root	1.00	0.63	1.8×10^3	209.9/288.8[h]	414.0[h]	>972[i]	>1 750.25[i]	...
LR-10	root	2.03	1.54	1.8×10^4	237.7/246.5	421.1	>85[j]	~1536	...
LS-8	surface	1.02	0.62	1.8×10^4	205.4/200.4	367.6	251[k]	4 501.30[k]	...

[a] $\Delta\varepsilon_t$ = total strain range, $\Delta\varepsilon_p$ = plastic strain range, $\Delta\sigma$ = total stress range, $\Delta\sigma_r$ = relaxed stress range, N_f = cycles to failure, and t_f = time to failure.

[b] Peak tensile stress/peak compressive stress.

[c] Test started in tensile direction.

[d] Tested at 584°C.

[e] First quarter cycle at $\dot{\varepsilon} = 4 \times 10^{-4}\,s^{-1}$.

[f] Failed at tool mark in transition region; no visible cracks in gage section.

[g] Machine shut down before failure; many large cracks in gage section.

[h] Values for cycle 252.

[i] Specimen removed unbroken; specimen overstrained at cycle 253 as a result of programmer malfunction.

[j] Test stopped before failure as a result of programmer malfunction; several cracks in gage section.

[k] Specimen removed unbroken; no visible cracks in gage section.

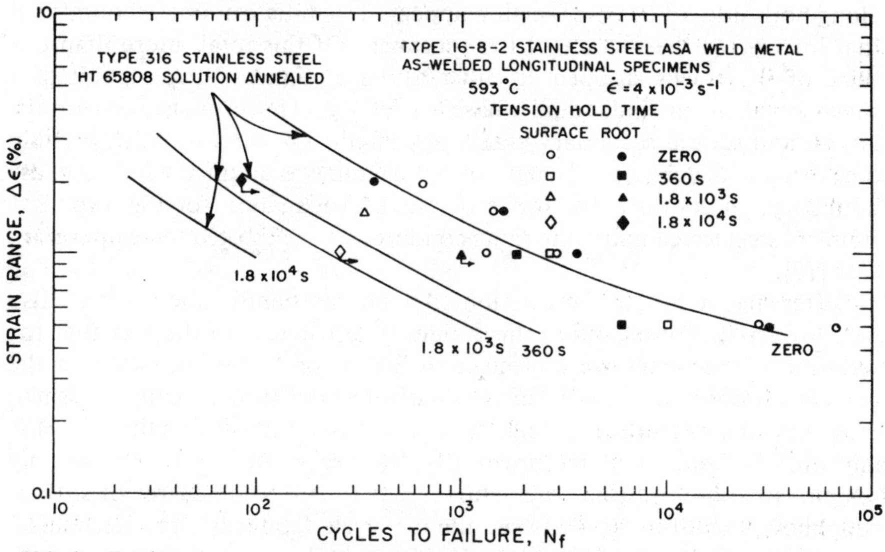

FIG. 5—*Total strain range versus cycles to failure for Type 16–8–2 stainless steel ASA weld metal compared with Type 316 stainless steel baseline data.*

icantly different, and in several cases specimens tested at 584°C had shorter fatigue lives than those tested at 593°C.

The results of the 360, 1.8 × 10^3, and 1.8 × 10^4–s tension hold-time tests are particularly significant because the fatigue life for the weld metal is from three to seven times longer than that for the base metal when tested under the same conditions. This was unexpected since the creep-rupture properties of the weld metal are significantly poorer than the average properties for Type 316 stainless steel at 566 and 650°C [*6*]. Moreover, it has been statistically established that heats of a similar austenitic stainless steel base metal, Type 304, with poor creep properties also tend to exhibit poorer than average creep-fatigue properties [*10*]. Thus, the correlation between creep and creep-fatigue behavior observed in base-metal specimens is apparently not valid for weld metals.

One mechanism that may explain this difference is the mode of fatigue crack propagation in these materials. The results of crack-growth-rate studies on stainless steel weld- and base-metal specimens indicate that the weld metal has significantly lower crack-growth rates than the base metal, particularly at 593°C [*15–17*]. This difference is attributed to the fine duplex delta ferrite-austenite microstructure in the weld metal which, with its many phase boundaries, inhibits the growth rate of cracks. Evidence that the fine microstructure of the weld metal is responsible for the observed differences in the fatigue lives of this material and the base metal was obtained from a specimen (LS–14) which was sectioned normal to the main crack and metallographically examined over the entire gage section. All 88 cracks observed along both sides of the gage section appeared to initiate at phase boundaries that intersected the surface of the specimen. Of this total, more than one third of the cracks stopped at either a phase-boundary triple point or a sharp break in an individual phase boundary. The propagation for the largest and several secondary cracks appeared to be almost entirely within the austenite matrix after a length of ~0.03 mm was attained within a phase boundary. In contrast, the mode of crack propagation for the Type 316 stainless steel tested under the same conditions was reported to be intergranular [*12*].

Differences in fatigue lives attributed to nonmechanistic factors must also be considered. For example, they might be attributed to the fact that the weld-metal specimens were electropolished prior to testing, whereas the base-metal specimens used for comparison were mechanically polished. However, the base-metal specimens were polished longitudinally to a 0.2-μm finish [*18*] and, as stated previously, the electropolished finish was only 0.4 μm. In addition, it has been shown that for the same degree of surface roughness, the difference between the low-cycle fatigue life for mechanical and electropolished specimens is insignificant [*19*].

Another potential source for the dissimilarity between these data may be the method of beginning the tests. All but one of the Type 316 stainless steel tests were begun with the first quarter cycle in tension. In contrast, all but one of the weld-metal tests were begun in the compressive direction. Although it has been reported that tests begun in tension would result in somewhat lower fatigue lives [*14*], recent studies [*10*] indicate that no statistically significant difference exists between tests begun in either manner.[4]

Finally, size effects also can be eliminated as a contributing factor to the observed differences in fatigue lives of the base and weld metal. A size effect based on the volume of the specimen has been shown to exist, but the volume differences must be larger than those in the present study [*20*]. Moreover, the specimens with the largest volume (the uniform-gage weld-metal specimens) would have the shortest fatigue lives, which was not observed.

Initial Monotonic Tensile Data and Cycle-Dependent Changes in Mechanical Properties

The monotonic 0.2 percent offset yield strengths and the elastic modulus for the Type 16–8–2 stainless steel ASA weld metal as determined from the initial portion of the first-cycle hysteresis loops are listed in Table 3. These results indicate that the metal from the root of the weld has a higher yield strength than that from the weld surface. This is consistent with results obtained at Oak Ridge National Laboratory for the material at 565 and 650°C [*6*]. A prominent feature in the present data is the large scatter associated

[4] Unpublished research by the author also supports this conclusion.

TABLE 3—*Type 16–8–2 stainless steel ASA weld-metal monotonic and cyclic tensile data from continuous-cycling fatigue tests on as-welded longitudinal specimens at 593°C and* $\varepsilon = 4 \times 10^{-3}$ *and* $4 \times 10^{-4}\ s^{-1}$.

Location in Weld	Monotonic 0.2% Offset Yield Strength (σ_{ys}), MPa	Elastic Modulus (E) × 10^3, MPa	Cyclic Properties[a] 0.2% Offset Yield Strength (σ_{ys}'), MPa	Strength Coefficient (K'), MPa	Strain Hardening Exponent (n')
Surface	214.1[b]	99.10[c]	231.9/270.3	688.0/475.3	0.175/0.094
Root	237.2[d]				

[a] Upper value from companion tests at constant strain ranges, lower value from monotonic tension tests after spectrum straining.
[b] Standard deviation = 40.90 MPa.
[c] Standard deviation = 5.75 MPa.
[d] Standard deviation = 18.88 MPa.

with the yield strength of the surface weld metal (more than twice that of the root weld metal). The scatter is probably due to differences in composition, structure, and local properties that result from the variety of thermal histories which occur in various locations of multipass weldments [2,4]. As expected, the elastic modulus exhibits relatively little scatter for these tests and is not statistically different for the surface and root weld metal.

The cyclic hardening and subsequent softening of the weld metal as a function of strain range and cycles are shown in Figs. 6 and 7. Note the widely differing values for the surface weld-metal stress range in Fig. 6. This is consistent with the large scatter in the monotonic yield-strength values discussed earlier. Beyond the last data point shown in these figures, the tensile stress amplitude drops to zero as failure of the specimen occurs.

The cyclic stress-strain curves shown in Fig. 8 were determined by two methods [21]. The lower curve (solid line) was obtained from the constant-amplitude continuous-cycling fatigue test data. A linear regression analysis was performed on the logarithms of the half-life stress $\Delta\sigma/2$ and plastic strain amplitudes from fatigue tests on both the surface and root weld metal. The results were of the form

$$\Delta\sigma/2 = K'(\Delta\varepsilon_p/2)^{n'} \tag{1}$$

where

K' = cyclic strength coefficient,
$\Delta\varepsilon_p$ = plastic strain range, and
n' = cyclic strain-hardening exponent.

Values of K', n', and the cyclic 0.2 percent offset yield strength are listed in Table 3. The cyclic stress-strain curves was subsequently calculated for the strain amplitude $\Delta\varepsilon/2$ using the relation [22]

$$\Delta\varepsilon/2 = \Delta\varepsilon_e/2 + \Delta\varepsilon_p/2 = \Delta\sigma/2E + (\Delta\sigma/2K')^{1/n'} \tag{2}$$

where

$\Delta\varepsilon_e$ = elastic strain range and

E = elastic modulus.

Initially, this exercise was performed for the surface and root weld metal individually. However, little difference exists between the calculated curves, particularly at low strains (<0.5 percent); thus, the data were combined to

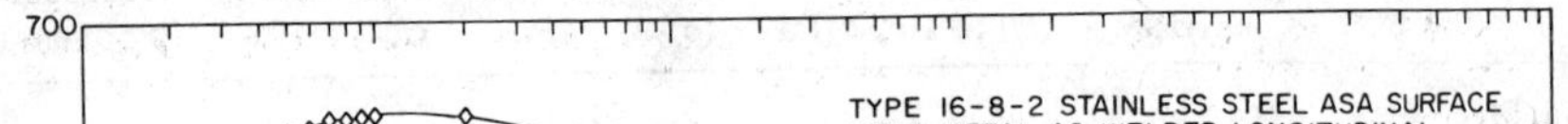

FIG. 6—*Changes in stress range during reversed strain cycling for surface weld metal (specimen numbers in parentheses).*

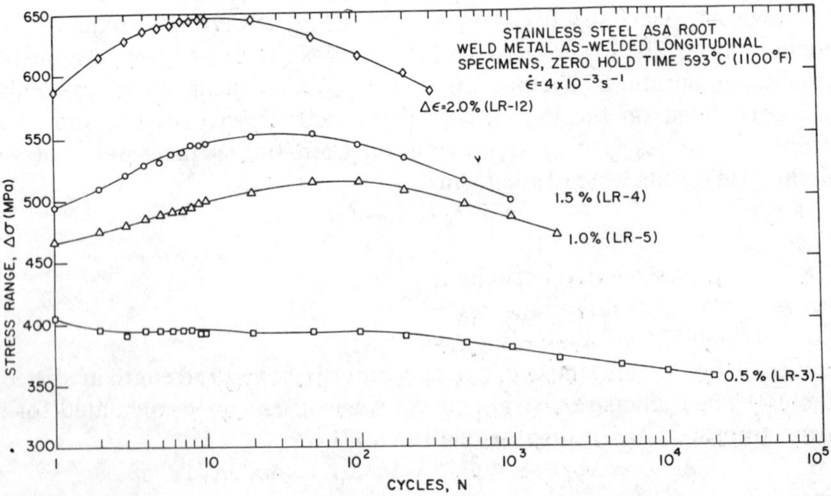

FIG. 7—*Changes in stress range during reversed strain cycling for root weld metal (specimen numbers in parentheses).*

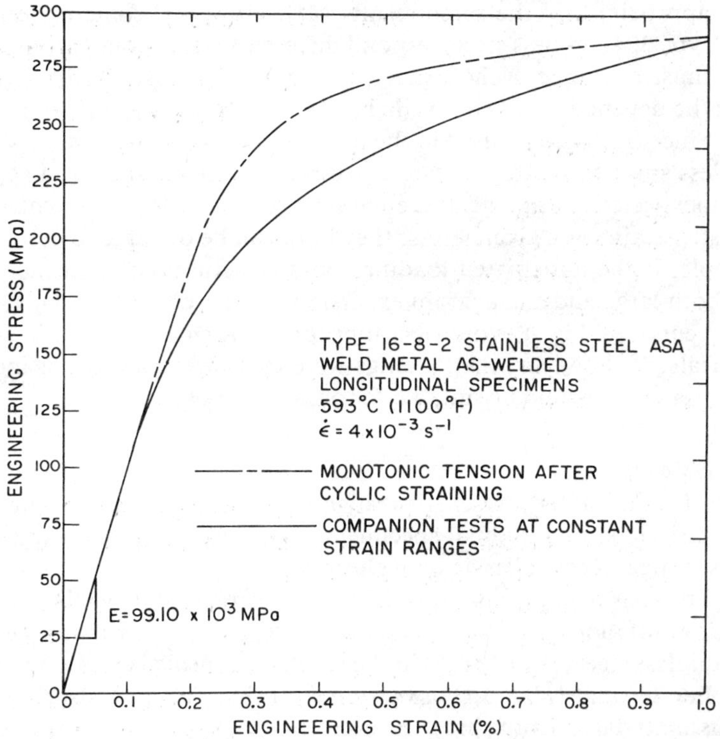

FIG. 8—*Cyclic stress-strain curves from constant-amplitude continuous-cycling fatigue tests and monotonic tension test after cyclic straining.*

produce a single curve. At strains >0.5 percent (strain ranges >1.0 percent), the surface weld metal has slightly higher cyclic strength values, but the difference is not sufficiently large nor do adequate data exist to justify separate cyclic stress-strain curves.

The upper curve (broken line) in Fig. 8 was obtained from a monotonic tension test after cyclic straining on a single root weld-metal specimen. The preliminary straining consisted of blocks of gradually increasing and then decreasing strain amplitudes, that is, incremental step tests interspersed with blocks of constant strain-range cycling. A total of ~2100 cycles at strain ranges between 0.2 and 2.0 percent and a constant axial strain rate of 4×10^{-3} s^{-1} were imposed on the specimen prior to the monotonic tension test. Values of K' and n' for this stress-strain curve were determined by back-fitting incremental values for the logarithms of stress and plastic strain from the curve in Fig. 8 to the form of Eq 1 by a linear regression analysis. These values, as well as the 0.2 percent offset yield strength, are also listed in Table 3.

Normally the cyclic stress-strain curve for a given material can be generated by several methods, including those just described, which usually produce approximately the same results [*21*]. However, some materials, notably Type 304 stainless steel, respond differently to various loading histories and, thus, a unique cyclic-strain curve does not exist [*9,23*]. This results from the dependence of the cyclic hardening of the material on loading history. Since the material used in the present investigation is also an austenitic stainless steel, the different cyclic stress-strain curves shown in Fig. 8 were not unexpected. Although these curves provide widely different values of inelastic strain for a given stress, they both can be of value to designers. For example, if the anticipated loadings on a structural component were such that both large and small strain excursions would occur randomly, then the upper curve of Fig. 8 would be appropriate for characterization of these materials. If, however, the loadings were cyclically constant or increasing, the lower curve would be suitable for design purposes.

Conclusions

The results of the present investigation on Type 16–8–2 stainless steel ASA as-welded longitudinal specimens tested in air at 593°C and $\dot{\varepsilon} = 4 \times 10^{-3} s^{-1}$ suggest the following conclusions.

1. The resistance of this material to axial strain-controlled constant-amplitude continuous-cycling fatigue test loadings is greater than that of Type 316 stainless steel (Heat 65808) in the solution-annealed condition.

2. For 360 and 1.8×10^4 – s tension hold-time fatigue tests, the cyclic life of this material is from three to seven times longer than that of Type 316 stainless steel. This is attributed to the fine duplex microstructure of the weld metal that inhibits the growth rate of cracks.

3. The correlation between resistance to static creep rupture and resistance to creep-fatigue failures previously observed in Types 304 and 316 stainless steel base metals is not observed for this weld metal.

4. The monotonic 0.2 percent offset yield strength is higher and more consistent in the root weld metal than in the surface weld metal. However, the elastic moduli for specimens from both locations in the weld are identical.

5. This material is strain-history dependent, and, therefore, a unique cyclic stress- strain curve does not exist. Monotonic tension tests after cyclic straining results in a different stress-strain curve than obtained from companion fatigue tests at various completely reversed constant strain ranges.

Acknowledgments

The author is grateful for the help of J. C. Tezak, C. F. Peterson, and M. D. Gorman in the test program and data reduction. Special appreciation is due to W. F. Burke for his assistance in resolving the many experimental difficulties encountered in this investigation. Appreciation is also due to R. T. King of Oak Ridge National Laboratory for aid in procuring the material and for helpful discussions on the properties of weld metals. I also wish to thank D. R. Diercks and R. W. Weeks for their support and encouragement.

This work was supported by the U.S. Department of Energy.

References

[1] Carpenter, O. R. and Wylie, R. D., *Metal Progress*, Vol. 70, No. 5, Nov. 1956, pp. 65-73.

[2] Ward, A. L., Irvin, J. E., Blackburn, L. D., Slaughter, G. M., Goodwin, G. M., and Cole, N. C., "Preparation, Characterization, and Testing of Austenitic Stainless Steel Weldments," HEDL-TME-71-118, Hanford Engineering Development Laboratory, Aug. 1971.

[3] Goodwin, G. M., Cole, N.C., and Slaughter, G. M., *Welding Journal*, Research Supplement, Vol. 51, No. 9, Oct. 1972, pp. 425s-429s.

[4] Ward, A. L., "Austenitic Stainless Steel Weld Materials—A Data Compilation and Review," HEDL-TME-74-25, Hanford Engineering Development Laboratory, May 1974.

[5] King, R. T., Cole, N. C., Berggren, R. G., and Goodwin, G. M., "Properties and Structure of 16-8-2 Stainless Steel Weld Metal," ASME Publication No. 75-PVP-45, presented at the Second National Congress on Pressure Vessels and Piping, San Francisco, 23-27 June 1975.

[6] King, R. T. and Edmonds, D. P., "Mechanical Properties Test Data for Structural Materials Quarterly Progress Report for Period Ending April 30, 1976," ORNL-5150, Oak Ridge National Laboratory, pp. 135-146.

[7] Ward, A. L. and Blackburn, L. D., *Journal of Engineering Materials and Technology*, Vol. 98, No. 3, July 1976, pp. 213-220.

[8] Brinkman, C. R., Sikka, V. K., and King, R. T., "Mechanical Properties of LMFBR Primary Piping Materials," ORNL-5199, Oak Ridge National Laboratory, Nov. 1976.

[9] Brinkman, C. R. and Korth, G. E., "Heat-to-Heat Variations in the Fatigue and Creep-Fatigue Behavior of AISI Type 304 Stainless Steel at 593°C," *Journal of Nuclear Materials*, Vol. 48, 1973, pp. 293-306; also ANCR-1097, Aerojet Nuclear Company, May 1973.

[10] Diercks, D. R. and Raske, D. T., "Elevated-Temperature Strain-Controlled Fatigue Data on Type 304 Stainless Steel—A Compilation, Multiple Linear Regression Model, and Statistical Analysis," ANL-76-95, Argonne National Laboratory, Dec. 1976.

[11] Maiya, P. S. and Majumdar, S., "Mechanical Properties Test Data for Structural Materials Quarterly Progress Report for Period Ending October 31, 1976," ORNL-5237, Oak Ridge National Laboratory, pp. 23-38.

[12] Brinkman, C. R., Korth, G. E., and Beeston, J. M., "Comparison of the Fatigue and Creep-Fatigue Properties of Unirradiated and Irradiated Type 304 and 316 Stainless Steel at 593°C (1100°F)," ANCR-1078, Aerojet Nuclear Company, Aug. 1972.

[13] Weeks, R. W., Diercks, D. R., and Cheng, C. F., "ANL Low-cycle Fatigue Studies—Program, Results, and Analysis," ANL-8009, Argonne National Laboratory, Nov. 1973.

[14] Slot, T., Stentz, R. H., and Berling, J. T. in *Manual for Low Cycle Fatigue Testing, ASTM STP 465,* American Society for Testing and Materials, 1969, pp. 100-128.

[15] James, L. A., *Welding Journal*, Research Supplement, Vol. 52, No. 4, April 1973, pp. 173s-179s.

[16] Shahinian, R., Smith, H. H., and Hawthorne, J. R., *Welding Journal*, Research Supplement, Vol. 51, No. 11, Nov. 1972, pp. 527s-532s.

[17] Raske, D. T. and Cheng, C. F., *Nuclear Technology*, Vol. 34, No. 1, June 1977, pp. 101-110.

[18] Beeston, J. M. and Brinkman, C. R. in *Irradiation Effects on Structural Alloys for Nuclear Reactor Applications, ASTM STP 484*, American Society for Testing and Materials, 1970, pp. 419-450.

[19] Maiya, P. S. and Busch, D. E., *Metallurgical Transactions A*, Vol. 6A, Sept. 1975, pp. 1761-1766.

[*20*] Raske, D. T., "Section and Notch Size Effects in Fatigue," T&AM Report No. 360, University of Illinois, Urbana, Ill., Aug. 1972.

[*21*] Landgraf, R. W., Morrow, JoDean, and Endo, T., *Journal of Materials*, Vol. 4, No. 1, March 1969, pp. 176-188.

[*22*] Raske, D. T. and Morrow, JoDean in *Manual on Low Cycle Fatigue Testing, ASTM STP 465*, American Society for Testing and Materials, 1969, pp. 1-25.

[*23*] Jaske, C. E., Mindlin, H., and Perrin, J. S. in *Cyclic Stress-Strain Behavior Analysis, Experimentation, and Failure Prediction, ASTM STP 519*, American Society for Testing and Materials, 1973, pp. 13-27.

Don Webber [1]

Evaluation of Possible Life Improvement Methods for Aluminum-Zinc-Magnesium Fillet-Welded Details

REFERENCE: Webber, Don, "**Evaluation of Possible Life Improvement Methods for Aluminum-Zinc-Magnesium Fillet-Welded Details,**" *Fatigue Testing of Weldments, ASTM STP 648,* D. W. Hoeppner, Ed., American Society for Testing and Materials, 1978, pp. 73–88.

ABSTRACT: Fillet welds have a low fatigue resistance compared with other types of weld detail, and it is difficult to design structures without using them. Results are presented of screening tests on methods of life improvement that have been applied successfully to fillet welds in steels under fluctuating tensile loading and are applicable to heat-treatable aluminum alloys. The welding method or shielding gas used did not affect fatigue life: (*a*) improvement of the weld profile during or after welding resulted in slightly increased lives; (*b*) local peening treatments had to be sufficiently intense to produce compressive residual stresses some depth below the surface to cause a worthwhile increase in life; and (*c*) specimen overloading treatments applied both initially and periodically generally were effective but the results for exclusion of the atmosphere by a conventional paint system were inconsistent. Results of tests on as-welded specimens, at various stress ratios which include compression in the loading cycle, corresponded with the results of treatments which include compressive residual stresses. The number of tests of each method was small, and further investigation of the more promising methods is recommended before they can be applied generally without qualifying tests.

KEY WORDS: fatigue tests, weldments, aluminium alloys, fillet welds, fatigue life, life improvement, welding, weld profile, peening, overloading, painting

It is well known that the fatigue strength of a welded structure containing fillet-welded joints in highly stressed regions is low in comparison with the static strength of the parent metal. This is a direct result of the relatively high stress concentration which such joints produce. In general, it is difficult to design structures without using fillet-welded joints and, therefore, it

[1] Principal scientific officer, Military Vehicles and Engineering Establishment, Christchurch, Dorset, England.

is of considerable value to devise methods of improving fatigue strength of such joints.

Measures often can be taken to improve the fatigue strength of a joint when the determining features have been established. Fatigue failures in welded joints generally occur by crack propagation from external or internal stress concentrations and defects, the most common being from the external stress concentration caused by the welded toe and any defects it may contain. Methods based on the following three principles have been found to produce improvements in fatigue strength for weld toe failures: (*a*) improvement of weld profile, (*b*) modification of the effective stress system by stress relief or preferably the introduction of compressive residual stresses, and (*c*) exclusion of the environment. In addition, it should be possible to improve fatigue strength against failure from internal stress concentrations and defects by using appropriate methods based on the second principle.

It should be noted that the effect of using a local improvement technique may be to change the location of failure, rather than to increase fatigue strength by a significant amount. For instance, a load-carrying weld with an improved weld profile may fail from an internal defect and give approximately the same fatigue strength as the as-welded joint [*1*].[2] A considerable amount of work on methods for improving the fatigue strength of fillet-welded joints in steel structures has been carried out in the past, and much of this work has been reviewed elsewhere [*2*]. However, few comparable tests appear to have been carried out on similar fillet-welded joints in aluminium alloys, and as far as is known no comparable data exist for aluminum-zinc-magnesium (Al-Zn-Mg) alloys. Therefore, the objective of this work was to "screen" a number of possible improvement methods by carrying out fatigue tests on fillet-weld joints for a variety of treatments with a view to define which methods are most likely to be worth further study.

Improvement Methods

Improvement of Weld Profile

A stress concentration results from the change of the cross section at the edge of a fillet weld on a stressed plate and may be expected to have a major influence on the fatigue strength of the joint. The magnitude of the stress concentration depends somewhat on the weld size compared with the plate thickness. However, a more important feature is the shape of the weld profile, a low contact angle resulting in a low stress concentration factor. This is illustrated by the results of experimental and theoretical stress analyses of transverse fillet welds, summarized in Ref *3*, which shows that the stress concentration factor at the weld toe can vary from 1.5 to 4.5 for contact

[2] The italic numbers in brackets refer to the list of references appended to this paper.

angles varying from 10 to 70 deg. The following methods for improving the weld profile and so reducing the stress concentration have been tried.

Toe Grinding—Many investigations have shown that an improvement with steel can be achieved by machining or grinding the weld to give a gradual transition from weld to plate surface [*2*]. This was repeated in the present investigation using a tungsten carbide burr described as "rounded tree" with an end radius of 1.2 mm rotating at 5 000 to 15 000 rpm. Care was taken to ensure that the grinding marks at the weld end were parallel to the direction of stressing so that additional stress concentrations were not introduced by machining marks. The typical shape of the weld after this treatment is shown in Fig. 1. Grinding had the combined effect of improving poor shapes and removing any sharp inclusions embedded in the weld toe as a result of the weld process. It also exposed porosity, but unlike dressed butt welds this was not observed to be a site for crack initiation.

Gas Tungsten Arc (GTA) Dressing—This method was explored since it has been used with success on steels, welded by other methods, to produce a significant modification to the weld profile and also to remove entrapped inclusions at the weld toe [*4*].

However, postweld treatments of this type are expensive and sometimes difficult to carry out if access to the weld toe is restricted. It would be useful if it were possible to produce welds with good profiles by suitable control during welding.

Welding Method—The gas metal arc (GMA) process is generally used on Al-Zn-Mg alloys, but, in addition, the GTA welding method was used. Also helium shielding gas was used with the GMA method instead of the more usual argon as shown elsewhere [5] for NS8 aluminium alloy (United States equivalent 5083) that it resulted in improved profiles and hence the possibility of improved fatigue life.

Weld Angle—Finally, because of analytical results, results presented by Maddox [*3*], and Christopher and Crabbe's previous experience with steel [*6*], welds were produced having the angles of 15 and 30 deg to the stressed plate in addition to the conventional 45 deg.

Introduction of Compressive Residual Stresses

Three methods of introducing compressive residual stresses are: mechanical working, loading to cause local yielding at stress concentrations, and local heating [*2*]. Mechanical working is useful where it can be applied locally to improve unavoidable details that have a low fatigue strength, but it tends to be expensive if a whole structure requires treatment. Loading, or perhaps more correctly overloading, is attractive because it can be done under controlled conditions and the whole structure is treated in one operation. Overloading has been shown to be particularly effective in improving the fatigue life of mechanical joints [*8*]. Local heating has not been used in view of the sensitivity of the alloy to heat treatment making temperatures above 180 °C impractical.

The following methods were used.

1. *Shot Peening*—This method has been shown to be effective on Al-Zn-Mg alloy butt welds [*1*]. Thirty-grade shot was used at an air pressure of 7

FIG. 1—*Typical weld shapes after local grinding of the weld toe.*

bar with the gun at right angles to the weld surface, and three passes were made to obtain complete coverage.

2. *Hammer Peening*—The efficiency of hammer peening has been demonstrated on steel fillet welds [*6*] and Al-Zn-Mg alloy butt welds [*9*]. It was carried out with a pneumatic hammer fitted with a solid tool which had a round end of approximately 6-mm radius. The piston diameter was 17.5 mm and the air pressure, 7 bar. The gun was traversed approximately 450 mm/min along the weld toes at the ends of the specimen only. This was repeated three times. The finished profile obtained is shown in Fig. 2.

3. *Prior Overloading*—This was carried out by applying either one or 10 "static" stress cycles before fatigue testing. A stress of 277 N/mm^2, applied for 5 s was based on an effective proof stress of the specimen taking account of the reduced strength of the heat affected zone.

4. *Periodic Overloading*—This was expected to reinforce the residual stress field and the stress concentration to give a greater effect than prior overloads. Single or 10 overloads of the same magnitude as the prior overloads were applied at every 1000 cycles. In all cases the tests began with the overload. The interval was chosen to correspond to an annual inspection at the anticipated rate of use of a particular equipment.

FIG. 2—*Typical weld shape after hammer peening of the weld toe.*

Exclusion of the Environment

Grit Blasting and Painting—In this test series the welds were finished in a way which is used in practice. Usually, fatigue specimens are tested in the as-welded condition and no account is taken of any possible influence of protective finish. The specimens were prepared according to Director General of Fighting Vehicles and Equipment (DGFVE) Specification 235/1, as used for actual Al-Zn-Mg alloy military bridges. This involved grit blasting at 3 bar with nonmetallic alumina grit, zinc spraying to a depth of 0.008 to 0.16 mm, and finally painting with primer, undercoat, and top coat. An improvement in fatigue life as a result of the treatment, of course, could be due to the introduction of compressive residual stresses by grit blasting rather than the exclusion of atmosphere by painting.

Method of Testing

All specimens were subjected to axial load within the frequency range of 5.5 to 11 Hz. In the main axial load fluctuating tension test series the minimum stress was maintained at 31 N/mm^2, and in the majority of cases stress ranges of 154, 92, and 46 N/mm^2 were used. This gave a life range for control specimens from just over 10^4 to beyond 10^6 cycles. However, other stress ranges were used when the treatments were effective and failure had not occurred in 2 x 10^6 cycles at 46 N/mm^2 stress range. The criterion of failure was taken as complete rupture of the specimen. After failure, the weld toe contact angle of the specimens that had not been given improvement treatments was measured and recorded.

In order to verify the effects of introducing compressive residual stresses, a series of zero mean stress ($R = -1$) and fluctuating compressive stress ($R = -\infty$) tests were carried out on untreated specimens. Extensometers were fitted to these specimens to ensure that the loading was truly axial under compressive loads. Failure under fully compressive loading was assumed when the crack had propagated about 25 mm on either side of the center line of the specimens. This condition was detected by cutout wires.

The majority of specimens was tested after a minimum of 15 days natural aging to enable the weld heat affected zone to recover strength; this condition will be described hereafter as as-welded. To investigate the effect of longer periods of natural aging of as-welded specimens in this alloy, some specimens were tested 6 to 7 weeks after welding; this condition will be described hereafter as aged 6 to 7 weeks.

The general form of the specimens used is shown in Fig. 3; it is essentially a 9.5-mm stressed plate with a single longitudinal nonload carrying stiffener or gusset fillet welded to one surface. The material was a weldable Al-Zn-Mg alloy to UK military specification DGFVE 232A, having a chemical composition shown in Table 1, which also shows the range of mechanical properties of the material in the solution and precipitation heat treated (TF) condition in which it was used. A similar United States alloy is 7005.

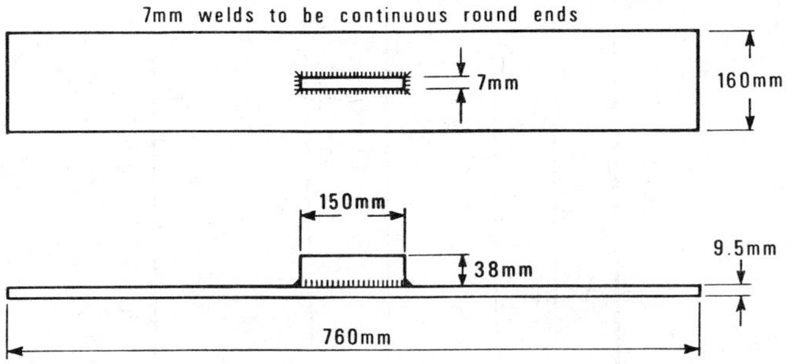

FIG. 3—*Details of nonload carrying fillet-weld specimen.*

This particular specimen type was selected as previous experience had shown the scatter of results to be low compared with those of other specimen types, even though only a small length of the weld is transverse to the stress field. Failure of untreated specimens initiated at the transverse weld toes.

The welds were made in the horizontal-vertical (HV) position with 1.6-mm-diameter NG61 wire (the equivalent United States wire is 5356), with the plate clamped so that the finished specimen was substantially straight. Additional straightening was applied to some specimens. This altered the residual stress level but ensured that consistent stress ranges were applied to the specimens.

The welds around the ends of the stiffener were completed first, and the remainder of the welds were made by welding towards the middle of the stiffener from the ends. This would tend to prevent fatigue cracks initiating from crater cracks in untreated specimens.

The GMA welds were made using the spray transfer technique and 180 A, 20 V direct current with argon flow rate of 14.2 litre/min. GTA welds were made at 250 A, 22 V alternating current and argon flow rate of 7.1 litre/min. Both spray and short arc techniques were used in the case of the helium shielded GMA welds.

A symmetrical double stiffener specimen was found to be necessary so that tests could be conducted wholly in compression in order to investigate the effects of the treatments that produce compressive residual stresses. The specimen was the same as that described with the addition of a second stiffener to the other face.

In view of the known influence of individual welder technique on the fatigue strength of welded joints [7], it was intended that all specimens would be made by the same man. Unfortunately, the original welder left the work and a control group, helium shielded and symmetrical specimens, were made by a different welder.

TABLE 1—*Properties of Al-Zn-Mg alloy to DGFVE specification 232A.*

Chemical composition (nominal)

Element	Al	Zn	Mg	Mn	Zr	Cr	Si	Fe	Ti	Cu
	93.5	4.0	2.0	0.35	0.15	<0.05	<0.2	<0.3	<0.05	<0.1

Mechanical properties

		0.2% Proof Strength, N/mm^2	Ultimate Tensile Strength N/mm^2	Elongation on $5.65\sqrt{S_o}$
From tension tests	range	326 to 380	384 to 437	—
	average	355	411	
From specification (minimum)		332	386	8%

Presentation of Results

The fluctuating tension results obtained at 154, 92, and 46 N/mm^2 stress ranges have been plotted in Fig. 4(*a*), (*b*), and (*c*), respectively, as bar charts. An analysis was then made of the original GMA and GTA argon shielded welds, ignoring weld angle, and a scatter band embracing these results was plotted on a conventional stress range versus life curve in Fig. 5 together with the remaining results. The measured weld toe angles for untreated welds have been plotted against fatigue life in Fig. 6. Finally, the results at different stress ratios are compared in Fig. 7, with the scatter band just established, and the upper limit obtained for the treatment methods. Comparison of the effectiveness of the treatments investigated has been made on a functional basis of welding variables, postweld treatment applied to the weld toes at the ends of the stiffener, and treatments that affect the whole specimen, in preference to grouping them according to the improvement methods described in the introduction.

Results and Discussion of Tests

Weld Method and Profile

All of the as-welded GMA and GTA specimens prepared by the first welder failed from the weld toe and the results spanned a range of endurances of approximately 3:1 at each stress range. There was a trend for the welds with low specified toe angles to be towards the upper end of the scatter band. However, differences are not considered to be sufficient or consistent enough under the present circumstances to warrant specifying a low toe angle in order to obtain an improved fatigue life. These results have been used to establish a scatter band, shown in Fig. 5, for comparison with the other specimens to which the various treatment methods were applied.

The results for the 45-deg GMA welds from the second welder fell at the lower end of the scatter band as did most of the welds made by the same man with helium shielding gas. Prolonged natural aging for 6 to 7 weeks of the 30 and 45-deg GMA welds also showed no discernible trend compared with as-welded specimens although there was an improvement in the static strength of the specimen. Consistently measured weld toe angles were obtained from the GTA welding process which were close to the specified angle, but there was considerable variation in the GMA welds. They ranged from 27 to 60-deg for 45-deg welds; 25 to 45-deg for 30-deg welds. This confirms the view that there is little likelihood of obtaining consistent improvements in fatigue life by control of weld angles unless GTA or automatic welding can be used. However, in order to estimate the potential of control of weld shape, Fig. 6 was plotted showing weld angle versus fatigue life for each stress range. It is seen that actual weld angle clearly has an effect; research on the control of weld angle therefore would seem to be worthwhile but the potential benefits are small. An examination of helium shielded and argon shielded GMA welds, all of 45-deg specified toe angle, made by the

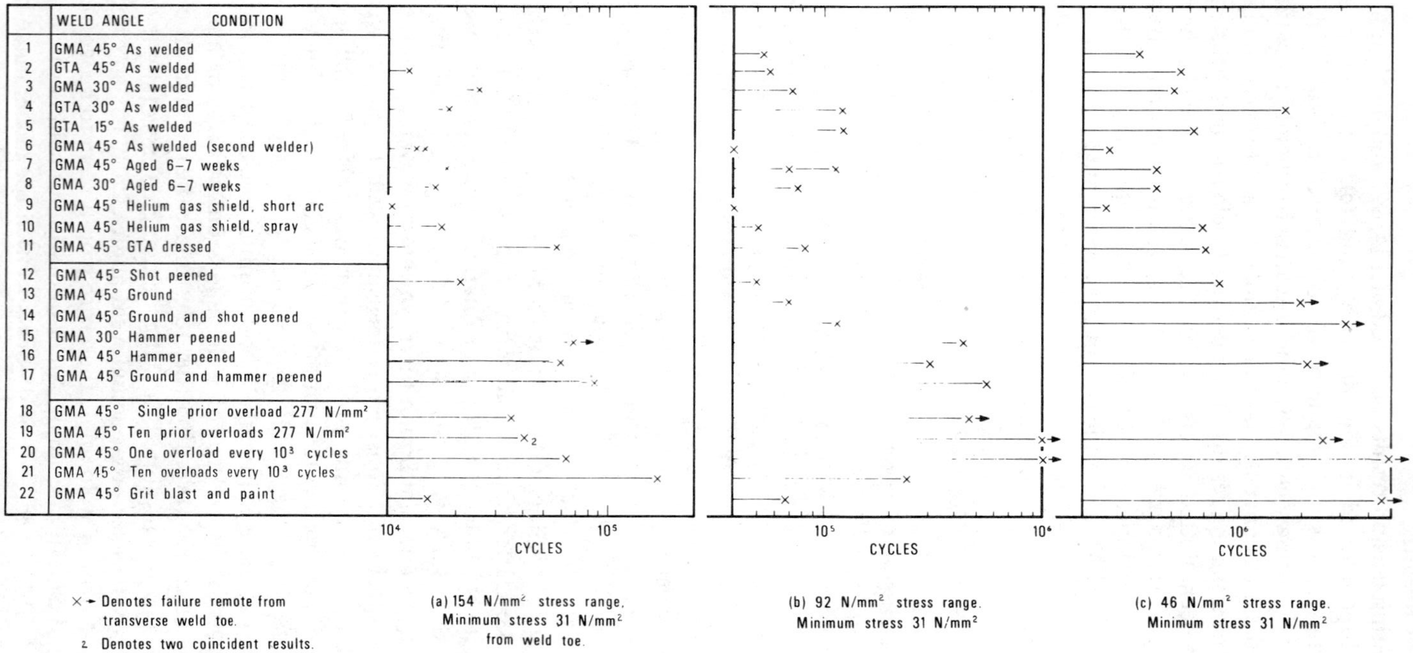

FIG. 4—*The effect of some variables on the fatigue life of nonload carrying fillet weld specimens. Minimum series 31 N/mm² denotes failure remote from transverse weld, 2 denotes number of coincident points. (a) 154 N/mm² stress range. (b) 92 N/mm² stress range.*

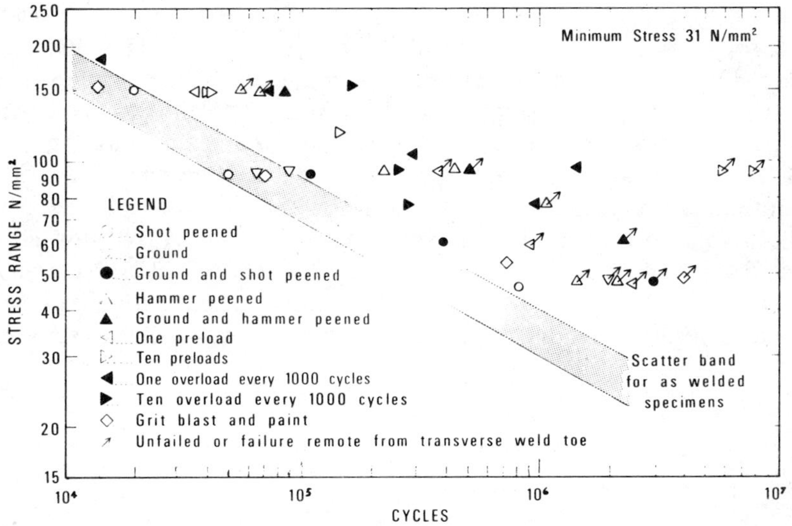

FIG. 5—*The effect of potential life improvement methods on the fatigue life of nonload carrying fillet welds.*

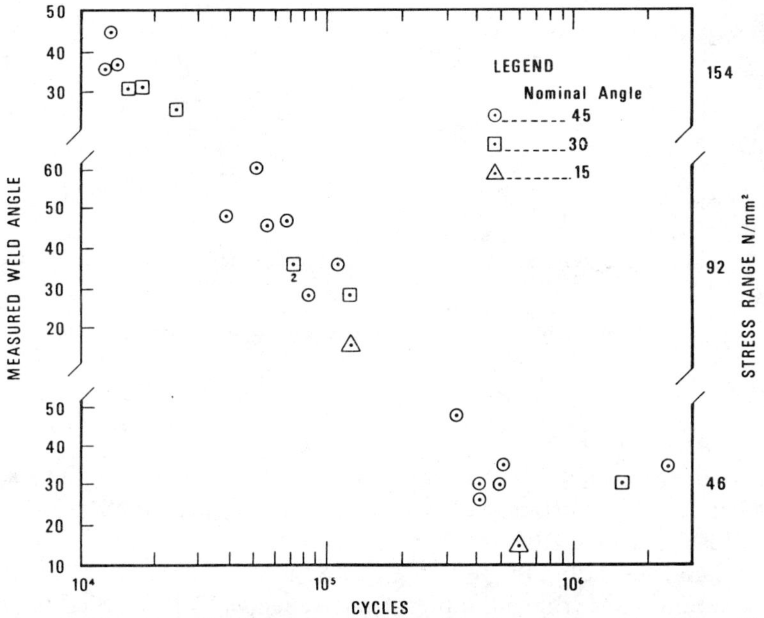

FIG. 6—*The effect of weld angle on the fatigue life of nonload carrying fillet welds.*

second welder, showed that the helium shielded welds made under spray transfer conditions had weld angles of 22 to 30 deg, argon shielded spray transfer welds 37 to 47 deg, while short arc helium shielded welds had toe angles of 40 to 58 deg and a bumpy appearance. There was little scatter except for a low angled helium shielded sprayed transfer weld at 46 N/mm^2 stress range and toe angle of 27 deg, which had a significantly longer life than the upper limit of the scatter band. Although the helium shielded weld showed considerable penetration, this also caused excessive dilution of the welds and the welds were very prone to cracking, so this method of welding had nothing to commend it for welding Al-Zn-Mg alloys.

Microscopic examination of the weld toe of GMA and GTA welds did not reveal sharp nonmetallic defects of the type which caused GMA welds in steel [*4*] to have shorter lives than GTA welds. It appears that both methods of welding result in similar toe conditions and that any inclusions are very small.

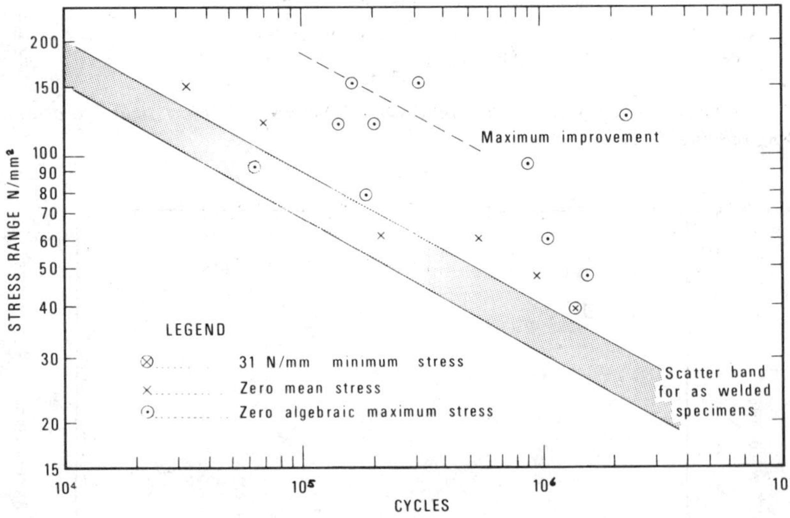

FIG. 7—*The effect of stress ratio on the fatigue life of nonload carrying fillet welds.*

Weld Treatments

GTA dressing of GMA welds was inconclusive except for the specimen tested at 154 N/mm^2 stress range which had a longer life than did any of the welds just presented, the lives for other GTA dressed welds fell within the scatter band. Grinding the weld toe had no effect on life at 92 N/mm^2, but at 46 N/mm^2 stress range the life of one specimen far exceeded the scatter band and eventually broke at the testing machine jaws. Shot peening did not appear to be effective and, not surprisingly, grinding followed by shot peening produced results similar to shot peening alone. Hammer peening, which also resulted in some improvement to weld profile, see Fig. 2, caused at least an eight-fold improvement in fatigue life in comparison with the minimum of the scatter band of results, and the improvement was such that only one out of three failures occurred at the ends of the weld that had been treated. The results of grinding and hammer peening were very similar to hammer peening alone. Figure 8 illustrates a typical failure away from the weld toe.

FIG. 8—*Failure of a hammer peened specimen away from the weld toe. Note the subsidiary crack away from the weld toe.*

Specimen Treatments

Grit blasting and painting specimens before testing had no effect on life compared with the scatter band at the higher stress ranges but at 46 N/mm^2 stress range a failure occurred from a weld pore away from the weld end, at a life beyond the limit of the scatter band indicating that there may be some improvement at low stress ranges. A single prior overload causing tensile stress of 277 N/mm^2 in the specimen brought about at least a three-fold increase in life, and at the lower stress ranges failures tended to be at locations other than the weld toe. Ten prior overloads had a similar effect to a single prior overload for a test at 154 N/mm^2 stress range, but at lower stress ranges the beneficial effect was much greater and exceeded that for hammer peening. When the overloads were repeated at intervals of 1000 cycles they also showed an improvement, the ten overloads being the most effective improvement method at 154 N/mm^2 stress range, but at lower fatigue stress ranges a single overload seemed to be more advantageous. As the fatigue stress range was reduced and the total number of overload cycles increased, the improvement became less than for prior overloads. The overload stress which was selected initially for the GMA welded specimen was so high that it was not possible to overload GTA welded specimens to the same level and they broke in the reduced strength region of the heat affected zone, which is more extensive than for GMA welds.

The effect of periodic overloading depends on the magnitude, number, and interval of applications of overload stress. At low test stress ranges the very process of periodic multiple overloading is fatigue damaging and will eventually limit the fatigue life. Alternatively, had a lower overload stress been used, the benefit due to the introduction of local compressive residual stresses would have been less, but the maximum possible endurance at lower test stress ranges might have been greater. If the interval between overloads had been greater so that the total number of overload cycles had been less, the beneficial effect of each overload cycle would have been similar, and it is again possible that the endurance would have been greater. The optimum condition, which would introduce greater benefit than the application of prior overloads only, would be the application of overloads when the retardation in crack growth due to the previous loads had disappeared. Such conditions would probably have to be derived by carrying out fatigue crack propagation tests and would only be appropriate to a particular type of operating conditions.

Prior overloading ten times is clearly the best general method of specimen treatment, but it would not be practical to use such high prior overloads on structures because of dimensional tolerance requirements and variations in material properties, and consequently the magnitude of improvement shown in these experiments could not be expected to be attained.

Stress Ratio

The zero mean stress results ($R = -1$) fell just above the upper end of the scatter band with the wholly compressive tests ($R = -\infty$) generally beyond these, Fig. 7, but exhibited large scatter in life and location of failure, most being from weld ripples and crater cracks away from the toes at the weld ends.

Measured residual stresses in parent plate at the weld toe ranged from 178 N/mm^2 tension to 157 N/mm^2 compression. The stresses tended to be lower for the specimens with the double gusset used for the stress ratio experiments and the greatest tensile stresses were in the direction of loading. Failure under cyclic compressive loading depends on the residual stresses causing an effective tensile stress range. Residual stresses are likely to be larger in structures than in specimens because of the bulk of the material and the remoteness of welds from free edges, so the lives would be lower than the present specimen lives suggest. It has been noted [*9*] that the weld would contain a much higher residual stress than exists in the heat affected zone at the toe of the weld, so although the nominal stress range would be lower in the section containing the weld than at the toe, due to the increased cross-sectional area, it is possible for it still to be wholly tensile, whereas at the toe the tensile portion of the stress range would be limited to the value of the tensile residual stress. This could account for the large amount of scatter of life and location of failure at surface ripples, crater cracks, and buried defects which would normally be harmless compared with the weld toe. At lives of less than the 10^6 cycles, Fig. 7 also shows that the results for compressive applied loads are consistent with the life improvement methods that involved treatments causing compressive residual stresses.

General

After the fatigue tests were completed, residual stress measurements were made in the toe of the welds of similar details. Grit blasting caused a peak compressive stress of at least 300 N/mm^2 at 0.25 mm below the surface which then reduced rapidly. Hammer peening caused a peak stress of 240 N/mm^2 at 0.6 mm below the surface which had only fallen to 75 percent of this value at 1.2 mm. The results for shot peening were similar to grit blasting but the depth extended to about 0.5 mm. The results indicate that any peening method must produce an adequate depth of sufficiently high residual compressive stress to be effective and further work is necessary to define the conditions under which this is so. Corresponding residual stress measurements are not available for the overloading specimens, but from the fatigue results they appear to be similar to hammer peening. The results for toe grinding were disappointing compared with the results for measured weld angle; as seen in Fig. 2, conditions at the weld toe were improved but there was still a rapid change in section sometimes and the reduction in stress concentration would not be as large as indicated by the toe angle. Unlike the case with steels, the method does not appear to be cost effective when applied to aluminium alloy fillet welds.

Concluding Remarks

Within the limited range of the present investigation under tensile loading, the following results of potential methods of increasing fatigue life were apparent.

1. The welding method (GMA and GTA) does not have a significant effect on life.
2. Decreasing weld angle caused a small increase in life, but the weld angle is not easily controlled, particularly with GMA welds.
3. The shielding gas (argon or helium) does not appear to have a significant effect on life, but helium shielding gas is not satisfactory on account of weld cracking.
4. Grinding weld toes does not appear to be effective at high stress ranges, but may be beneficial at low stress ranges if the whole weld profile is improved.
5. Light shot peening or grit blasting does not improve fatigue life, but a more severe hammer peening was effective, a certain depth of relatively high compressive residual stress appears to be necessary to obtain an improvement in fatigue life.
6. Combined grinding and peening was no more effective than peening alone.
7. Prior overloading was effective; the improvement was not as great as

with hammer peening at high stress ranges, but became more effective at the lower stress ranges. Ten prior overloads were better than one at low stress ranges.

8. Periodic overloading improved life at high stress ranges, but the effect was not so great at lower stress ranges.

There was little that could be done during welding to produce a substantial improvement in fatigue life, but weld treatments and specimen treatments that induced compressive residual stress fields of sufficient magnitude were extremely effective. There is little to choose between severe peening and 10 prior overloads in effectiveness and the choice depends on individual circumstances. The control of treatment clearly needs some care and the comparison had not been extended beyond the constant amplitude loading.

Acknowledgments

The author wishes to thank the staff of the Welding Institute, Cambridge, who cooperated in the several phases of this investigation.

References

[*1*] Brine, F. E., Webber, D., and Baron, H. G., *British Welding Journal*, Vol. 15, No. 11, Nov. 1968, pp. 541-546.

[*2*] Gurney, T. R., *Fatigue of Welded Structures*, Cambridge University Press, Cambridge, 1968.

[*3*] Maddox, S. J., *International Journal of Fracture*, Vol. 11, No. 2, 1975, pp. 221-243.

[*4*] Watkinson, F., *The Fatigue of High Strength Steels in Fatigue Performance of Welded High Strength Steels*, The Welding Institute, 1974.

[*5*] Dye, S. A., *Welding Institute Bulletin*, Vol. 13, No. 10, Oct. 1972.

[*6*] Christopher, P. R. and Crabbe, D. R., *Ocean Engineering*, Vol. 1, July 1969, pp. 497-520.

[*7*] Gurney, T. R., *Welding Institute Research Bulletin*, Vol. 12, No. 1, 1971.

[*8*] Heywood, R. B. in *Designing Against Fatigue*, Chapman and Hall, London, 1962, Chapter 15.

[*9*] Nordmark, G. E., *Metal Progress*, Nov. 1963, pp. 101-103.

W. H. Munse[1]

Fatigue of Weldments—Tests, Design, and Service

REFERENCE: Munse, W. H., **"Fatigue of Weldments—Tests, Design, and Service,"** *Fatigue Testing of Weldments, ASTM STP 648*, D. W. Hoeppner, Ed., American Society for Testing and Materials, 1978, pp. 89–112.

ABSTRACT: For many years the fatigue behavior of welds and weldments has been studied in terms of the geometry of the members, the stresses to which they are subjected, and the materials of which they are fabricated. Numerous laboratory tests have been conducted and have served as a basis for the development of current fatigue design specifications. More recently, research on crack initiation and propagation, expanded data on loadings, and the development of new concepts of structural reliability have provided information that should help to make possible in the future greatly improved fatigue design specifications for welded structures. However, to achieve this goal, extensive further detailed research and study are necessary.

KEY WORDS: fatigue tests, welded joints, weldments, design, crack initiation, crack propagation, reliability, random loads

Although fatigue has been studied for nearly 150 years, research on the fatigue behavior of weldments and welded structures is more recent. This is primarily because welded structures that are subjected to fatigue loadings, such as bridges, were introduced only about 50 years ago. However, since then numerous laboratory fatigue tests of welds and weldments have been conducted, the results of which have been used in the development of fatigue design requirements.

Obtaining behavioral fatigue data is only one step in providing structures with adequate resistance to fatigue: the quality of the structure must be comparable to that assumed in design, and the maintenance during the life of the structure adequate. Unfortunately the quality of welding is sometimes difficult to control. Welding is an art, so the care and skill of the welder is significant; the welder controls the quality of the weld and it's

[1] Professor, Civil Engineering Department, University of Illinois, Urbana-Champaign, Ill. 61801.

placement in the proper manner, both of which can be expected to affect a weld's fatigue resistance.

Thus, a designer to consider properly fatigue in a welded structure must have: (*a*) adequate information concerning the fatigue behavior of welds and weldments and the factors that affect this behavior, (*b*) suitable design criteria, and (*c*) knowledge of the service conditions to which the structure will be subjected during its full expected lifetime. It is these various aspects of the fatigue problem in welds and weldments that will be considered here in rather general terms.

Fatigue, Laboratory Tests

During the nearly 50 years that laboratory studies have been conducted on weldments, numerous papers, conference or seminar proceedings, and books have provided detailed fatigue data for welds and weldments [*1–9*].[2] Thousands of tests have been conducted in many parts of the world. One of the principal objectives of this research has been to provide basic engineering data for the development of fatigue design criteria. However, because of the many variables involved and the large degree of scatter found in the tests, empirical relationships generally have been developed for design. In recent years the use of fracture mechanics and the results of crack growth studies have shown signs of leading to a more basic understanding of the mechanics of fatigue in welded structures, although much more must be done before adequate relationships can be developed for fully effective fatigue design based on fracture mechanics.

Laboratory investigations have demonstrated that numerous factors affect the fatigue behavior of welds and weldments, factors that can generally be separated into four categories.

1. The geometry of the member or detail. This includes both the general configuration and the local geometry of the member.
2. The stresses or loading conditions to which the member or detail is subjected. These may include constant cyclic stresses, residual stresses, stresses from random loadings, frequency of loading, etc.
3. The materials from which the members are fabricated. For structural purposes the steels generally have yield strengths ranging from 248 to 689 MPa (36 to 100 ksi).
4. The environmental conditions existing during the life of the member.

Geometry

Welding is a powerful tool that can be used to produce continuity in the joints and members of a welded structure. However, because of the manner in which such members are assembled, discontinuities are introduced and produce increased local stresses when loads are applied. These stress con-

[2] The italic numbers in brackets refer to the list of references appended to this paper.

centrations result from the general configuration of the members, the local configuration of the weld details, and the discontinuities that occur within the welds. The internal discontinuities may consist of such defects as porosity, slag inclusions, lack of fusion, lack of penetration, or cracks.

An example of the effect on fatigue of the general configuration of two specific types of members is shown in Fig. 1; the fatigue resistances of the members are seen to differ by a factor of approximately three. Similarily, the addition of a partial length cover plate to a rolled I-beam reduces the fatigue resistance of the beam by a factor of about 3. These are extreme examples, but they clearly demonstrate the important role played by the configuration of the members.

The importance of the local geometry of weldments can be demonstrated by examining in more detail the behavior of a butt-welded splice of the type shown in Fig. 1(*a*). The introduction of the butt weld reduces the fatigue resistance of the basic plate by as much as 30 to 50 percent. The magnitude of this reduction will depend upon the local configuration or geometry of the weld and a variety of other factors [*10*].

In several investigations of weld geometry the shape has been defined in terms of the parameters shown in Fig. 2 [*10–13*]. These parameters (the radius at the toe of the weld, the angle the reinforcement makes with the surface of the plate, the height of the weld reinforcement, and the width of

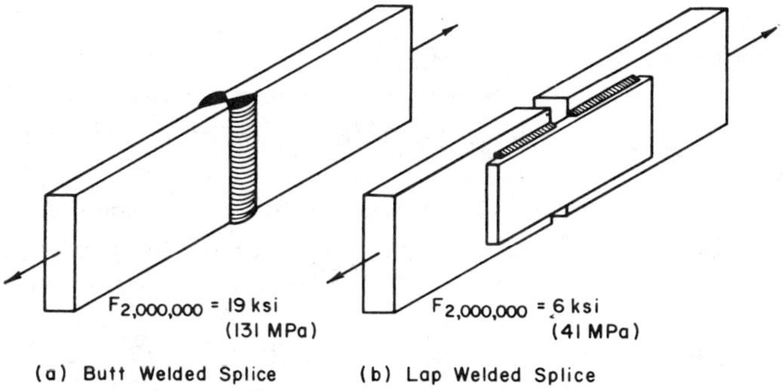

FIG. 1—*Welded splices of two different configurations.*

the weld reinforcement) are the factors that affect the local stress at the toe of the weld and consequently the fatigue resistance of the member. Decreasing the height of the reinforcement from 3.18 to 1.59 mm (1/8 to 1/16 in.), for example, increased the fatigue life for a tensile stress cycle of 13.8 to 241 MPa (2 to 35 ksi) by a factor of 2. However, for a tensile stress cycle of 13.8 to 186 MPa (2 to 27 ksi) the fatigue life was increased only by a factor of 1.5. Apparently the effects of geometrical parameters also vary with the magnitude of the loading to which a member is subjected: at longer lives the

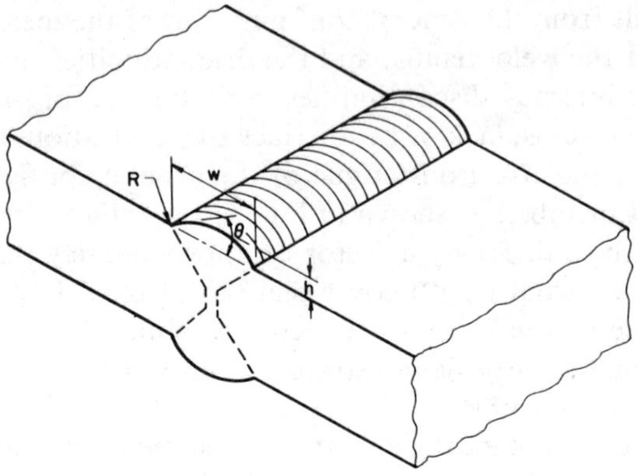

FIG. 2—*Geometrical parameters for transverse butt welds.*

effect of the external geometry of the weld was not as significant as at the shorter lives. This is typical of the effects observed for many of the other factors that have been studied; there are many interrelated effects of the various parameters involved in fatigue. A further indication of the interrelated effects can be seen in Fig. 3 wherein the reduction in fatigue strength is shown as a function of the angle of the weld reinforcement and radius at the weld toe for two different stress cycles [*10*]. The effect of the flank angle and the weld toe radius was greater for the 0 to 241 MPa (0 to 35 ksi) stress cycle than for the ± 172 MPa (± 25 ksi) stress cycle.

In other studies, the weld geometry has been shown to affect, in a similar manner, the propagation of fatigue cracks in butt-welded joints [*11*]. Thus, the important role played by the local configuration of a weld on its fatigue behavior is evident.

The third type of geometrical parameter that may affect the behavior of a welded joint is the internal weld geometry (defects). Weld defects generally

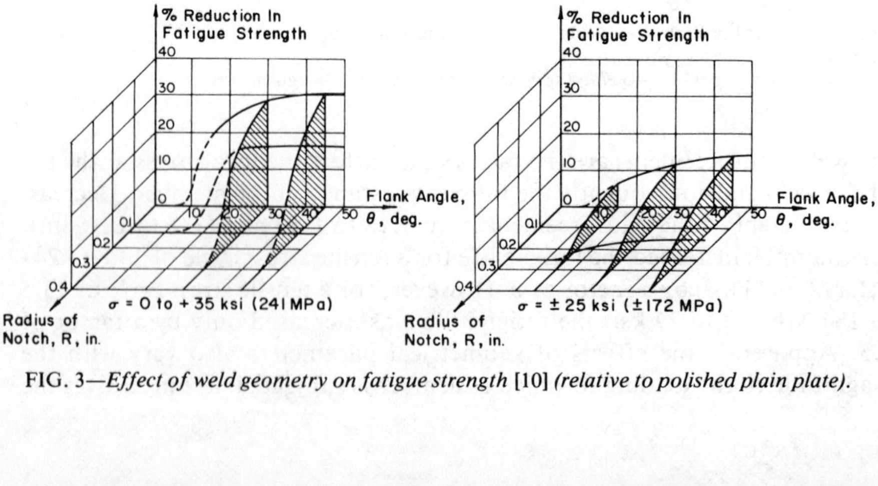

FIG. 3—*Effect of weld geometry on fatigue strength* [10] *(relative to polished plain plate).*

appear to have a greater effect at long lives than at short lives. This is just the opposite of what has been observed in the case of the external weld geometry. Small amounts of porosity appear to have a relatively minor effect on the fatigue resistance of a sound weld [*14*]. However, large clusters, such as shown in Fig. 4 (*a*), produce a significant reduction in fatigue strength. This gross porosity produced approximately a 40 percent reduction in the fatigue resistance of the joint. A more severe internal discontinuity is the lack of penetration shown in Fig. 4(*b*). Such a defect can be expected to reduce the fatigue strength to 50 percent of its original value, even when the reduction in cross-sectional area produced by the lack of penetration is taken into account. Clearly, internal weld defects can have a significant effect upon the fatigue resistance of a weld. Such defects are no doubt responsible for much of the scatter observed in laboratory tests of welds.

Stresses

Numerous tests have been made to evaluate the effects of stress cycles on the fatigue of welds and welded members. However, to generalize on the effects is extremely difficult because of the many interrelated variables that affect a weldment's fatigue behavior. Nevertheless, there are a number of general observations that can be made.

Because of the limited capabilities of much of the equipment used to conduct the early fatigue tests, most studies were conducted under constant-cycle conditions. Nevertheless, by conducting tests at various stress ratios, fatigue diagrams of the type shown in Fig. 5 were often developed. Such diagrams provide individual curves for a constant life and an indication of the effect on fatigue behavior of the various constant cycle stress parameters. For members that contain severe geometrical stress concentrations, the fatigue diagram curves, particularly for long lives, tend to be very low and almost parallel to the mean stress axis, thereby indicating that the fatigue resistance is essentially a function of the alternating stress or stress range. However, for other members and details with only minor stress concentrations and for shorter lives, there often appears to be an increased effect of mean stress on the fatigue behavior: the stress range increases somewhat as the mean stress decreases, particularly under a reversal of stress. Apparently the compressive stresses do not do as much fatigue damage as the tensile stresses. Nevertheless, it is still obvious from the various tests that the stress range is the dominant factor controlling the fatigue life of welds and weldments.

Another way of examining the effect of the stress parameters is in terms of the stress ratio, the ratio of minimum to maximum stress. One such comparison of the variation in fatigue resistance with stress ratio is made in Table 1 for numerous tests on several different types of members. In general, under a reversal of axial stress ($R = -1$), the stress range is about 20

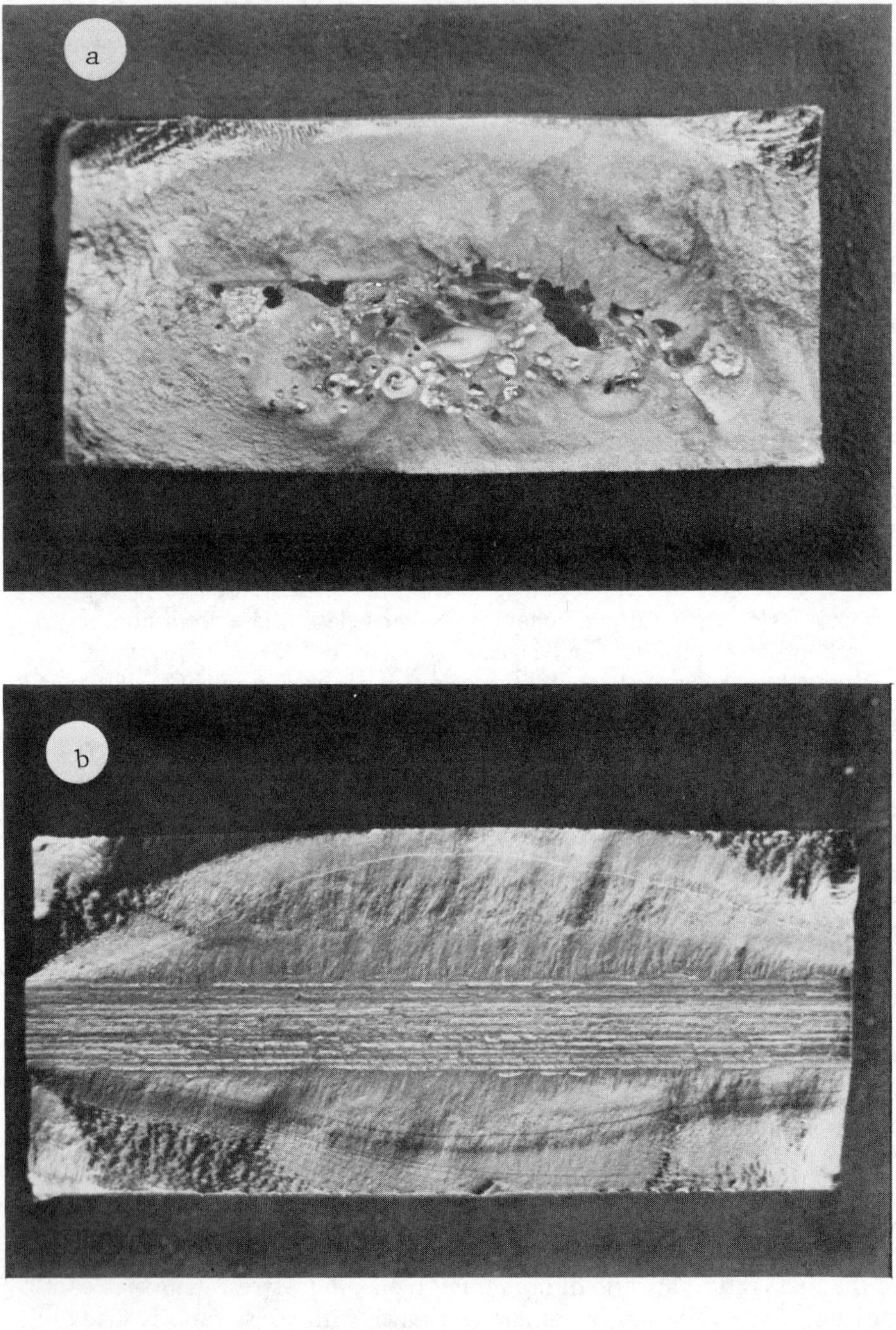

FIG. 4—*Internal defects in transverse butt welded joints* [14]. (a) *Large cluster of porosity, and* (b) *lack of penetration at midthickness.*

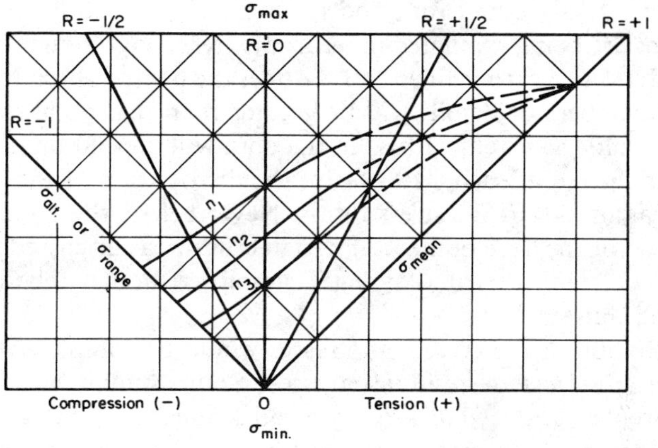

FIG. 5—*Fatigue diagram of constant life curves and stress parameters.*

percent greater than that for a zero-to-tension axial loading ($R = 0$). Under a stress cycle in which the stress varies from one-half tension to tension, the stress range is approximately 90 percent of the stress range for a zero-to-tension cycle.

TABLE 1—*Fatigue resistance of welded joints at various stress ratios.*

Type of Joint Steel	Stress[a] Ratio, R	Stress Range MPa (ksi) for n = 100 000	Stress Range[b] Ratio, r	Stress Range MPa (ksi) for n = 2 000 000	Stress Range[b] Ratio, r
1. Transverse full	−1	474(68.8)	1.26	188(27.20)	1.27
penetration	0	376(54.6)	1.00	148(21.40)	1.00
butt welds in	+ ½	343(49.8)	0.91	131(19.05)	0.89
quenched and					
tempered steel [*15*]					
2. Fillet welded	−1	169(24.58)	1.14	78(11.36)	1.18
plates of	0	149(21.64)	1.00	66(9.64)	1.00
A-7 steel [*16*]	+ ½	142(20.57)	0.95	56 (8.13)	0.84
3. Plug welded	−1	196(28.36)	1.08	74(10.68)	1.05
joints of	0	181(26.30)	1.00	70(10.13)	1.00
A-7 steel [*16*]	+ ½	153(22.20)	0.84	72(10.40)	1.03
4. Welded beams with	−1	290(42.0)	1.25	131(19.0)	1.09
butt-welded splices,	0	231(33.5)	1.00	121(17.5)	1.00
A-7 and A-373	+ ½	...	...	107(15.5)	0.89
steel [*17*]					
5. Plain flat plates of	−1	855(124.0)	1.37	310(45.0)	1.20
quenched and	0	622(90.2)	1.00	258(37.4)	1.00
tempered steel [*15*]					
6. Transverse butt welds	−1	...	...	...	1.24
Reported by	0	...	...	...	1.00
Gurney [2]	+ ½	...	...	...	0.82

[a] Stress ratio, the ratio of minimum to maximum stress in the stress cycle.
[b] Stress range ratio, the ratio of the stress range to the stress range for $R = 0$.

In many instances, and for good reason, fatigue data from a variety of stress ratios are being evaluated in terms of stress range alone. It must be remembered, however, that when this is done, the degree of scatter in the data for some members or details will be greater than that obtained at a single stress ratio and the extent of bias in the data will depend upon the number of tests conducted at each stress ratio, the distribution of the stress ratios, and the magnitudes of the stress ratios. Nevertheless, the use of a constant stress range for the development of design criteria, as discussed later, makes possible the establishment of greatly simplified design relationships and design procedures.

No discussion of the effects of stresses would be complete without comment on the effects of residual stresses. Some studies have shown that residual stresses produced in welding or subsequent to welding may alter the life of a member; other studies have shown no effect. In one recent investigation, residual stresses associated with periodic overloads were found to provide a significant increase in the fatigue life of a weldment [*18*]. In other studies [*19*], the importance of the magnitude and type of residual stress as well as the applied cyclic stresses are examined and suggest that the effects depend upon the relationship between the residual and applied stresses. Under relatively high applied tensile stresses (short lives), the effects of residual stresses can be quickly relaxed, and these effects would tend to be relatively small, whereas at long lives the effects become much more significant.

Another important aspect of the loading or stress-cycle problem concerns the effects of variable or random loadings. In recent years, as a result of the availability of more versatile testing equipment, increased consideration has been given to the effects of such loadings. The large and important effect of variations in amplitude of loading on the fatigue behavior of one type of weldment can be seen in Fig. 6 [*6*]. Another study, in which a systematic variation in the loadings had been used, indicated that changes in the resulting residual stresses can affect the behavior [*18*]. However, when the variation in amplitude is applied in a random fashion, the effects of residual stresses appear to be greatly diminished.

This brief discussion summarizes the effects of some of the principal stress parameters involved in fatigue. Other factors which may affect the behavior, but generally to a lesser extent, include frequency of loading, the possibility of extended rest between applications of loading, etc.

Material

Structural weld and weldment fatigue tests have been conducted with a variety of structural steels having tensile strengths ranging from approximately 414 to 827 MPa (60 to 120 ksi). Just as in the case of mean stress or stress ratio, structural fatigue design for weldments in these materials is generally based on the stress range and the combination of all steels. Neglecting

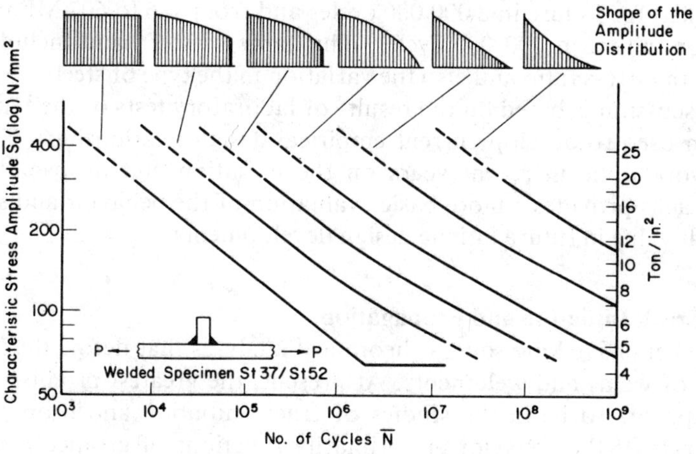

FIG. 6—*Effect of shape of amplitude distribution on cycles to failure* [6]

the effect of type of steel simplifies the design criteria, but again produces an increase in degree of scatter in fatigue life. For transverse butt welds, the resulting mean stress range for all steels is greater than that obtained for ordinary structural steels and is below that obtained from the higher strength quenched and tempered steels (see Table 2). The scatter band for the 517 tests reported in Table 2 ranges from approximately 76 to 248 MPa

TABLE 2—*Fatigue resistance of transverse full-penetration butt-welded joints in various steels (517 tests of as-welded joints)* [*15*].

		Stress Range, MPa (ksi)			
Item	Tests	n = 100 000	Ratio to Item 1	n = 2 000 000	Ratio to Item 1
1. Mean fatigue resistance, ordinary structural steels (yield approximately 207 to 248 MPa (30 to 36 ksi))	219	286(41.5)	1.00	117(16.9)	1.00
2. Mean fatigue resistance, high strength, low alloy steels (yield approximately 345 MPa (50 ksi))	43	350(50.8)	1.22	143 (20.7)	1.22
3. Mean fatigue resistance, quenched and tempered steels (yield approximately 620 MPa (90 ksi))	255	363(52.7)	1.27	148 (21.4)	1.27
4. Mean fatigue resistance, various steels (yield approximately 248 to 620 MPa (36 to 90 ksi))	517	317(46.0)	1.11	131(19.0)	1.12

(11 to 36 ksi) for failure in 2 000 000 cycles and from 186 to 607 MPa (27 to 88 ksi) for failure in 100 000 cycles. This large scatter band includes the effects of the stress ratio and also the variation in the type of steel.

This discussion is based on the results of laboratory tests of the type that have been used to develop current empirical design relationships. The extensive work done in recent years on the initiation and propagation of fatigue cracks provides a more basic evaluation of the behavior and should be of great value in future fatigue design developments.

Fatigue Crack Initiation and Propagation

Researchers have long sought theoretical analyses that define the fatigue behavior of welds and weldments. At present, the greatest promise in this respect appears to be in the studies of crack initiation and propagation. Both aspects of the behavior are primarily functions of geometry and the stress to which a member is subjected, although they are also affected by such other factors as material, residual stresses, environment, etc.

Initiation

Since there will be no crack to propagate if one does not initiate, the level of stress below which cracks will not initiate is of great importance. Rolfe and Barsom [*20*] note that for steels, fatigue crack initiation thresholds exist that are dependent upon the tensile properties of the steels and the acuity of the geometrical discontinuity at which initiation might occur. In the case of a weld, the initiation may also be affected by residual weld stresses [*18*].

The percentage of total fatigue life devoted to initiation appears to vary considerably, depending to some extent on the definition of initiation. Tests of butt-welded joints with a lack of penetration at mid-depth have provided initiation lives ranging from 40 to 60 percent of the total life when failure occurred at approximately 100 000 to 150 000 cycles [*14*]. In a study of flat plates with a polished hole in the center [*21*], approximately 50 percent of the life also appeared to be involved with initiation when failure occurred at approximately 200 000 cycles. However, when sharp notches are subjected to high stresses, cracks appear to initiate almost immediately. Thus, at short lives (high stresses) the percentage of life devoted to initiation may be small, whereas at longer lives (lower stresses) the percentage of life for initiation may become large. This has been verified in the study by Mattos [*19*] for weldments subjected to a full reversal of stress (see Fig. 7). The study also suggests that the type of material may have a significant effect upon the initiation life.

Propagation

Many crack propagation relationships are presented in the literature [*22*]. Although much of the research on which these relationships are based has

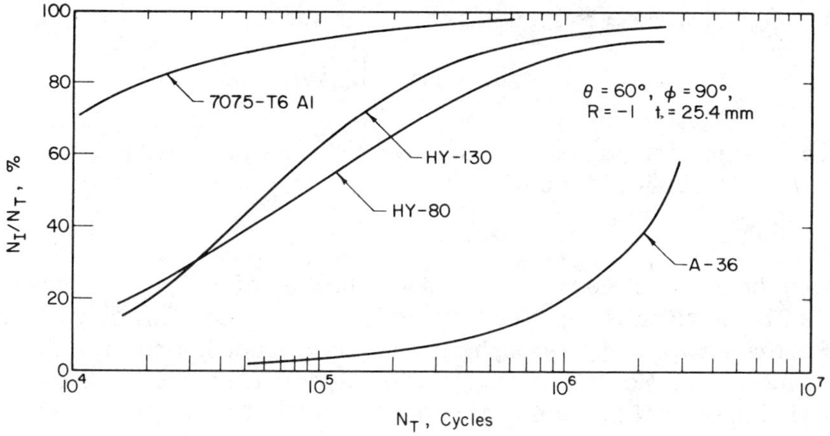

FIG. 7—*Predicted proportion of fatigue life for crack initiation in butt welds in various materials* [19]. *(0.05 in. = 1.27 mm, and 1 in. = 25.4 mm.)*

been conducted on plain plate-type specimens, many studies also have been conducted on weldments. In such studies [*23–26*], the reliability of the theoretical propagation relationships depends upon the accuracy with which the stress intensity for the flaws can be established. For weldments in "real" structures, the magnitude of the stress intensity has been difficult to define because of the complexity of most weldments, the fact that the stress intensity is a function of the size and shape of the weld as well as the weld discontinuities, and is affected by "real" loadings whereby overloads may produce residual stresses, crack tip blunting, crack closure, etc. [*18*]. Nevertheless, a great deal can be learned from even approximate crack propagation relationships.

For example, a fatigue crack propagation threshold is thought to exist for steels subjected to tensile stress cycles and to be a function of the stress ratio to which the member is subjected and also the acuity of the geometrical discontinuity from which a crack may propagate. Rolfe and Barsom [*20*] suggest that the threshold stress intensity range be given by

$$(\Delta K)_{th} = 6.4\,(1 - 0.85\,R) \text{ for } R \geq 0.10 \qquad (1)$$

where

$$R = \text{stress ratio, } \sigma_{min}/\sigma_{max}$$

When and if verified, relationships of the type given in Eq 1 may help to provide an indication of the magnitudes of stress to which a member with a flaw can be subjected without the flaw propagating to failure. For example, if the stress intensity[3] can be given by

$$(\Delta K) = \Delta\sigma\sqrt{\pi a} \qquad (2)$$

where

$$2a = \text{length of crack}$$

[3] Equation 2 is for a central through-thickness crack in a plate.

the range in stress corresponding to the threshold stress-intensity range of Eq 1 will be

$$\Delta\sigma = \frac{6.4\,(1 - 0.85\,R)}{\sqrt{\pi a}} \tag{3}$$

The maximum stress corresponding to this stress range for a given stress ratio can be obtained from

$$\sigma_{max} = \frac{\Delta\sigma}{1 - R} \tag{4}$$

From these relationships, (Eqs 3 and 4) a diagram of the type shown in Fig. 8 can be obtained for specified values of flaw size, $2a$. This diagram suggests that a plate with a through-thickness central notch, represented by the relationship of Eq 2, can sustain without propagation a relatively high maximum stress if the stress ratio is large and the flaw is sufficiently small. However, this stress level decreases rapidly as the size of the flaw is increased and the stress ratio decreased. This same general type of behavior can be expected from weldments too; however the appropriate relationships

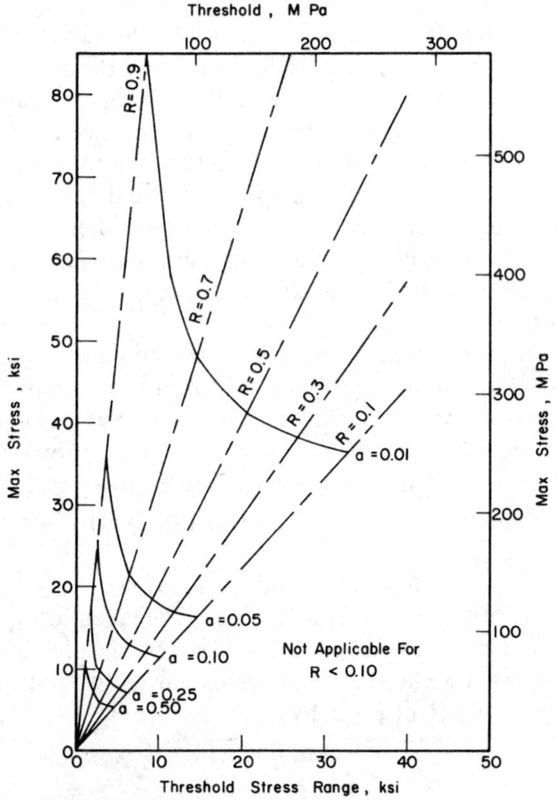

FIG. 8—*Threshold values of stress range for various flaw sizes in center-crack flat plate. (0.01 in. = 0.25 mm and 0.05 in. = 12.7 mm.)*

for such members need to be developed.

Although the information on crack initiation and propagation has helped greatly to define the mechanics of fatigue behavior and shows promise of greatly clarifying our understanding of this phenomenon, much more research is needed to define adequately the relationships necessary for the development of design and inspection criteria to protect welded structures adequately against failure.

Design

The results of laboratory fatigue tests and diagrams of the type shown in Fig. 5 have provided the basic information on which many fatigue design relationships have been developed. For example, the 1947 edition of the American Welding Society's specifications [*27*] introduced straight-line relationships to approximate the curves of Fig. 5 and factors of safety that provided maximum allowable design stresses in the form of Eq 5 for a variety of design details.

$$F_R = \frac{f_0}{1 - kR} \tag{5}$$

where

f_0 = the fatigue resistance under a zero-to-tension loading reduced by the desired factor of safety (also range of stress),

R = the ratio of minimum to maximum stress, and

k = a constant representing the slope of the straight line design relationship in terms of the mean stress and alternating stress.

Equation 5 was used for a number of years but, in many instances, proved to be rather difficult to apply. A constant range of stress criteria which relates the fatigue behavior to the live-load stress is much simpler to use. Consequently, the stress range design criteria that has been introduced into most of the current structural fatigue design specifications [*27–30*] has been readily accepted. This corresponds to a value of $k = 1.0$ in Eq 5. However, as noted previously, use of a constant stress range neglects some of the stress and material factors that affect fatigue.

There is apparently considerable differences of opinion around the world as to the appropriate fatigue design criteria. Figure 9 [*6*], for example, shows for several countries the marked variation in fatigue design criteria used for a single type of welded joint and a life of 2 000 000 cycles. With the apparent difference in maximum allowable design stress of 3.5, markedly different basic assumptions must have been used in developing the design criteria. Greatly different levels of reliability will also result from such criteria.

Although fatigue design provisions are already included in many specifications, studies [*31–33*] are under way to develop more realistic design criteria. The developments in some of these new studies are based on a philosophy of separating the problem into two functions, resistance and

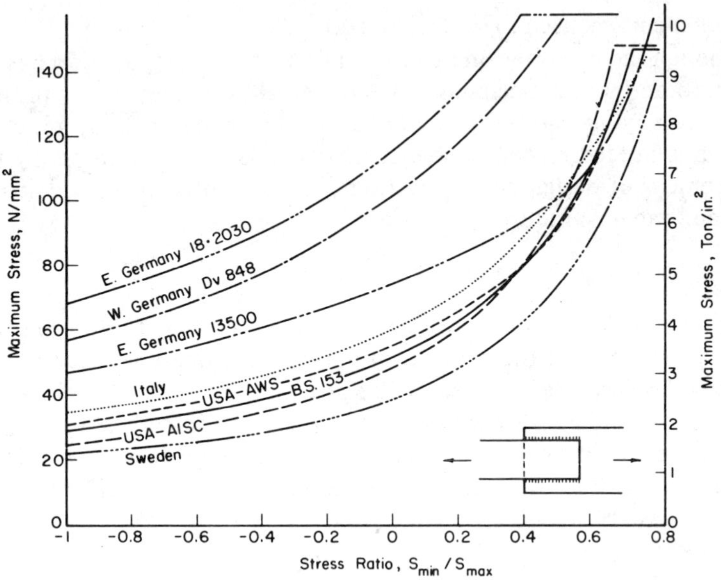

FIG. 9—*Comparison of design stresses at 2 000 000 cycles for longitudinal load-carrying fillet welded joints* [6].

loading, and in introducing consideration of the level of structural reliability. The degree of variability in the fatigue behavior of many welded details (a factor required for a reliability analysis) can be obtained from laboratory data. However, to this must be added the effects of the other variables: errors in analysis, variations in fabrication, the effects of corrosion, and the effects of any other many variables that may be involved.

The variability in loading must also be taken into account. For bridges, some loading information in the form of histograms and traffic data are available [*5,34*]; however, a great need exists for more information covering urban and rural conditions as well as predictions of loadings that can be expected in the future. With improved fatigue information and better loading predictions, greater design economy in both lives and dollars can be achieved.

The structural fatigue reliability design philosophy suggested by Ang and Munse [*31*] makes possible the design of many different types of members and details that are subjected to random loading conditions and for a preselected level of reliability. The variety of details covered by the procedure is presented in Fig. 10. The constant-cycle fatigue resistance of these details, based on a constant stress range philosophy, was determined and the variability in the data as well as estimates of the effects of errors in analysis, the uncertainty in the slope of the fatigue regression relationships, and errors in the mathematical modeling and in the use of linear damage rule were introduced. With such values and estimates, the values of fatigue

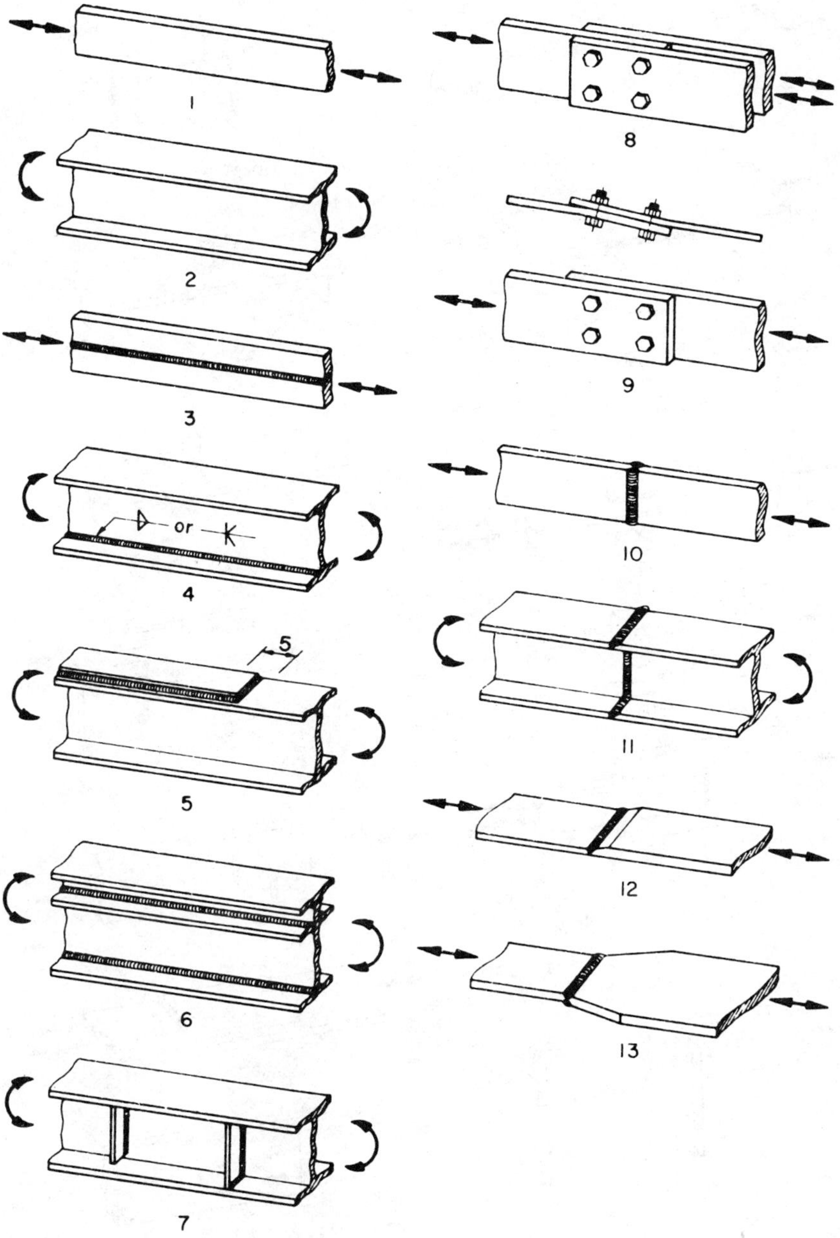

FIG. 10—*Structural details for fatigue design* [28].

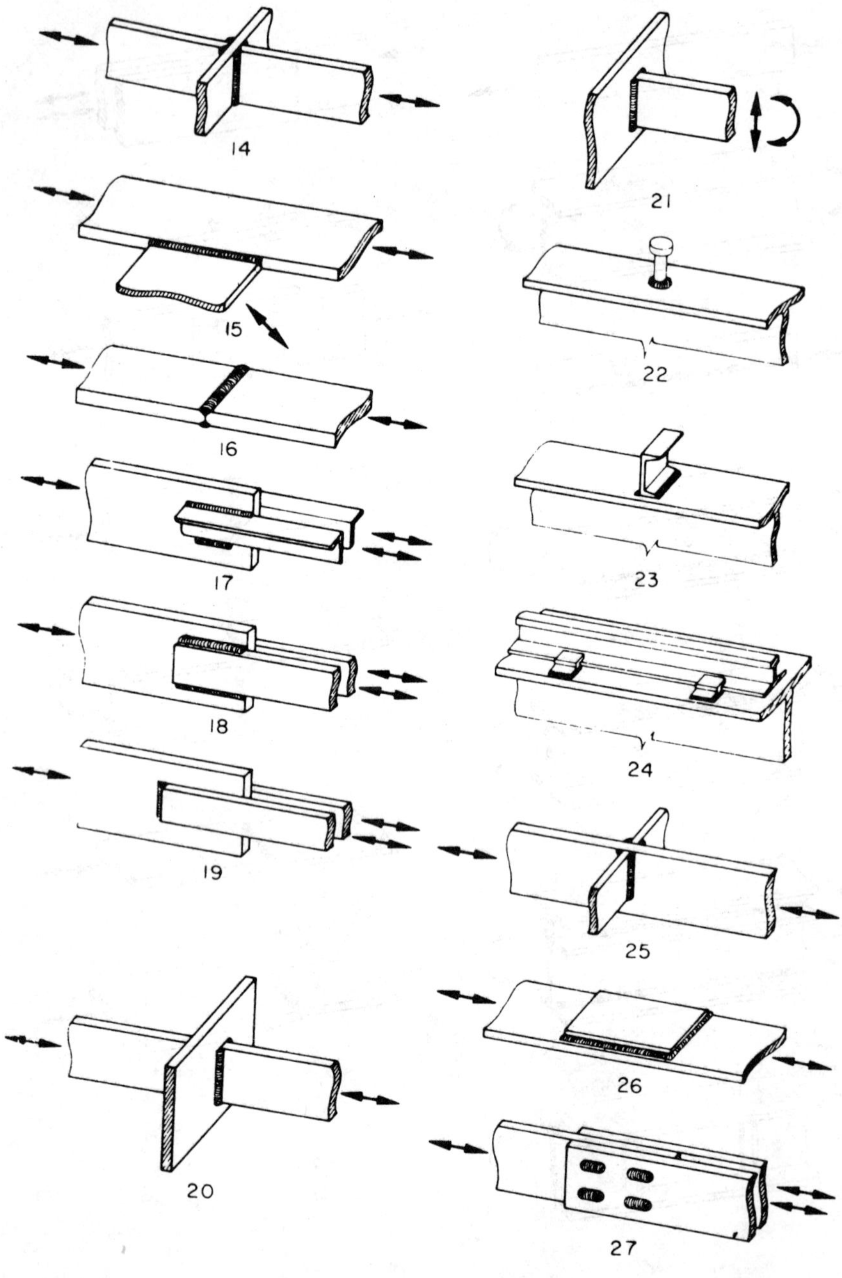

FIG. 10—*Continued.*

stress range for constant cycle loading and a 95 percent level of reliability were determined (see Table 3).

A selection must be made next in the loading function. Field data in the form of histograms (see Fig. 11) can be of help in choosing the appropriate probability density function that best represents the expected loading history for a particular design. Various distributions such as the Rayleigh [*33*] or beta distribution [*31*] can be used. An indication of three variations in beta distributions to represent markedly different loading histograms is presented in Fig. 12. Changing the coefficients of the beta distribution makes possible an infinite variety in such loading functions.

Using the beta distributions shown in Fig. 12, the linear damage rule, and the method presented by Ang and Munse [*31*] mean loading coefficients can be established which provide approximately and very simply for a beta probability density function of loading and a desired level of reliability. The appropriate design stress range then may be found from the following equa-

TABLE 3—*Basic fatigue stress range, for details in Fig. 10 (constant cycle - 0.95 reliability).*

Detail[a] (See Fig. 10)	Stress Range, MPa (ksi), for n cycles			
	n = 100 000 S_{r1}	n = 500 000 S_{r2}	n = 2 000 000 S_{r3}	n = 10 000 000 S_{r4}
1	236(34.2)	200(29.0)	174(25.2)	148(21.4)
2	228(33.0)	178(25.8)	143(20.8)	112(16.2)
3	203(29.5)	152(22.0)	118(17.1)	88(12.8)
4	223(32.4)	124(18.0)	75(10.9)	42(6.1)
5	105(15.2)	63(9.2)	41(5.9)	25(3.6)
6	223(32.4)	124(18.0)	75(10.9)	42(6.1)
7	155(22.5)	98(14.2)	66(9.6)	42(6.1)
8	239(34.6)	193(28.0)	161(23.4)	131(19.0)
9	135(19.6)	109(15.8)	90(13.1)	73(10.6)
9(S)	187(27.1)	150(21.8)	125 (18.1)	101(14.6)
10	160(23.2)	99(14.4)	66 (9.6)	41(6.0)
11	177(25.7)	117(16.9)	81(11.8)	54(7.8)
12	142(20.6)	81(11.8)	50(7.3)	29(4.2)
13	191(27.7)	134(19.4)	99(14.3)	69(10.0)
14	145(21.0)	91(13.2)	61(8.8)	38(5.5)
15	106(15.4)	67(9.7)	45(6.5)	28(4.1)
16	130(18.9)	84(12.2)	58(8.4)	37(5.4)
17	107(15.5)	67(9.7)	45(6.5)	28(4.1)
18	66(9.5)	34(5.0)	20(2.9)	10(1.5)
19	114(16.6)	88(12.7)	70(10.1)	53(7.7)
19(S)	113(16.4)	86(12.5)	68(9.9)	52(7.6)
20	150(21.7)	88(12.8)	56(8.1)	33(4.8)
20(S)	76(11.0)	53(7.7)	39(5.7)	28(4.0)
21	177(25.7)	139(20.2)	113(16.4)	89(12.9)
22	177(25.7)	98(14.2)	59(8.5)	32(4.7)
23	150(21.8)	92(13.3)	60(8.7)	37(5.3)
24	150(21.8)	92(13.3)	60(8.7)	37(5.3)
25	157(22.8)	83(12.1)	48(7.0)	26(3.7)
26	116(16.8)	75(10.9)	52(7.5)	34(4.9)
27	92(13.3)	65(9.4)	48(7.0)	34(5.0)
27(S)	99(14.3)	69(10.0)	50(7.3)	35(5.1)

[a] (S) Indicates shear stress on fasteners or welds.

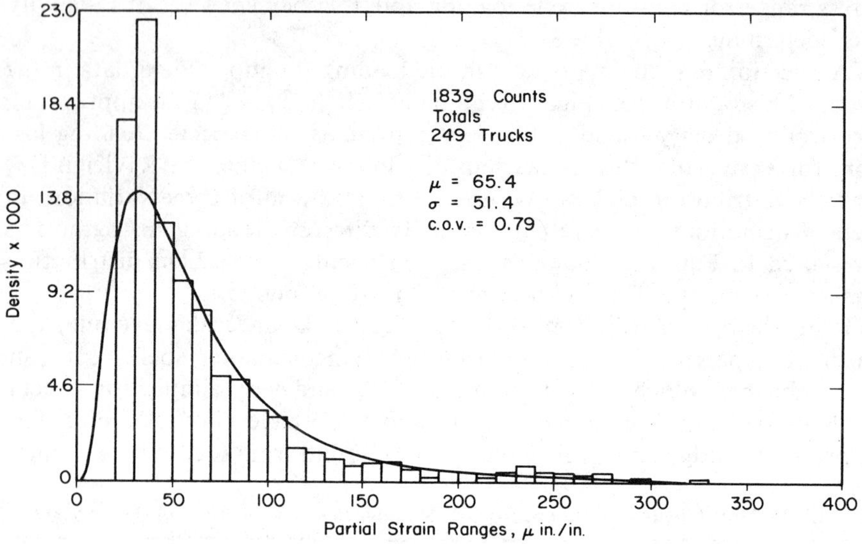

FIG. 11—*Histogram for bottom flange strain range for Shaffer Creek Bridge* [34].

tion

$$S_R = S_r \times C_L \tag{6}$$

where

S_r = the basic fatigue stress range for constant cycle loading and a 0.95 level of reliability (values from Table 3), and

C_L = the fatigue loading coefficient obtained from Table 4 for the appropriate loading shown in Fig. 12 and the desired level of reliability.

Other loading functions and other levels of reliability can be added readily to Table 4.

To evaluate the procedure, at least in part, use can be made of the random tension-biased loading shown in Fig. 13 and employed in a recent

TABLE 4—*Fatigue loading coefficients*, C_L.

Load Type	Load Description (See Fig. 12)	Loading Coefficient, C_L $R^* = 0.90$	$R^* = 0.95$	$R^* = 0.99$
I	Primarily light loading cycles, mean range of stress 30 percent of maximum	2.87	2.50	1.87
II	Medium loading cycles, mean range of stress 50 percent of maximum	2.07	1.80	1.35
III	Primarily heavy loading cycles, mean range of stress 70 percent of maximum	1.53	1.33	1.00
IV	Constant loading cycles, stress range constant and equal to 100 percent of maximum	1.15	1.00	0.75

NOTE—R^* is the level of reliability.

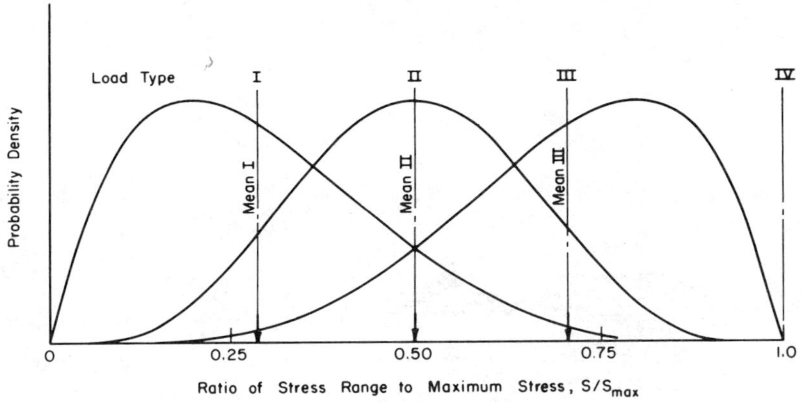

FIG. 12—*Beta distributions to model loading frequency distributions.*

study of butt welded joints [*35*]. Tests were conducted under the random loading and also under constant cycle loading (see Fig. 14 for the results). The predicted fatigue resistance of the members determined on the basis of the reliability procedure just described is also shown in Fig. 14. The agreement is excellent, indicating that realistic design procedures are certainly possible for random loads and should be sought for more effective design of welds and weldments in such structures as welded bridges.

In the design process, the designer of a structure that is subjected to repeated loads must classify the details in his structure that will be subjected to fatigue conditions. Although fatigue information is available on many details (such as shown in Fig. 10), in many instances, data are not available or details must be classified in terms of similar details for which the behavior is known [*36*]. This judgment is an important part of the process and requires a basic knowledge and understanding of the fatigue behavior of the various types of welds and weldments and of the factors that affect their behavior.

Also of importance for proper design is a knowledge of: (*a*) the great range in fatigue resistance possible in welds and welded details, (*b*) the locations in structures that are susceptible to fatigue cracking (for example, see Fig. 15 [*37*]), and (*c*) an appreciation of the need for quality control and adequate inspection to control defects at acceptable levels.

A variety of methods are available for the inspection of structures that are subjected to repeated loadings. However, the selection must be made with care since the sensitivity and costs of the methods vary greatly. Furthermore, the value of the various inspection methods is no better than the operator who is conducting the inspection [*38*].

Summary and Recommendations

Through fifty years of fatigue research on welds and weldments, many

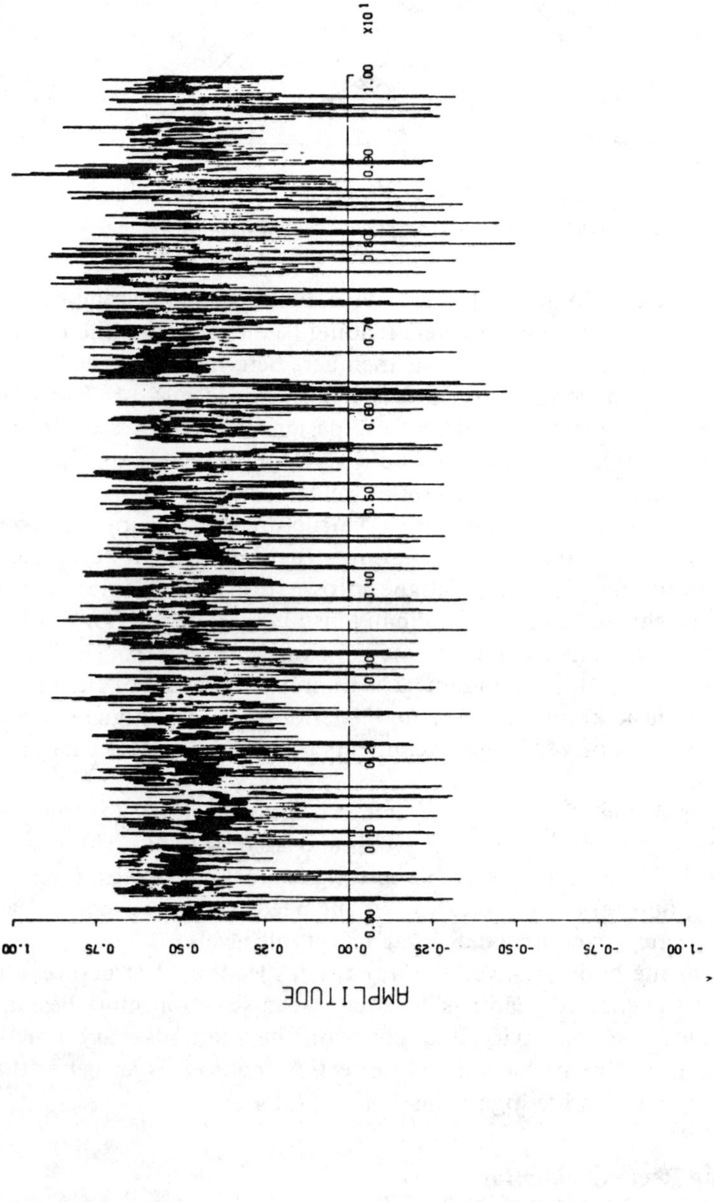

FIG. 13—*1710-cycle tensile biased random loading history from computer-controlled test of welded joint* [35].

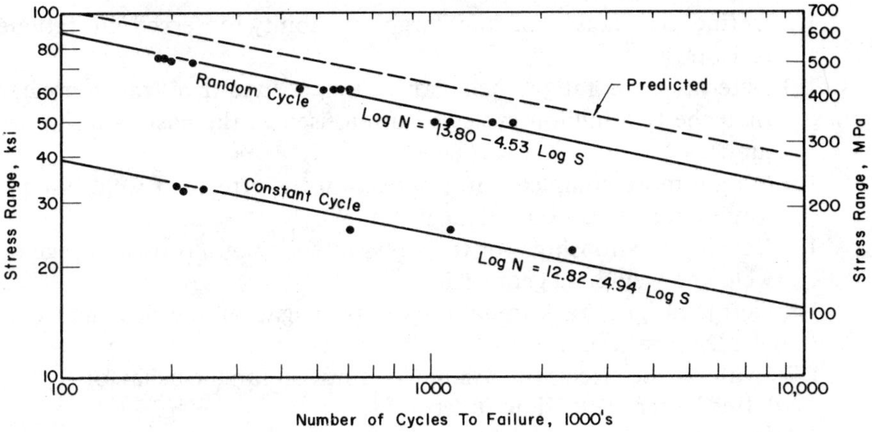

FIG. 14—*Comparison of fatigue tests under constant-cycle and random loadings* [35].

thousands of tests have been conducted to evaluate the variables that affect their fatigue resistance. The resulting data have been used to develop the current fatigue design relationships for welded structures. However, much more information is needed to provide for more accurate and more effective design. Some of these needs are as follows:

1. To establish better fatigue resistance data
   1.1 Define more accurately the fatigue strengths of welds and weldments and their variability

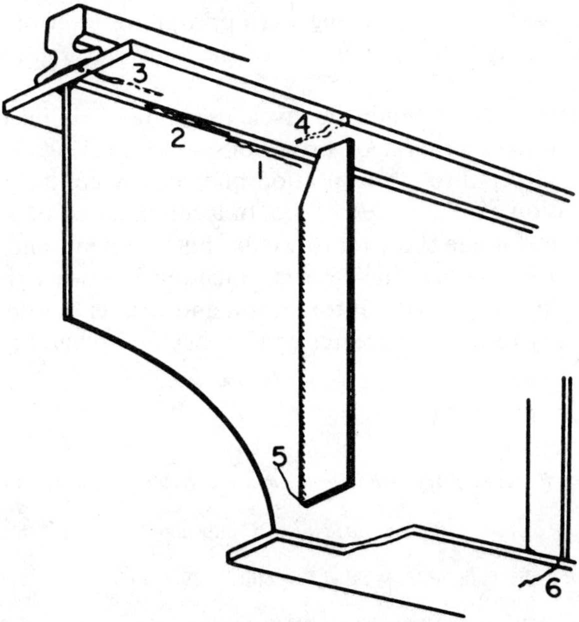

FIG. 15—*Location of fatigue cracks reported from the field* [37].

1.2 Define the causes of the large variability observed in fatigue resistance
1.3 Determine the fatigue behavior at longer lives (lives much greater than the two million cycles so often used in the past as an upper limit)
1.4 Obtain more complete information on the effects of weld discontinuities or defects on fatigue behavior
1.5 Develop relationships for the stress intensity of a broader range of weld and weldment geometries
1.6 Define fatigue crack initiation and propagation in welds and weldments more accurately
1.7 Evaluate the effects of variable loadings on fatigue behavior

2. To define better the loading functions
   2.1 More realistic loading histories for the many types of structures that are subjected to repeated loadings are needed (This is probably one of the most important areas in which further information is required.)
   2.2 Develop methods whereby predictions of future loading requirements might be established
   2.3 Evaluate in detail the various probabilistic loading density functions usable for fatigue design
3. To develop more effective design criteria
   3.1 Evaluate the various design criteria in existence and under development
   3.2 Consider ways of establishing the appropriate level of reliability and of defining better the effects of the many variables associated with fatigue
   3.3 Consideration also should be given to the possible introduction into the fatigue criteria or provisions to control against brittle fracture, either through inspection and quality control, through the limitation of allowable stresses, or a combination of these

Knowledge of the fatigue behavior of welds and weldments and of the design of welded structures has increased tremendously during the past 50 years; however, much more basic information and further developments in design are necessary to provide greater consistency and reliability in safety and economy.

References

[1] Munse, W.H., *Fatigue of Welded Steel Structures*, Welding Research Council, New York, 1964.

[2] Gurney, T.R., *Fatigue of Welded Structures*, Cambridge University Press, England, 1968.

[3] "Symposium on the Fatigue of Welded Structures, 29 March-1 April 1960," *British Welding Journal*, Mar.-Sept., 1960.

[4] Munse, W.H., Stallmeyer, J.E., and Drew, F.P., "Structural Fatigue and Steel Rail-

road Bridges," Proceedings of American Railway Engineering Association Seminar, 1968.

[5] "Symposium on Structural Fatigue," *Journal of the Structural Division,* American Society of Civil Engineers, Vol. 44, No. ST12, Dec. 1968, pp. 2663–2797.

[6] "Proceedings of the Conference on Fatigue of Welded Structures," 6-9 July 1970, The Welding Institute, Cambridge England, 1971.

[7] Pollard, B. and Cover, R.J., *Welding Journal,* Vol. 51, No. 11, Nov. 1972, pp. 544s–554s.

[8] Fisher, J.W., Frank, K.H., Hirt, M.A., and McNamee, B.W., "Effect of Weldments on the Fatigue Strength of Steel Beams," NCHRP Report 102, Highway Research Board, 1970.

[9] Fisher, J.W., Albrecht, P.A., Yen, B.T., Klingerman, D.J., and McNamee, B.M., "Fatigue Strength of Steel Beams with Welded Stiffeners and Attachments," NCHRP Report 147, Transportation Research Board, 1974.

[10] Sanders, W.W., Derecho, A.T., and Munse, W.H., *Welding Journal,* Vol. 44, No. 2, Feb. 1965, pp. 49s–55s.

[11] Lawrence, F.V. and Mainali, P.C., "Fatigue Crack Propagation Life Predictions for Butt and Fillet Welds," FCP Report No. 11, College of Engineering, University of Illinois, Urbana-Champaign, March 1974.

[12] Williams, H.E., Óttsen, H., Lawrence, F.V., and Munse, W.H., "The Effects of Weld Geometry on the Fatigue Behavior of Welded Connections," SRS No. 366, Department of Civil Engineering, University of Illinois, Urbana-Champaign, Aug. 1970.

[13] Derecho, A.T. and Munse, W.H., "Stress Concentration at External Notches in Members Subjected to Axial Loadings," Engineering Experiment Station Bulletin 494, University of Illinois, Urbana-Champaign, 5 Jan. 1968.

[14] Ekstrom, D.H. and Munse, W.H., "The Effect of Internal Weld Defects on the Fatigue Behavior of Welded Connections," SRS No. 395, Department of Civil Engineering, University of Illinois, Urbana-Champaign, Feb. 1973.

[15] Radziminski, J.B., Srinivasan, R., Moore, D., Thrasher, C., and Munse, W.H., "Fatigue Data Bank and Data Analysis Investigation," SRS No. 405, Department of Civil Engineering, University of Illinois, Urbana-Champaign, June 1973.

[16] Wilson, W.M., Munse, W.H. and Bruckner, W.H., "Fatigue Strength of Fillet-Weld, Plug-Weld, and Slot-Weld Joints Connecting Steel Structural Members," Bulletin Series No. 380, University of Illinois, Engineering Experiment Station, May 1949.

[17] Munse, W.H. and Stallmeyer, J.E., "Fatigue in Welded Beams and Girders," Bulletin 315, Publication No. 961, Highway Research Board, 1962, pp. 45–62.

[18] Abtahi, A., Albrecht, P., and Irwin, G.R., *Journal of the Structural Division,* Proceedings of the American Society of Civil Engineers, Vol. 102, No. ST11, Nov. 1976, pp. 2103–2119.

[19] Mattos, R.J. and Lawrence, F.V., "Estimation of the Fatigue Crack Initiation Life in Welds Using Low Cycle Fatigue Concepts," FCP Report No. 19, College of Engineering, University of Illinois, Urbana-Champaign, Oct. 1975.

[20] Rolfe, S.T. and Barsom, J.M., *Fracture and Fatigue Control in Structures—Applications of Fracture Mechanics,* Prentice-Hall, N.J., 1977.

[21] Herman, P.M., "Initiation and Propagation of Fatigue Cracks in Notched Steel Plates," Department of Civil Engineering, University of Illinois, Urbana-Champaign, Oct. 1974 (Special Research Study).

[22] Hoeppner, D.W. and Krupp, W.E., *Engineering Fracture Mechanics,* Vol. 6, 1974, pp. 47–70.

[23] Maddox, S.J., *Welding Journal,* American Welding Society, Vol. 53, No. 9, Sept. 1974, pp. 401s–409s.

[24] Lawrence, F.V., *Welding Journal,* American Welding Society, Vol. 52, No. 5, May 1973, pp. 212s–220s.

[25] Burk, J.D., and Lawrence, F.V., "Influence of Bending Stresses on the Fatigue Crack Propagation Life in Butt Welds," FCP Report No. 21, College of Engineering, University of Illinois, Urbana-Champaign, March 1976.

[26] Lawrence, F.V. and Radziminski, J.B., *Welding Journal,* Vol. 49, No. 10, Oct. 1970, pp. 445s–452s.

[27] "Structural Welding Code," American Welding Society, D1.1–75, 1975.

[28] "Specification for the Design, Fabrication and Erection of Structural Steel For Buildings," American Institute of Steel Construction, 12 Feb. 1969.

[29] "Standard Specifications for Highway Bridges,"American Association of State Highway and Transportation Officials, 1975.

[30] "Steel Structures-Specifications for Steel Railway Bridges," American Railway Engineering Association, 1975.

[31] Ang, A.H-S. and Munse, W.H., "Practical Reliability Basis for Structural Fatigue," American Society of Engineers' National Structural Engineering Conference, 14–18 April 1975 (Preprint No. 2494).

[32] Moses, Fred, "New Procedure for Fatigue Design of Highway Bridge Girders," No. 507, Transportation Research Record, Washington, D.C., 1974, pp. 58–67.

[33] Schilling, C.G., Klippstein, K.H., and Reilly, R.J., "Simulated Traffic Fatigue Loading of Steel Bridges," *Proceedings,* American Society of Engineers' Specialty Conference on Metal Bridges, St. Louis, Mo., 12–13 Nov., 1974, pp. 379–410.

[34] Walker, W.H. and Ruhl, J.A., "Investigation of Dynamic Stress in Highway Bridges-An Interim Report," SRS No. 404, Department of Civil Engineering, University of Illinois, Urbana-Champaign, Oct. 1973.

[35] Stefaniuk, A.R., "Fatigue Behavior of Butt Welded Joints Under a Random Loading." Department of Civil Engineering, University of Illinois, Urbana-Champaign, May 1975 (Special Research Study).

[36] "Classification of Welded Bridge Details for Fatigue Loading," Highway Research Board, National Cooperative Highway Research Program, Research Results Digest, No. 59, March 1974.

[37] Gorenc, B.E., *Journal of the American Institute of Steel Construction,* Steel Construction, Vol. 10, No. 4, 1976, pp. 2–18.

[38] "Nondestructive Methods of Fatigue Crack Detection in Steel Bridge Members," Highway Research Board, National Cooperative Highway Research Program, Research Results Digest, No. 66, Jan. 1975.

P. J. Haagensen [1]

Effect of Tungsten Inert Gas Dressing on Fatigue Performance and Hardness of Steel Weldments

REFERENCE: Haagensen, P. J., **"Effect of Tungsten Inert Gas Dressing on Fatigue Performance and Hardness of Steel Weldments,"** *Fatigue Testing of Weldments, ASTM STP 648*, D. W. Hoeppner, Ed., American Society for Testing and Materials, 1978, pp. 113–133.

ABSTRACT: The fatigue performance of steel weldments can be improved significantly by remelting the toe region using standard tungsten inert gas (TIG) equipment. While the cost of TIG dressing is much lower than for comparable methods, the TIG process causes an increase in maximum hardness in the base metal adjacent to the remelted material. This may preclude the use of the TIG method for some applications. The hardness distribution in TIG dressed T- joints of St 52.3N and NVE 36 steel plates with thickness ranging from 20 to 38 mm was studied. A modified TIG dressing technique involving a second (tempering) run was developed, and a substantial reduction in maximum hardness as compared with conventional TIG dressing was obtained.

Fatigue tests were performed on specimens with load carrying fillet welds in the as-welded and TIG-dressed conditions. The material was a quenched and tempered steel with a yield strength of 880 MN/mm^2 (128 ksi). The increase in fatigue strength at the endurance limit due to TIG dressing amounted to approximately 65 percent.

KEY WORDS: fatigue tests, weldments, hardness in welded joints, fatigue (materials), fatigue strength at N cycles, gas tungsten arc weldiing

The fatigue strength of steel weldments, in particular, fillet welds, is known to be largely unrelated to the ultimate strength of the base material. Thus the fatigue strength of welded high strength steel, in general, does not exceed the value of the fatigue strength of mild steel. Consequently, the use of high strength steels in welded structures has few advantages when fatigue is a limiting factor.

There are several reasons for the poor fatigue performance of steel weldments, the most important ones being the stress concentration caused by the weld bead and the presence of crack-like slag intrusions at the toe of the

[1] Senior research engineer, The Foundation of Scientific and Industrial Research at the University of Trondheim (SINTEF), Norway.

weld, as shown by Signes et al [*1*].[2] Fatigue cracks are initiated very shortly after the start of dynamic loading and the major portion of the fatigue life consists of the growth of fatigue cracks to final failure. Since the crack growth rate varies little with the ultimate strength, the fatigue strength of welded high strength steels cannot be expected to exceed the fatigue strength of mild steel weldment by any large amount.

The obvious way of improving the fatigue strength of steel weldments in general is to prolong the crack initiation stage. Several techniques have been developed for this purpose [*2-5*], for example, the introduction of favorable compressive residual stresses to the toe area of the weld by different techniques such as prestraining, peening (by hammer or shot blasting), or by spot heating.

For industrial applications, however, the most important methods are grinding and tungsten inert gas (TIG) dressing. In both methods, the increase in fatigue strength is achieved through improvement of the weld geometry and removal of weld defects in the toe region. The TIG dressing technique consists of remelting the toe of the weld, using standard equipment for manual welding without the addition of filler material.

While grinding with a small-diameter rotating file and TIG dressing give comparable results with respect to increase in fatigue life [*6*], TIG dressing offers some important advantages in terms of economy [*7*], improved working environment (less noise and dust), and ease of inspection.

The major drawback of the TIG method is the increase in hardness that occurs in the heat affected zone (HAZ) in the base material adjacent to the remelted metal. For several types of structures, for example, pressure vessels and offshore platforms, maximum allowable hardness values are usually specified in the welding procedure qualification tests that are required by controlling or classifying agencies such as American Society of Mechanical Engineers (ASME), Lloyds, and Det norske Veritas, a typical value for welds in offshore structures [*8*] being 300 VHN_{10}[3]. Since TIG dressing of welded carbon steels may give hardness values well in excess of this number, this is a major obstacle to the application of the TIG method to high performance welds.

This paper reports some early results from an investigation in which the feasibility of adapting the TIG method for use in the fabrication of offshore structures was studied. The investigation is part of a current research project which is also aimed at quantifying the effect of TIG dressing on fatigue strength under various conditions; results from some introductory fatigue tests are given.

[2] The italic numbers in brackets refer to the list of references appended to this paper.
[3] Vickers hardness number obtained with a 10-kg load.

Materials and Specimens

The materials used for the hardness tests were St.52.3N and NVE 36 steel. The specified chemical compositions of these steels are identical, see Table 1. The fatigue tests were performed on a high strength steel, KF 23 M 14B, heat treated to a yield strength of 880 MN/m^2. The chemical composition and mechanical properties of the three steels are given in Table 1 and 2, respectively.

Specimens for hardness testing were cut from the manually welded joints shown in Fig. 1. All welds were full penetration (*K*) welds with a bevel angle of 45 deg. The total number of passes per weld was approximately 18. All welding was performed at a preheat temperature of 150°C. Further details of the welding procedures are given in Refs *9* and *10*. The dimensions of the joints are shown in Fig. 1.

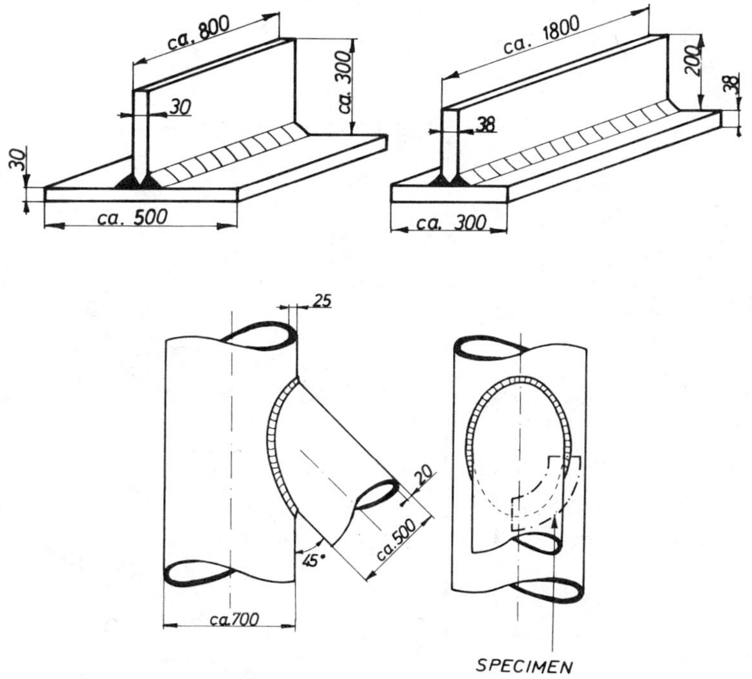

FIG. 1—*Dimensions of welded joints for the hardness investigation, all measurements in millimetres.*

The fatigue specimens were cut by a cold saw from long welded strips. The welds were single-pass fillet welds. The specimens were heat treated after welding and TIG-dressing. Only the fatigue specimens were heat treated. The geometry of the fatigue specimen is shown in Fig. 2.

TABLE 1—*Chemical composition of test materials, weight percent, maximum values.*

Steel	C	Si	Mn	P	S	Cr	Ti	Al	B
St.52.3N	0.18	0.50	1.6	0.04	0.04	. . .	. . .	. . .	. . .
NVE 36	0.18	0.50	1.6	0.04	0.04	. . .	. . .	. . .	. . .
KF23M14B	0.20 to 0.25	0.20 to 0.30	2.70 to 0.90	max 0.03	max 0.004	0.010 to 0.020	0.020 to 0.050	0.020 to 0.050	0.0025 to 0.0070

TABLE 2—*Mechanical properties of test materials (nominal values).*

Steel	Tensile strength, MN/m^2	Yield strength, MN/m^2	Elongation, %
St.52.3N	510	350	...
NVE 360	510	350	...
KF23M14B	950	880	16

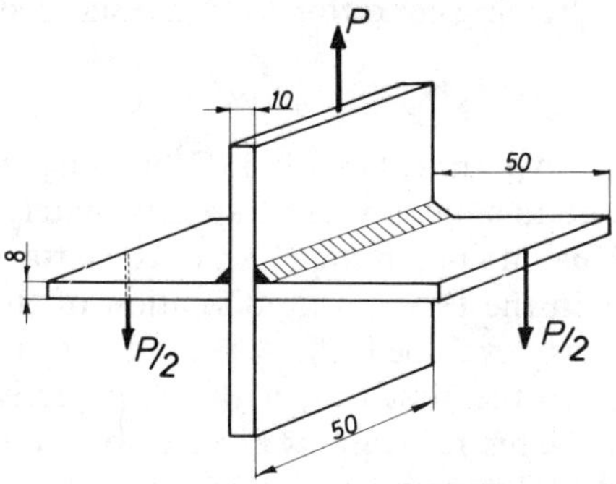

FIG. 2—*Fatigue specimen, all measurements in millimetres.*

TIG Dressing Conditions

All TIG dressing was performed manually using a standard direct current TIG welding machine of 400-A capacity with a 3.2-mm-diameter 2 percent thorium tungsten electrode. Argon shielding gas of 99.95 percent purity was used at flow rates of 10 to 12 litres/min.

The TIG dressing was performed in accordance with recommendations given by Millington [*11*] and Kado et al [*12*]. The position of the electrode relative to the weld was as indicated in Fig. 3; this was found by Kado et al [*13*] to provide an optimum weld bead profile.

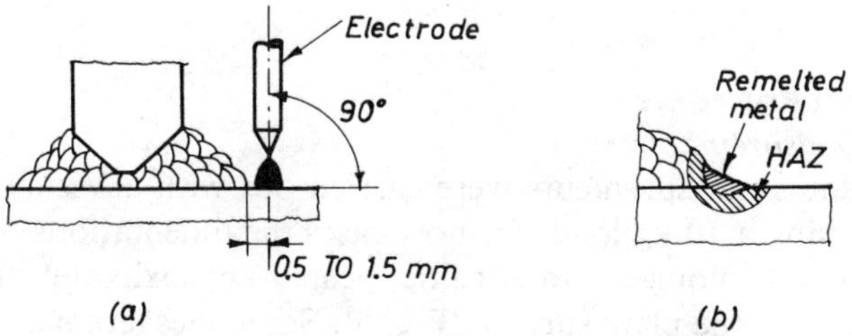

FIG. 3—*TIG dressing:* (a) *position of electrode relative to the weld, and* (b) *resulting bead profile, schematic.*

The various forms of TIG dressing were performed on portions of the welds which in no cases were less than 60 mm long. All dressing of 30-mm-thick T-joints and the tubular joint were done on the horizontal (2F) welding position while some of the dressing of the 38-mm-thick T-joints were dressed in the vertical (3F) position. The majority of the 38-mm joints were dressed under production conditions outdoors in a shipyard.

All four toes were dressed except for the 38-mm-thick joints and the fatigue specimens, only the toes of the horizontal plate were dressed on these specimens. The TIG dressing conditions are summarized in Table 3.

Modified TIG Dressing

In all introductory experiments with TIG dressing, the maximum hardness values were found in a narrow zone, approximately 1 mm wide, located 0.5 mm outside the fusion line, in the HAZ of the base material. In order to reduce the hardness in the HAZ, a modification of the conventional TIG dressing technique was developed [9]. A second TIG run was made on the weld metal adjacent to the first run, with the objective of tempering the HAZ of the first run. This principle is illustrated in Fig. 4. The conditions used in the modified TIG dressing method are summarized in Table 4.

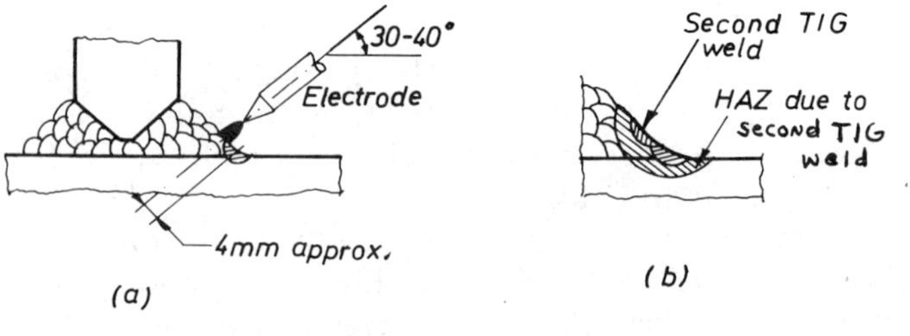

FIG. 4—*Modified TIG dressing technique:* (a) *position of electrode relative to the weld and* (b) *resulting weld bead profile, schematic.*

Experimental Procedure

Hardness Measurements

The hardness measurements were carried out with a Vickers-type hardness tester using a 10-kg load. In most cases the indentations were made at 0.5-mm intervals along a straight line located approximately 1 mm below and parallel with the plate surface (Fig. 5). Some measurements were made with a 0.5-kg load ($VHN_{0.5}$) at 0.2-mm intervals. There were no significant hardness variations in the fatigue specimens due to the fact that these specimens were heat treated after fabrication.

TABLE 3—*TIG dressing conditions.*

Joint Type	Toe[a]	Weld portion	Dressing position[b]	Voltage, V	Current, A	Welding Speed, mm/min	Preheat Temperature, °C	Heat Input,[c] kJ/mm	Remarks
T-joint, 30-mm-thick plates	1	all	2F	9.0	150	200	150	0.41	dressing performed in laboratory
	2	all	2F	9.5	160	110	150	0.83	
	3	all	2F	9.5	160	140	150	0.65	
	4	all	2F	9.5	165	180	150	0.52	
Tubular joint, 20 and 25-mm-thick plates	1	all	2F	9.5	160	140	150	0.65	dressing performed in laboratory
	2	all	2F	9.5	160	140	150	0.65	
	3	all	2F	9.5	160	140	150	0.65	
	4	all	2F	9.5	160	140	150	0.65	
T-joint, 38-mm-thick plates	1,2	I	2F	9.0	160	150	none	0.58	dressing performed in open air in shipyard
	1,2	I	3F	9.0	160	160	none	0.54	
	1,2	II	2F	9.0	160	210	150	0.41	
	1,2	II	3F	9.0	160	190	150	0.46	
	1,2	III	2F	9.0	225	160	150	0.76	
	1,2	III	3F	9.0	225	240	150	0.51	
	1,2	IV	2F	9.0	236	200	none	0.61	dressing performed in laboratory
	1,2	IV	3F	9.0	232	170	none	0.78	
Fatigue specimens	1,2	all	2F	12	200	350	none	0.41	dressed in fabrication shop

[a] 1 and 2 refer to the toes on the horizontal plate, and 3 and 4, toe on the vertical rib.
[b] 2F is the horizontal (downhand) position, and 3F is the vertical position.
[c] The heat input is computed from 60 VA/1000S (kJ/mm), where V = volts, A = current in amperes, and S = welding speed in mm/min.

TABLE 4—*Conditions of the modified TIG dressing process.*

Joint Type	Toe[a]	Weld Portion	Dressing Position[b]	Voltage, V	Current, A	Welding Speed, mm/min	Preheat Temperature, °C	Heat input,[c] kJ/mm	Distance between TIG Welds, mm	Remarks
T-joint, 30-mm-thick plates	all	all	2F	10	165	140	150	0.71	4	
	four	all	2F	10	170	150	150	0.68	4	
T-joint, 38-mm-thick plates	1,2	IV,B+C	2F	9.0	230	170	none	0.73	4 to 6	
	1,2	IV,E+F	2F	9.0	230	80	none	1.55	4 to 6	weaving in a 4-mm-wide band
	1,2	IV,B+C	3F	9.5	230	140	none	0.94	4 to 6	
	1,2	IV,E+F	3F	9.5	230	60	none	2.07	4 to 6	weaving in a 10-mm-wide band
	1	I,E	2F	9.0	230	176	150	0.70	2 to 2.5	
	2	I,E	2F	9.0	230	176	150	0.70	2 to 2.5	

[a] 1 and 2 refer to the toes on the horizontal plate, and 3 and 4, toe on the vertical rib.
[b] 2F is the horizontal (downhand) position, and 3F is the vertical position.
[c] The heat input is computed from 60 VA/1000 S (kJ/mm where V = volts, A = current in amperes, and S = welding speed in mm/min.

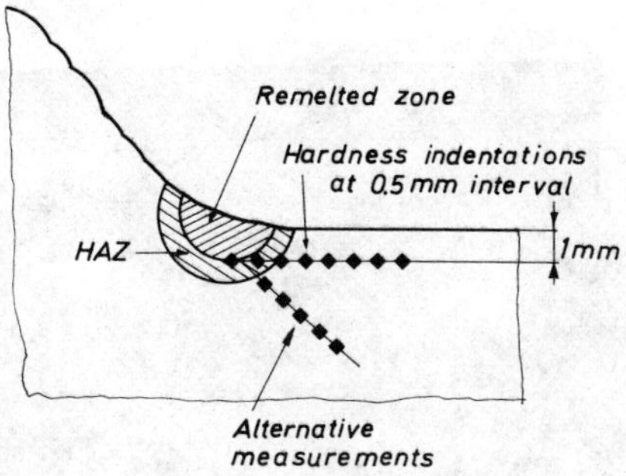

FIG. 5—*Position of hardness indentations.*

Fatigue Testing

The fatigue tests were conducted in a room-temperature environment using a 10-kN electrohydraulic closed loop testing machine. The fillet-welded arms of the specimens were subjected to constant-amplitude loading in bending, a load ratio (R) of 0.1 being used in all tests. The testing frequency was 30 Hz.

Results

Profile Improvements

The geometry of the welds were clearly improved by TIG dressing as evidenced by Fig. 6.

Hardness Distribution

The results of the hardness measurements are summarized in Table 5. For each portion of the weld, corresponding to a given set of welding and TIG dressing conditions, several sections were cut and the maximum hardness at each toe was determined. The maximum values were averaged; only these average values are listed in Table 5. The increases in hardness caused by the different TIG dressing techniques as compared with the as-welded hardness are listed in Table 6.

In Fig. 7(a), the hard zone adjacent to the TIG weld is shown; Fig. 7(b) shows the microstructure in the zone of maximum hardness just outside the fusion line.

Fatigue Test Results

The results of the fatigue test are given in the *S-N* diagrams in Figs. 8 and 9. The straight lines representing average fatigue life in the finite life region of Fig. 8 were fitted by linear regression analysis assuming a stress-life relationship of the form

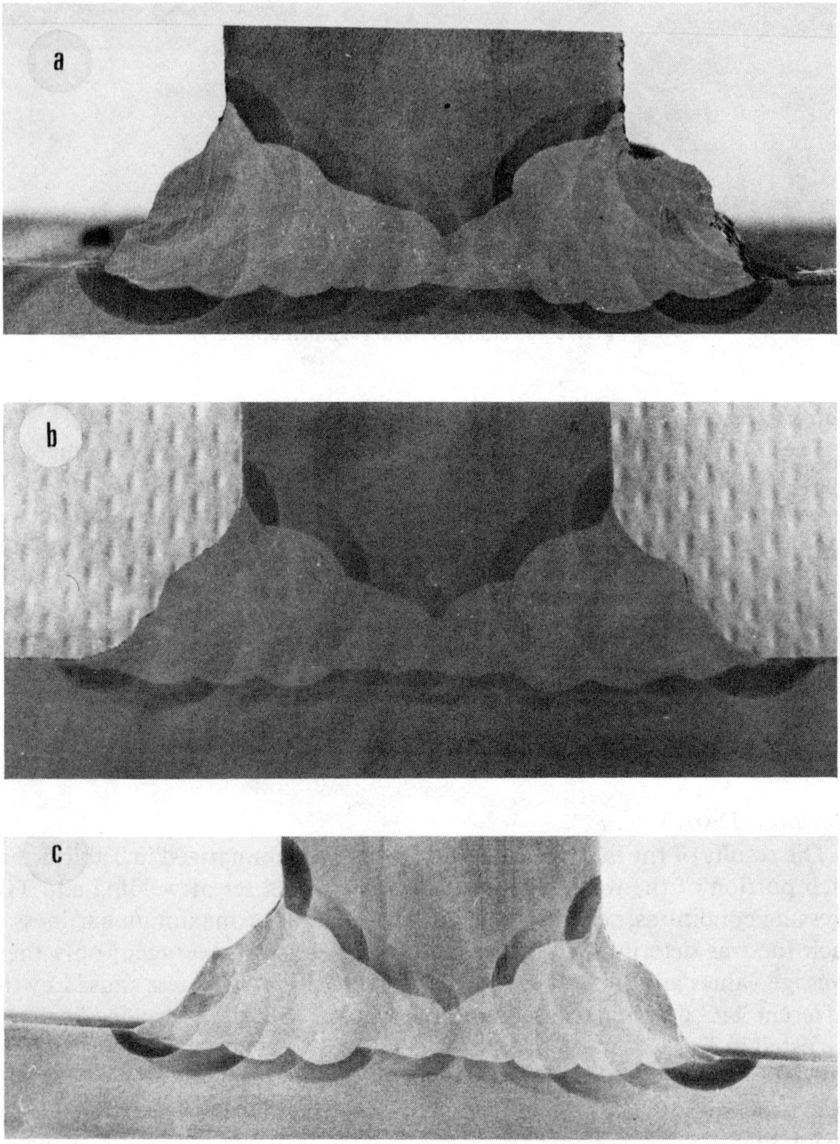

FIG. 6—*Examples of improvement of weld profiles due to TIG dressing.* (a) *As welded, 30-mm plate.* (b) *Conventional TIG dressing, 30-mm plate.* (c) *Modified TIG dressing, 30-mm plate.*

$$N(\Delta\sigma)^b = C$$

where b and C are constants, and N and $\Delta\sigma$ are cycles to failure and stress range, respectively.

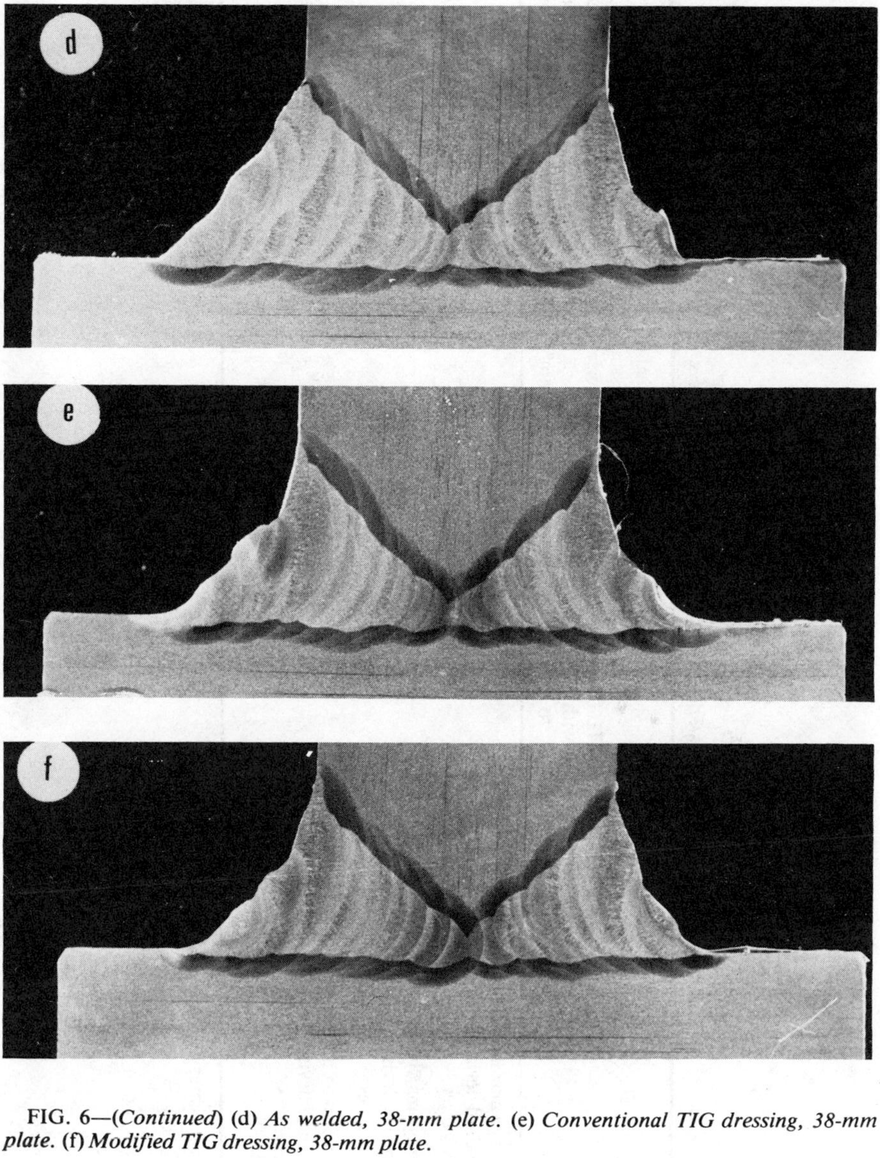

FIG. 6—(*Continued*) (d) *As welded, 38-mm plate.* (e) *Conventional TIG dressing, 38-mm plate.* (f) *Modified TIG dressing, 38-mm plate.*

The horizontal parts (endurance limits) of the *S-N* curves in Fig. 8 were fitted by eye. Figure 8 shows that the endurance limit is raised from 200 to 330 MN/m^2 or 65 percent due to TIG dressing.

The increase in fatigue life varies with stress level in the finite life region; this is shown in Fig. 10.

In Fig. 9, the scatter bands bounded by the upper and lower 95 percent

TABLE 5—*Average values and standard deviation of maximum hardness.*

Joint Type		As Welded	TIG Dressed without Preheat	TIG Dressed with Preheat to 150°C	Modified TIG Dressing without Preheat	Modified TIG Dressing with Preheat to 150°C, Δ = 4 to 6 mm[a]	Modified TIG Dressing with Preheat to 150°C, Δ = 2 to 2.5 mm[a]	Hardness of Base Material	Hardness of Weld Material
T-joint, 30-mm plates	average max hardness	349		431		396		210	257
	standard deviation	45		14		25		12	13
	max values	8		8		4		17	19
T-joint, 38-mm plates	average max hardness	376	476	447	437		322	193	248
	standard deviation	58	11	12	16			10	25
	max values	6	6	8	8		2	28	22
Tubular joint, 20 and 25-mm plates	average max hardness	321		454		396		185	261
	standard deviation			42		25			13
	max values	2		4		4		2	4

[a] Δ = distance between the center lines of first and second TIG runs.

TABLE 6—*Summary of hardness increase due to TIG dressing over the as-welded hardness.*

Joint Type	Hardness due to Welding	Increase due to TIG Dressing without Preheat	Increase due to TIG Dressing with Preheat to 150°C	Increase due to Modified TIG Dressing without Preheat	Increase due to Modified TIG Dressing with Preheat to 150°C, Δ = 4 to 6 mm[a]	Increase due to Modified TIG Dressing with Preheat to 150°C, Δ = 2 to 2.5 mm[a]
T-joint, 30-mm plates	349		431 − 349 = 82		396 − 349 = 47	
T-joint, 38-mm thick plate	376	476 − 376 = 100	447 − 376 = 71	437 − 376 = 61		322 − 376 = −54
Tubular joint	321		454 − 321 = 133		395 − 321 = 74	

[a] Δ = distance between center lines of first and second TIG runs.

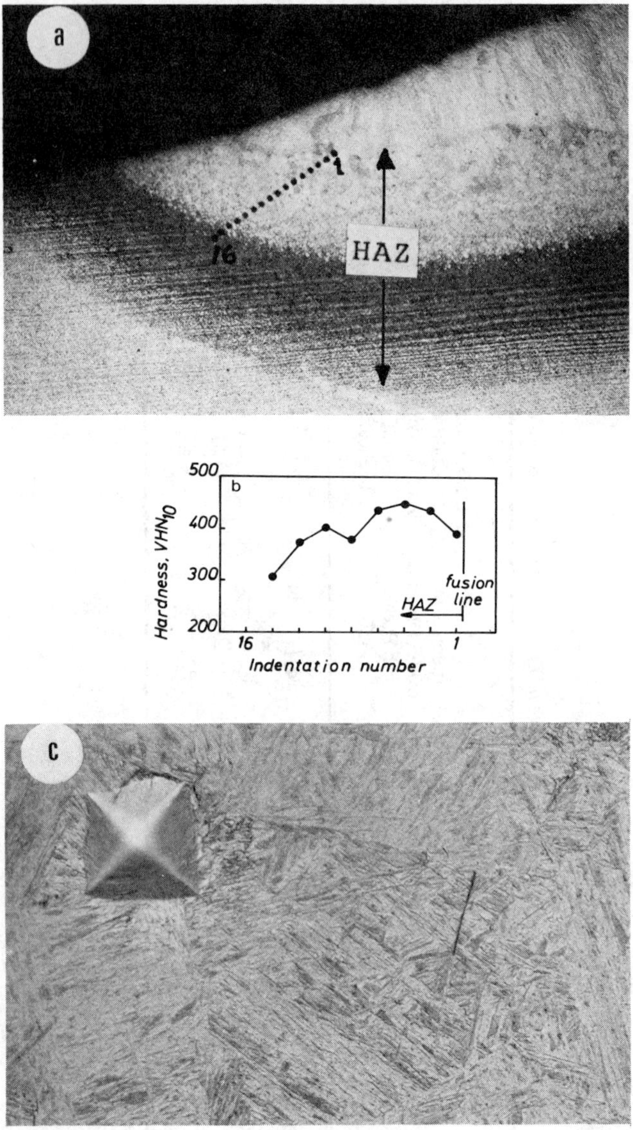

FIG. 7—(a) *Hardness measurements in the hard (bright) zone just outside a TIG weld, indentation 1 at the fusion line, Vickers hardness load, 0.5-kg (×13).* (b) *Hardness distribution in the hard zone in* (a). (c) *Indentation 2 in* (a) *($VHN_{0.5}$ = 450); martensite (×400).*

confidence limits are shown. Details of the regression analysis are given in Table 7.

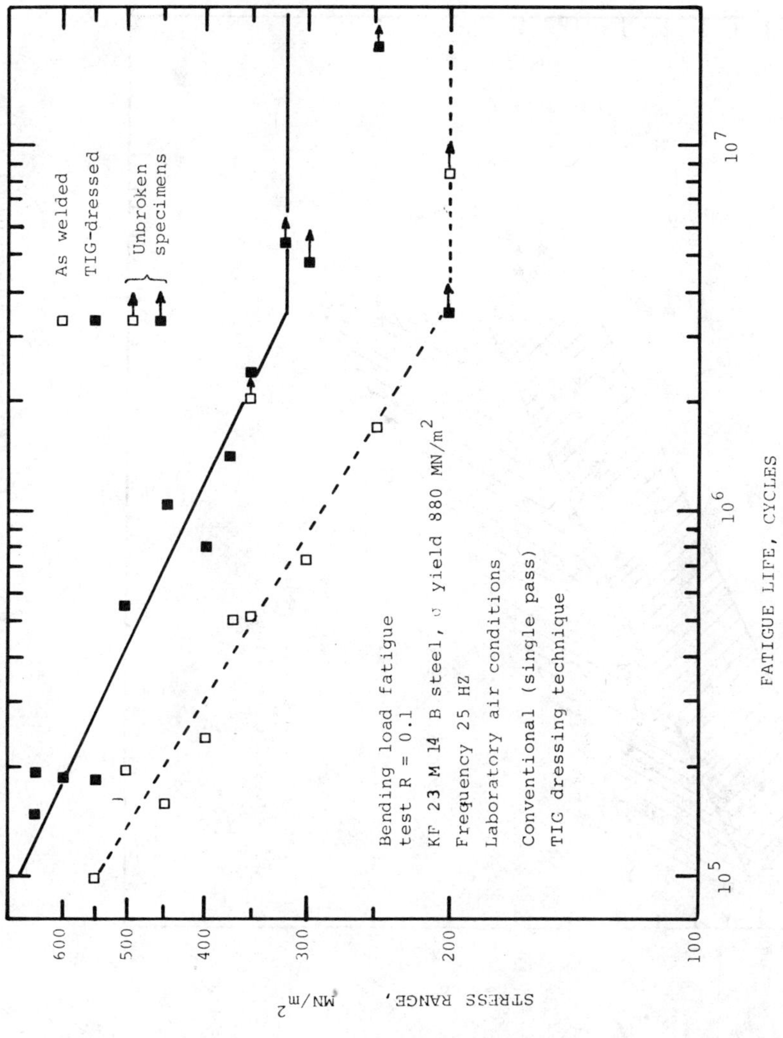

FIG. 8—S-N *curves.*

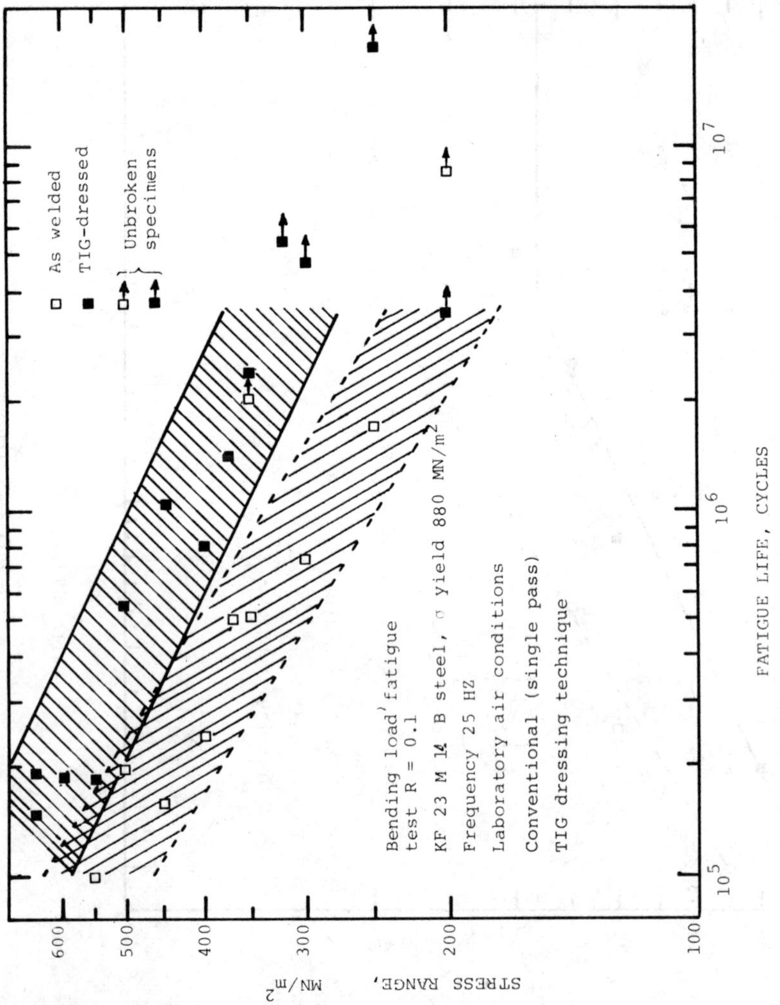

FIG. 9—*Scatterbands of S-N curves in Fig. 8 (95 percent confidence limits).*

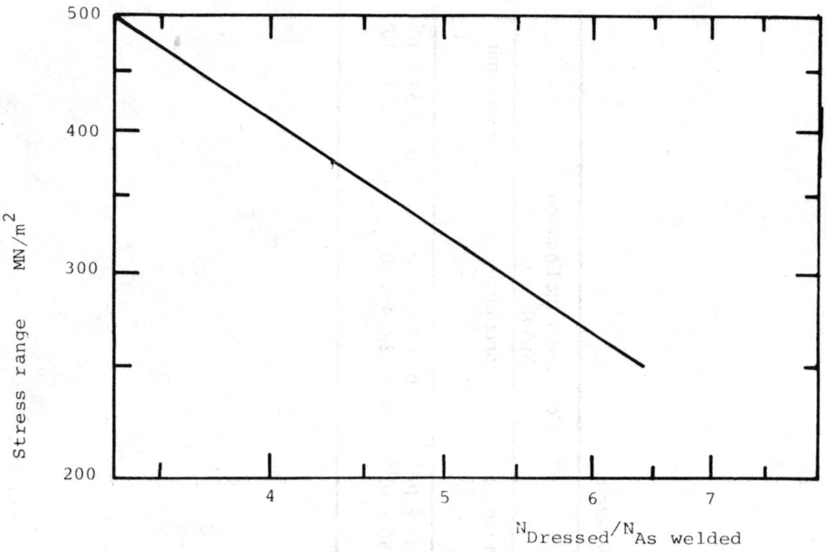

FIG. 10—*Variation in increase in fatigue life due to TIG dressing.*

Discussion

Hardness Investigation

The maximum hardness of all specimens in the as-welded condition exceeded the 300-VHN limit given in Ref *8*. Furthermore, the scatter of these measurements was much larger than for TIG dressed specimens. A possible reason for the large scatter is inaccurate positioning of the last weld passes relative to the HAZ; if the last passes are placed too far from the HAZ, the temperature will not be high enough to temper the martensite in the HAZ and large variations in hardness may be expected.

The measurements show that conventional TIG dressing gives a considerable increase in maximum hardness as compared with the as-welded hardness.

Variations in heat input between 0.41 and 0.76 kJ/mm produced no significant changes in maximum hardness for conventional TIG dressing. Preheating to 150°C, however, gave a small (~30 VHN) but distinct reduction in hardness. The small effects of variations in heat input and preheating on hardness are consistent with the results obtained by Kanazawa et al [*6*].

Modified TIG dressing gave a substantially smaller hardness increase than the conventional TIG technique when the second TIG pass was positioned on the weld metal at a distance of 4 to 6 mm from the first pass. Moreover, with a distance of 2 to 2.5 between the TIG welds, a hardness *reduction* was obtained, compared with the as-welded hardness. However, the

TABLE 7—*Details of regression analysis of fatigue results.*

Test Series	Number of Results (Fractured Specimens)	Standard Deviation of $\log_{10} N$	Correlation Coefficient	Constants in the Equation $N(\Delta\sigma)^b = C$					
				Mean Line		Upper Limit		Lower Limit	
				b	C	b_u	C_u	b_1	C_1
As-welded specimens	8	0.25	0.95	3.70	9.837×10^{14}	3.70	1.746×10^{15}	3.70	5.54×10^{14}
TIG dressed specimens (conventional dressing technique)	9	0.34	0.92	4.55	1.196×10^{18}	4.55	2.614×10^{18}	4.55	5.473×10^{17}

positioning of the second TIG weld is critical; too large a distance gives too low a temperature for tempering of the martensite in the HAZ. On the other hand, to avoid martensite formation, temperatures in excess of ~750°C must be avoided.

In a theoretical study of the effects of TIG dressing on hardness in carbon steels, Christensen [*14*] has shown that the conditions used in the present investigation are likely to produce hardnesses well in excess of 300 VHN. As an alternative to TIG dressing, he suggests tempering with a "soft" flame, for example, from a multinozzle propane torch, rectangular in shape and somewhat wider than the weld [*14*].

Further studies are necessary to ascertain the relative merits of the modified TIG method and the proposed flame tempering technique. In particular, the economics of the latter should be investigated.

Another aspect of TIG dressing which is deemed worthy of further study is the applicability of the hardness limits to TIG dressed welds. Such limits are imposed to avoid problems of brittle fracture and stress corrosion. A third problem is HAZ cracking which is related to the amount of trapped hydrogen which is present. Since the TIG welding process does not introduce any additional hydrogen to the weld, but is likely to reduce the amount of hydrogen that may be present, TIG dressing is likely to improve the resistance of welding joints to brittle fracture and stress corrosion cracking. In an investigation [*6*] of the impact properties of TIG dressed weldments, an increased resistance to brittle fracture was attributed to the reduced stress concentration. For these reasons a relaxation of the present limitations on maximum hardness should be possible.

Fatigue Tests

The test results indicate a consistent increase in fatigue life due to TIG dressing. The increase in endurance limit is somewhat smaller than reported for other high strength steels [*6*]; this is probably due to the fact that the transition (toe) region of the as-welded specimens in the present investigation was quite smooth. The increase in fatigue life is larger for the lower stresses (Fig. 10); the overlapping scatter bands in Fig. 9 show that the difference in fatigue life is not statistically significant on the 95 percent confidence level for stress ranges larger than approximately 550 MN/m^2. This is consistent with the well known fact that fatigue cracks are initiated at an early stage at high alternating stresses, regardless of the severity of the notch; thus *S-N* curves for notched and smooth specimens tend to converge in the low-cycle region.

Conclusions

1. The TIG dressing gives an appreciable increase in hardness in the HAZ of the base material, the increase amounting to up to 130 VHN_{10} in the present investigation.

2. Variations in heat input between 0.4 and 0.75 kJ/mm and preheating to 150°C have only a small effect on maximum hardness.

3. Modified TIG dressing with a second TIG pass gives a smaller hardness increase than the standard TIG method, the difference amounting to approximately 40 VHN_{10}. For a second TIG pass located 2 to 2.5 mm from the first pass, the modified TIG method gives a lower hardness than the original welding process. However, more tests are necessary to determine whether such results can be obtained consistently.

4. From a limited number of fatigue tests, the following results were obtained: the endurance limit was raised by 65 percent, from 200 to 330 MN/m^2 by TIG dressing; the fatigue life was increased 5 to 6 times at low stress levels.

Acknowledgments

The work described here is part of a continuing research project which is carried out to investigate various aspects of the TIG dressing method. The project is sponsored by the following companies and institutions: Statoil, Norsk Hydro, Gränges Oxelösund Järnverk, Domnarvets Jernverk, Norsk Jernverk, Brødr. Søyland, Kvernelands Fabrikk, Norgas, The Aker Group, The Kvaerner Group, Vigor, and The Norwegian Council for Scientific and Industrial Research. The author is happy to acknowledge the contributions to the project made by representatives of these companies through discussions. Special thanks are due to J. Drugli, N. Christensen, and A. Baggerud for their participation and continued interest in the project, and to Mr. H. O. Knagenhjelm who carried out the fatigue testing.

References

[*1*] Signes, E. G., Baker, R. G., Harrison, J. D., and Burdekin, F. M., *British Welding Journal,* Vol. 14, No. 3, 1967, pp. 108-116.

[*2*] Gurney, T. R., *British Welding Journal,* Vol. 7, No. 6, 1960, pp. 415-431.

[*3*] Harrison, J. D., *British Welding Journal,* Vol. 13, No. 11, 1966, pp. 642-649.

[*4*] Harrison, J. D., "Techniques for Improving the Fatigue Strength of Welded High Strength Steels," Contract Report C215/7/68, The Welding Institute, May 1968.

[*5*] Gurney, T. R., *British Welding Journal,* Vol. 15, No. 12, 1968, pp. 601-609.

[*6*] Kanazawa, S., Ishiguro, T., Hanazawa, M., and Yokota, H., "The Improvement of Fatigue Strength in Welded High Strength Steel", Document XIII-735-74, International Institute of Welding.

[*7*] Harrison, J. D. and Hale, N. P., "A Summary of Techniques for Improving the Fatigue Strength of Welded Structures and an Estimate of Their Costs," Contract Report C215/12/69, The Welding Institute, Sept. 1969.

[*8*] Valland, G. et al, "Guidance for Welding Procedure Qualification Tests—Structural Steel Parts," Fixed Offshore Structures, Technical Note C1/2, Det norske Veritas, 1 June 1975.

[*9*] Haagensen, P. J., "Hardness in TIG-Dressed Weldments, Introductory Investigation," SINTEF Report STF18 F76068, The Foundation of Scientific and Industrial Research at the University of Trondheim. June 1976 (in Norwegian)

[*10*] Drugli, J., "Hardness in TIG-Dressed Weldments," SINTEF Report STF16 F77008, The Foundation of Scientific and Industrial Research at the University of Trondheim, Feb. 1977 (in Norwegian).

[11] Millington, D., "TIG Dressing for the Improvement of Fatigue Properties in Welded High Strength Steels," Contract Report C12/22/71, The Welding Institute, July 1971.
[12] Kado, S. et al, "The Method of TIG Dressing," Document XIII-773-75, International Institute of Welding.
[13] Kado, S. et al, "Influence of the Conditions in TIG Dressing on Fatigue Strength in Welded High Tensile Strength Steels," Document 771-75, International Institute of Welding.
[14] Christensen, N., "Hardness in TIG Dressed Weldments, A Theoretical Analysis," SINTEF Report STF34 77029, The Foundation of Scientific and Industrial Research at the University of Trondheim, March 1977 (in Norwegian).

F. V. Lawrence, Jr.,[1] *R. J. Mattos,*[2] *Y. Higashida,*[3] *and J. D. Burk*[1]

Estimating the Fatigue Crack Initiation Life of Welds

REFERENCE: Lawrence, F. V., Jr., Mattos, R. J., Higashida, Y., and Burk, J. D., **"Estimating the Fatigue Crack Initiation Life of Welds,"** *Fatigue Testing of Weldments, ASTM STP 648*, D. W. Hoeppner, Ed., American Society for Testing and Materials, 1978, pp. 134–158.

ABSTRACT: The fatigue crack initiation life (cycles to obtain a 0.25-mm fatigue crack) was estimated for butt welds using strain-controlled fatigue concepts. Key developments which facilitated these estimates were the assumption of $(K_f)_{max}$ conditions (the largest value of K_f possible for a given weld shape), the use of computer simulation methods which modeled cyclic hardening and softening as well as mean stress relaxation effects, and the use of the fatigue properties of the actual weld zone in which the initial notch was located. The initiation life was found to be very sensitive to changes in K_f, but rather insensitive to strength level. The importance of residual stresses and mean stress varied with material as did the fraction of total life devoted to crack initiation. Mild steel (ASTM Specification for Structural Steel (A 36)) high strength, low alloy (ASTM Specification for High-Yield Strength, Quenched and Tempered Alloy Steel Plate, Suitable for Welding (A 514)) steel and aluminum alloy welds were considered.

KEY WORDS: fatigue tests, life estimations, weldments, crack initiation, weld defects

Fatigue Crack Initiation and Propagation in Welds

Although the many factors which influence the fatigue life of a weld have been identified and are qualitatively understood[*1*],[4] it is difficult to gage the relative importance of each in determining the total fatigue life of a weld without a rational computation scheme which correctly deals with all

[1] Professor, Departments of Civil Engineering and of Metallurgy and Mining Engineering, and research assistant, Department of Metallurgy and Mining Engineering, respectively, University of Illinois at Urbana-Champaign, Urbana, Ill. 61801.

[2] Quality control and research manager, Siderurgica de Medellin, Simesa, Medellin, Columbia, South America.

[3] Research engineer, Hiratsuka Research Laboratory, Sumitomo Heavy Industries, Ltd., Tokyo, Japan.

[4] The italic numbers in brackets refer to the list of references appended to this paper.

known variables. It is assumed that the total fatigue life (N_T) is composed of a fatigue crack initiation period (N_I) and a fatigue crack propagation period (N_P) and that N_T can be calculated as

$$N_T = N_I + N_P \tag{1}$$

if N_I and N_P can be properly separated and independently estimated.

At present, there is no standard or practical method of separating the initiation and propagation portions of life, although, there are two standard types of fatigue information and attendent predictive methods: those based upon strain-controlled testing of smooth specimens [*2*] and those based on measurement of fatigue crack growth rates [*3*]. In the work to be described, the total fatigue life of welds was estimated by calculating a crack initiation period using strain-controlled fatigue concepts and by adding to this a fatigue crack propagation life which was calculated arbitrarily assuming that in the absence of any preexisting large defect that initiation ended and propagation began at a crack size of 0.25 mm [*4-9*].

Fatigue Crack Initiation Life Estimation

Fatigue cracks initiate at highly strained regions (notches) such as those shown for a butt weld in Fig. 1. According to Topper [*2,10*], the fatigue crack initiation life of a notched member can be reproduced by a smooth specimen subjected to the notch-root strain history. It is thereby assumed that the notch root stresses may be taken as uniaxial and that the behavior of the smooth specimen is like that of a filament of material at the notch root.

The first step in the estimation of the fatigue crack initiation life of a weld is to relate the remote stresses to the local (notch-root) stresses and strains. The expression derived by Neuber[*11*] and modified for fatigue by Topper[*2*] is used to relate the range in remote stress (ΔS) to the product of local stress and strain ranges ($\Delta\sigma$,$\Delta\varepsilon$). For elastic conditions except at the notch root

$$\Delta\sigma\Delta\varepsilon = \frac{(K_f\,\Delta S)^2}{E} \tag{2}$$

where

K_f = fatigue notch factor, and
E = Young's modulus.

The fatigue notch factor(K_f) may be determined from long life fatigue tests as the ratio of unnotched to notched fatigue strength or may be estimated by several empirical relationships among which is the simple and popular Peterson's equation[*12*]

$$K_f = 1 + \frac{K_t - 1}{1 + \dfrac{a}{r}} \tag{3}$$

where

K_t = theoretical (elastic) stress concentration factor,

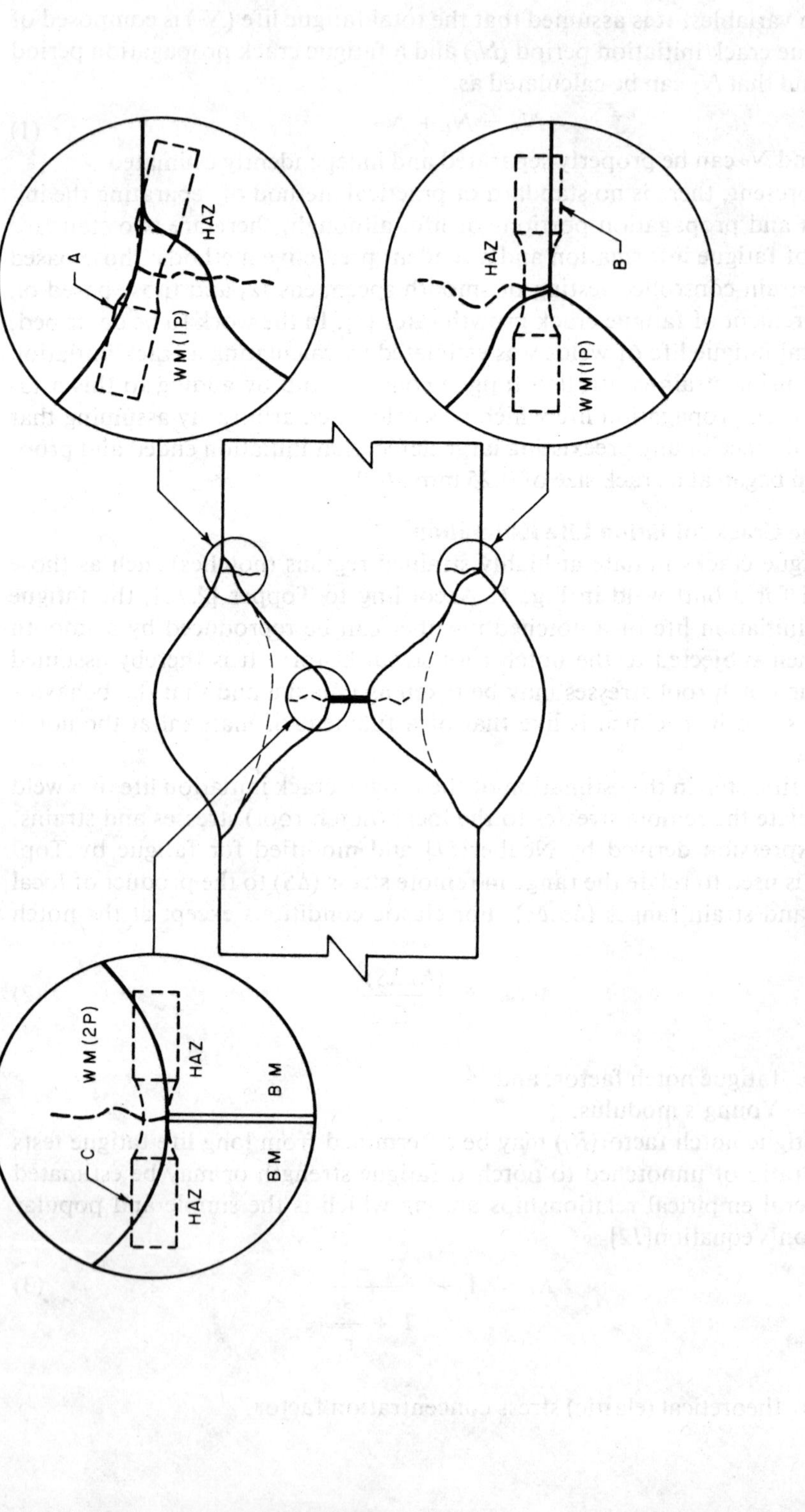

FIG. 1—*Typical fatigue crack initiation sites for a butt weld: Location A—Initiation in untempered weld metal (WM 1P); Location B—Initiation in HAZ; Location C—Initiation in tempered weld metal (WM 2P)).*

r = notch root radius, and
a = an experimentally determined material parameter.

The second task is to model the cyclic stress-strain response of the material at the notch root. Morrow[13] has shown that the cyclic stress-strain curve can be fitted by the following expression

$$\frac{\Delta\varepsilon}{2} = \frac{\Delta\sigma}{2E} + \left(\frac{\Delta\sigma}{2K'}\right)^{1/n'} \quad (4)$$

where

$\Delta\varepsilon$ = total true strain range,
$\Delta\sigma$ = total true stress range,
K' = cyclic strength coefficient, and
n' = cyclic strain hardening exponent.

Alternatively, the cyclic stress-strain behavior can be computer simulated using a multielement rheological model such as the one developed by Martin[14,15], Jhansale and Topper[16], and Plummer[17]. Each element of the model consists of a spring and frictional slider, the stiffness and yield stress of which are set and periodically readjusted to model the cyclic stress-strain curve, the effects of previous stress-strain excursions, cyclic hardening and softening, and mean stress relaxation effects. Such models have been widely used to simulate the cyclic stress-strain response under complex, variable load histories. In the much simpler, constant (remote) stress amplitude case at hand, this model allows one to cycle between stress and strain limits set by the Neuber-Topper relation (Eq 2) and to include the effects of mean stress relaxation and cyclic strain hardening and softening. Simulated and measured histories are compared in Fig. 2.

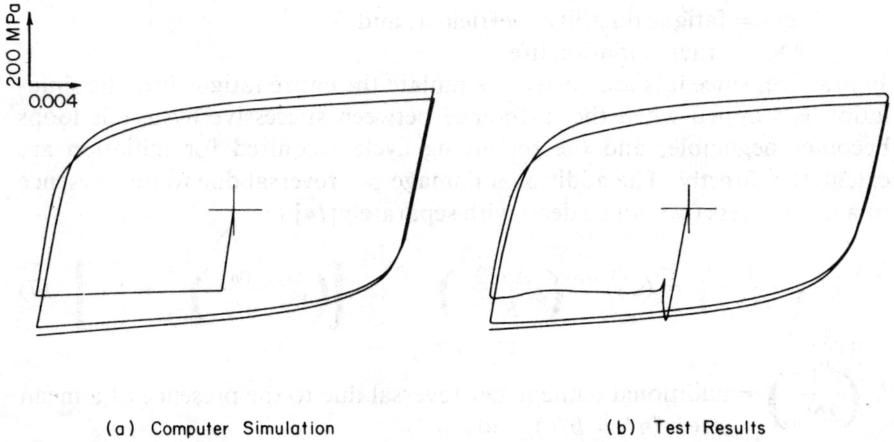

FIG. 2—*Comparison of computer simulated* (a) *and experimentally determined* (b) *cyclic stress-strain response for ASTM Specification A 36 steel base metal* $\Delta e_t/2 = \pm 0.015$.

With the ability to simulate the notch root stresses and strains on a cycle-by-cycle basis for any pattern of remotely applied load, one can calculate and total the fatigue damage caused by each successive stress reversal. Fatigue crack initiation is considered to have occurred when the Palmgren-Miner[*18*] rule of linear cumulative damage is satisfied

$$\sum_{1}^{2N_f} \left(\frac{1}{2N_f}\right) = 1 \qquad (5)$$

where $\left(\frac{1}{2N_f}\right)$ = damage in a given reversal, that is, either

$$\left(\frac{1}{2N_{f_e}}\right) = \left(\frac{\Delta\sigma/2}{\sigma_f{}'}\right)^{-1/b} \qquad (6)$$

or

$$\left(\frac{1}{2N_{f_p}}\right) = \left(\frac{\Delta\varepsilon p/2}{\varepsilon_f}\right)^{-1/c}$$

where

$\left(\frac{1}{2N_{f_e}}\right)$ = damage per reversal associated with the stress amplitude,

$\left(\frac{1}{2N_{f_p}}\right)$ = damage per reversal associated with the plastic strain amplitude,

$\Delta\sigma$ = stress range,

$\Delta\varepsilon_p$ = plastic strain range,

b = fatigue strength exponent,

c = fatigue ductility exponent,

$\sigma_f{}'$ = fatigue strength coefficient,

$e_f{}'$ = fatigue ductility coefficient, and

$2N_f$ = crack initiation life.

In practice, since it is too costly to simulate the entire fatigue life, the simulation is stopped when the difference between successive hysteresis loops becomes negligible, and the remaining cycles required for initiation are calculated directly. The additional damage per reversal due to the presence of a mean stress (σ_0) can be dealt with separately[*14*]

$$\left(\frac{1}{2N_{f_0}}\right) = (\varepsilon'_f)^{1/c} \left(\frac{\Delta\sigma/2}{K'}\right)^{-1/n'c} \left[\left(1 - \frac{\sigma_0}{\sigma'_f}\right)^{1/n'c} - 1\right] \qquad (7)$$

where

$\left(\frac{1}{2N_{f_0}}\right)$ = additional damage per reversal due to the presence of a mean stress ($n' = b/c$), and

σ_0 = current value of mean stress.

The mean stress was allowed to relax using two different schemes; Martin's relaxation constant method [*14*] was used initially. More recently, the power law relationship suggested by Jhansale and Topper [16] was found to be more useful

$$\frac{\sigma_0}{\sigma_{0s}} = (2N_i - 1)^k \tag{8}$$

where

σ_{0s} = initial (stabilized) mean stress,
σ_0 = current mean stress at $2N_i$, and
k = relaxation exponent.

Estimation of K_f for Welds

Of the many types of information required for the analysis of fatigue crack initiation life of welds, good estimates of K_f are perhaps the most critical and difficult to obtain. As Peterson's equation (Eq 3) implies, K_f depends upon the elastic stress concentration factor (K_t), the notch root radius (r), and a parameter (a) which must be determined for the material in question. The parameter (a) is thought to be related to the tensile strength of the material (S_u), and the relationship[19] below permits rough estimates

$$a = 2.5 \times 10^{-5} \left(\frac{2068}{S_u}\right)^{1.8} \quad \text{(SI units)}$$

K_t can be estimated in a variety of ways. For complex geometries such as welds, finite element methods (FEM) are most useful. The variation in K_f with each value of K_t can be calculated using Peterson's equation (Eq 3). Although K_t becomes steadily larger with decreasing notch-root radius, Peterson's equation predicts that K_f will become maximum at some critical value of r, r_{crit}. Assuming a general form for K_t

$$K_t = A \left(\frac{C}{r}\right)^B + 1 \tag{10}$$

where

A,B,C = constants, and
r = notch-root radius.

Substituting Eq 10 into Peterson's equation (Eq 3), differentiating K_f with respect to r, and setting that expression equal to zero leads to the result that K_f is maximum when

$$r = r_{crit} = \frac{a}{B}(1 - B)$$

so that

$$(K_f)_{max} = 1 + A \frac{\left[\dfrac{BC}{a(1-B)}\right]^B}{1 + \dfrac{B}{1-B}} \tag{11}$$

which, for an elliptical flaw ($K_t = 2\sqrt{c/r} + 1$), reduces to

$$r_{crit} = a$$
$$(K_f)_{max} = 1 + \sqrt{c/a} \quad (12)$$

where

c = semimajor axis of ellipse.

Thus, while K_f varies with both the shape of the body (macrogeometry) and the notch root radius (microgeometry), there is only one value of $(K_f)_{max}$ for a given macrogeometry: that predicted by $r_{crit} \cong a$ (see Fig. 3). This maxi-

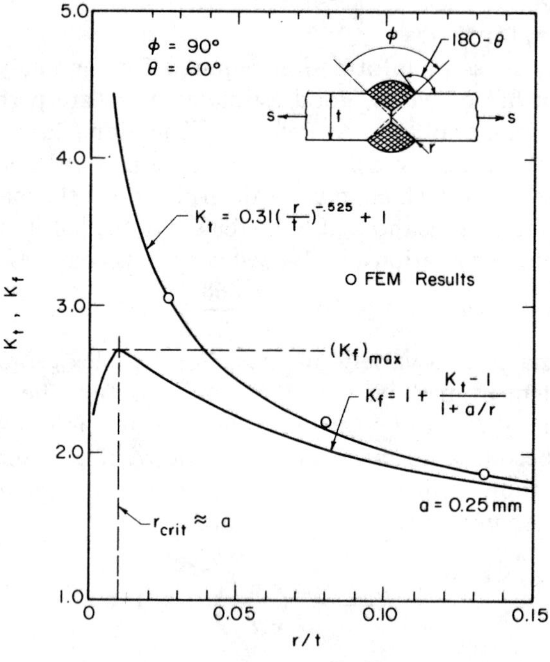

FIG. 3—*Variation of elastic stress concentration factor* (K_t) *and fatigue notch factor* (K_f) *with notch root radius* (r) *for a butt weld* (ϕ = *90 deg,* θ = *60 deg).* $(K_f)_{max}$ *occurs when* $r = r_{crit} \cong a$.

mum value of K_f, $(K_f)_{max}$, is the largest possible value of K_f and leads to the most conservative estimates of N_I. Since practically any value of r can exist at some location along a notch such as a weld toe, it is believed that the use of r_{crit} and $(K_f)_{max}$ is also realistic.

Fatigue life predictions using the methods described and using $(K_f)_{max}$ are compared with experimental results in Fig. 4. The experimental values of N_I were derived from strain measurements at the weld toe which permitted the presence of 0.25-mm-deep fatigue cracks to be detected[*20*]. The propagation portion of life was calculated[*9*] and added to the initiation life estimate to give the total life prediction shown in Fig. 5. The concept of $(K_f)_{max}$ has also been applied with good results to elliptical notches found in welds such

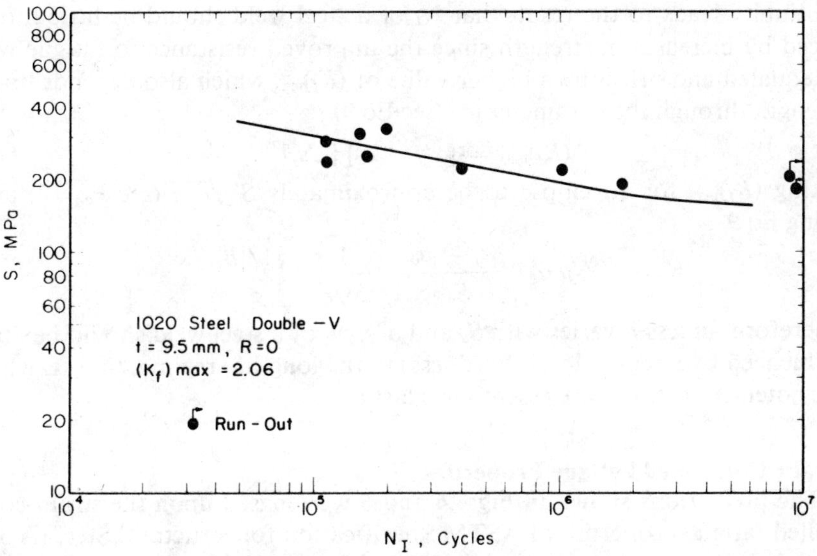

FIG. 4—*Comparison of predicted and experimentally determined fatigue crack initiation lives* (N_I) *for 9.5-mm-thick, 1020 steel, double-V butt welds* (R = 0).

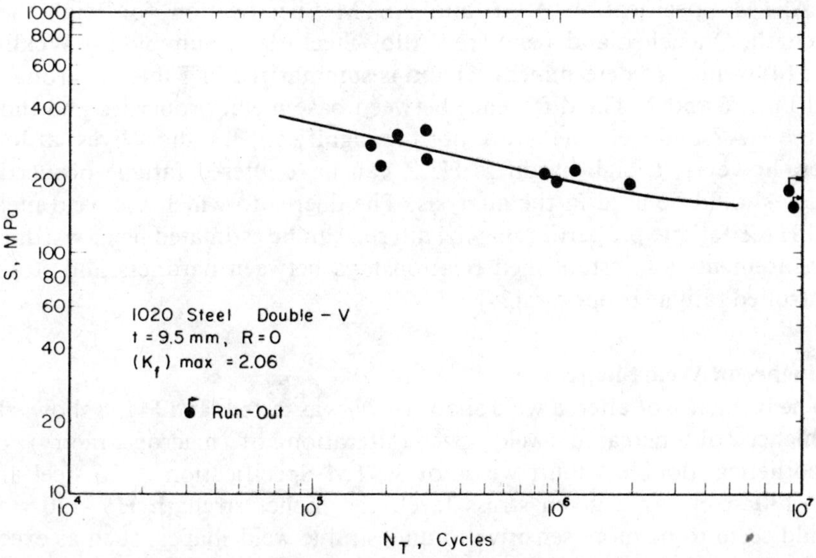

FIG. 5—*Comparison of predicted and experimentally determined total fatigue life* ($N_T = N_I + N_P$) *for 9.5-mm-thick, 1020 steel, double-V butt welds* (R = 0).

as incomplete joint penetration[21,22].

The assumption of $(K_f)_{max}$ conditions at the root of an elliptical weld discontinuity leads to the result that N_I of a steel weld should be little influenced by increases in strength since the improved resistance to fatigue will be equaled and offset by a higher value of $(K_f)_{max}$ which also depends upon strength through the parameter (a) (see Eq 9)

$$S\,(K_f)_{max} = [\sigma'_f - \sigma_0]\,[2N_f]^b \qquad (13)$$

taking $(K_f)_{max}$ for an ellipse to be approximately $S\sqrt{c/a}$ (see Eq 12) and using Eq 9

$$N_I \propto \left[\frac{\sigma'_f - \sigma_0}{S_u} \cdot \frac{1}{S\sqrt{c}} \right]^{1/|b|}$$

Therefore, unless b varies with S_u and σ'_f, N_I of a steel weld should be little influenced by strength level (hardness) in the long life regime, that is, when the notch-root strains are essentially elastic.

Strain-Controlled Fatigue Properties

The predictions shown in Figs. 4 and 5 were based upon the strain-controlled fatigue properties of ASTM Specification for Structural Steel (A 36) steel *base metal*. Fatigue crack initiation in welds actually occurs in the grain-coarsened heat affected zone (HAZ), tempered or untempered weld metal (WM (2P) or WM (1P)) rather than base metal (BM), see Fig. 1. The strain-controlled fatigue behavior of the important microstructures in ASTM Specification A 36 and ASTM Specification for High-Yield Strength, Quenched and Tempered Alloy Steel Plate, Suitable for Welding (A 514) welds was determined[*23*] and is summarized in Tables 1 through 6 and Figs. 6 and 7. The difference between base metal properties and those of the HAZ and weld metal are not too significant for short lives; at long lives, however, a high hardness HAZ can have altered fatigue properties which should be used in the analysis. The degree to which the weld metal and HAZ fatigue properties may be altered can be estimated using hardness measurements and established relationships between hardness and strain-controlled fatigue properties[*24*].

Influence of Weld Shape

The influence of altered weld shape on N_I was tested, and Fig. 8 shows the influence of increased weld size (alteration of macrogeometry) on hypothetical double-V butt welds of ASTM Specification A 36 steel and HY-130 steel. At a given stress level, the higher strength HY-130 steel would seem to be more sensitive to undesirable weld shapes, such as excessive weld reinforcement, than the ASTM Specification A 36 mild steel.

Figure 9 shows the predicted effect of altering the notch-root radius (microgeometry). One could imagine this as having been the result of tungsten inert gas (TIG) dressing the weld toe. Again, the HY-130 material seems to be most affected; and, on this basis, TIG dressing should be most beneficial to higher strength metals. In general, reducing the fatigue notch

TABLE 1—*Mechanical properties of base, weld, and heat-affected materials for ASTM Specification A 36 steel welds.*

Material	ASTM A 36–BM	ASTM A 36–HAZ	E60S–WM(1P)	E60S–WM(2P)
Hardness, DPH/HB	168/160	255/243	245/233	211/201
Modulus of elasticity, E, $\times 10^3$ (MPa)	190	190	190	190
0.2% offset yield strength, (MPa)	224	535	580	408
Ultimate tensile strength, Su, (MPa)	414	667	711	580
Percent reduction in area, %RA	69.7	52.5	44.6	60.7
True fracture strength, σ_f, (MPa)	800	821	890	870
True fracture ductility, ε_f	1.19	0.745	0.590	0.933
Strain hardening exponent, n	0.0146/0.258	0.102	0.098	0.130
Strength coefficient, K, (MPa)	778	980	987	849

TABLE 2—*Cyclic and fatigue properties of base, weld, and heat-affected materials for ASTM Specification A 36 steel welds.*

Material	ASTM A 36–BM	ASTM A 36–HAZ	E60S–WM(1P)	E60–WM(2P)
Cyclic yield strength, 0.2% offset, (MPa)	232	402	385	363
Cyclic strain hardening exponent, n'	0.249	0.215	0.155	0.197
Cyclic strength coefficient, K' (MPa)	1097	1490	1007	1235
Fatigue strength coefficient, σ_f', (MPa)	1014	725	904	1028
Fatigue ductility coefficient, ε_f',	0.271	0.218	0.607	0.602
Fatigue strength exponent, b	−0.132	−0.066	−0.075	−0.090
Fatigue ductility exponent, c	−0.451	−0.492	−0.548	−0.567
Transition fatigue life, $2N_t$, reversals	200 000	13 234	28 022	19 259

TABLE 3—*Mechanical properties of base, weld, and heat-affected materials for ASTM Specification A 514 welds.*

Material	ASTM A 514–BM	ASTM A 514–HAZ	E110–WM(1P)	E110–WM(2P)
Hardness, DPH/HB	320/303	496/461	382/362	327/310
Modulus of elasticity, E, $\times 10^3$ (MPa)	209	209	209	209
0.2% offset yield strength, (MPa)	890	1180	835	759
Ultimate tensile strength, Su, (MPa)	938	1408	1035	911
Percent reduction in area, %RA	63.0	52.7	57.6	59.3
True fracture strength, σ_f, (MPa)	1270	1988	1911	1428
True fracture ductility, ε_f	0.994	0.750	0.857	0.899
Strain hardening exponent, n	0.060	0.092	0.092	0.085
Strength coefficient, K, (MPa)	1187	2111	1559	1290

TABLE 4—*Cyclic and fatigue properties of base, weld, and heat-affected materials for ASTM Specification A 514 welds.*

Material	ASTM A 514–BM	ASTM A 514–HAZ	E110–WM(1P)	E110–WM(2P)
Cyclic yield strength, 0.2% offset, (MPa)	604	938	649	603
Cyclic strain hardening exponent, n'	0.091	0.103	0.177	0.166
Cyclic strength coefficient, K', (MPa)	1090	1766	2022	1670
Fatigue strength coefficient, σ_f', (MPa)	1304	2001	1890	1408
Fatigue ductility coefficient, ε'_f	0.975	0.783	0.848	0.595
Fatigue strength exponent, b	−0.079	−0.087	−0.115	−0.079
Fatigue ductility exponent, c	−0.699	−0.713	−0.734	−0.590
Transition fatigue life, $2N_t$, reversals	3461	1138	1536	6448

TABLE 5—*Mechanical properties of base, weld, and heat-affected materials for 5083 aluminum welds.*

Material	5083–BM	5183–WM
Hardness, DPH/HB	106/93	105/92
Modulus of Elasticity, E, $\times 10^3$ (MPa)	71	71
0.2% offset yield strength, (MPa)	131	142
Ultimate tensile strength, Su, (MPa)	300	305
Percent reduction in area, %RA	30	33
True fracture strength, σ_f, (MPa)	422	430
True fracture ductility, ε_f	0.36	0.40
Strain hardening exponent, n	0.129	0.133
Strength coefficient, K, (MPa)	306	314

TABLE 6—*Cyclic and fatigue properties of base and weld materials for 5083 aluminum welds.*

Material	5083–BM	5183–WM
Cyclic yield strength, 0.2% offset, (MPa)	292	275
Cyclic strain hardening exponent, n'	0.114	0.072
Cyclic strength coefficient, K', (MPa)	594	500
Fatigue strength coefficient, σ_f', (MPa)	725	638
Fatigue ductility coefficient, ε_f'	0.405	0.581
Fatigue strength exponent, b	−0.122	−0.107
Fatigue ductility exponent, c	−0.692	−0.890
Transition fatigue life, $2N_t$, reversals	640	205

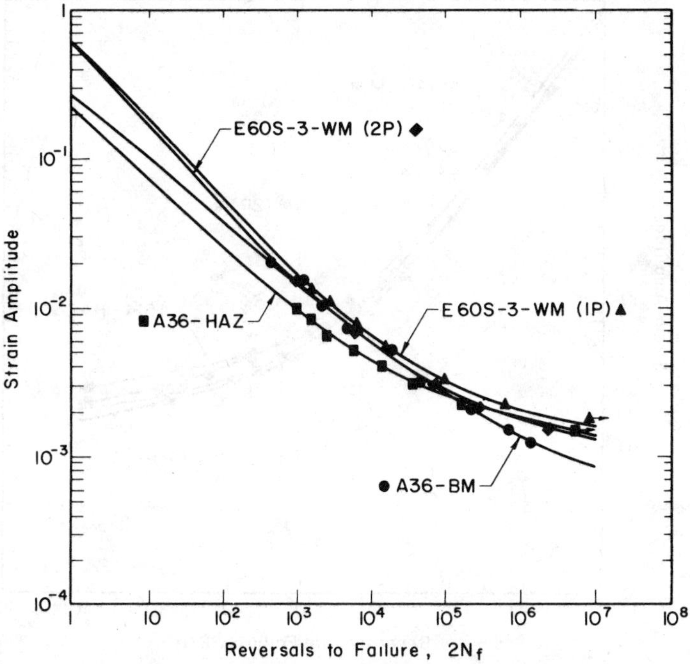

FIG. 6—*Strain-controlled fatigue behavior of ASTM Specification A 36 steel weld microstructures: ASTM Specification A 36 base metal (A36–BM); grain coarsened HAZ (A36–HAZ); untempered weld metal (E60S-3-WM (1P)); tempered weld metal (E60S-3-WM (2P)).*

factor (K_f) would seem to be one of the most effective means of lengthening the fatigue crack initiation period of a weld.

Behavior of Different Metals

The fatigue crack initiation life was estimated for hypothetical butt welds having the base metal properties of HY-130, HY-80, ASTM Specification A 36 and 7075-T6 (aluminum alloy) at the weld toe for the case of $R = -1$ ($R = S_{min}/S_{max}$). As seen in Fig. 10, HY-130 exhibits the longest N_I followed by HY-80, ASTM Specification A 36 steel, and 7075-T6. Estimates of N_P were made assuming the range in stress intensity (ΔK) to be truncated at zero (an assumption which should overestimate N_P). As seen in Fig. 11, ASTM Specification A 36 steel had the longest predicted N_P followed by the HY steels and 7075-T6. The predicted total fatigue lives (N_T) which are plotted in Fig. 12 were calculated by adding the N_I and N_P estimates of Figs. 10 and 11. As has been often observed in tests on welds[*8*], the total fatigue life of high strength steel welds and mild steel welds are about the same, but for quite different reasons: ASTM Specification A 36 steel has a short N_I but long N_P; the HY steels have a long N_I but short N_P.

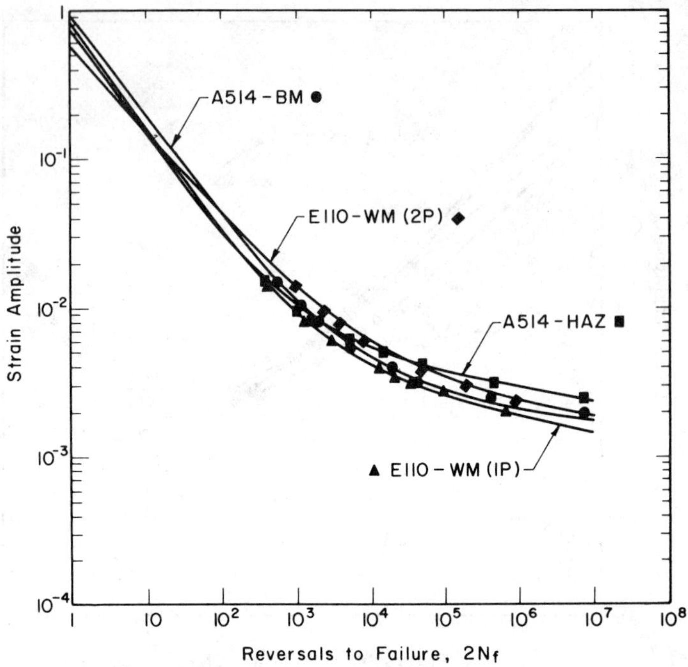

FIG. 7—*Strain-controlled fatigue behavior of ASTM Specification A 514 steel weld microstructures: ASTM Specification A 514 base metal (A514–BM); grain coarsened HAZ (A514–HAZ); untempered weld metal (E110–WM (1P)); tempered weld metal (E110–WM (2P)).*

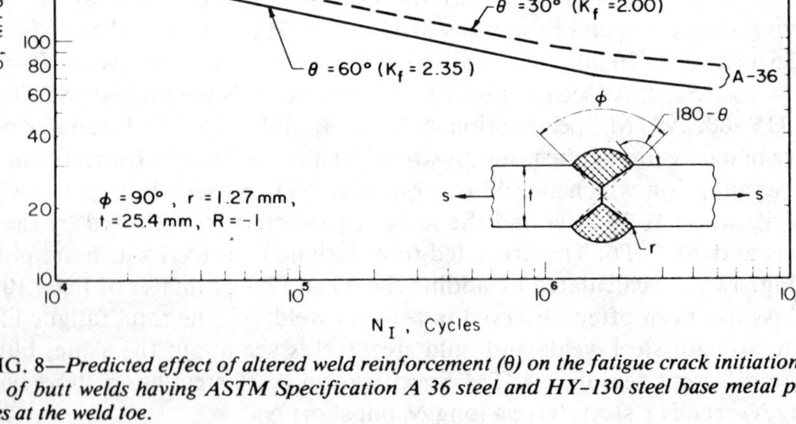

FIG. 8—*Predicted effect of altered weld reinforcement (θ) on the fatigue crack initiation life* (N_I) *of butt welds having ASTM Specification A 36 steel and HY–130 steel base metal properties at the weld toe.*

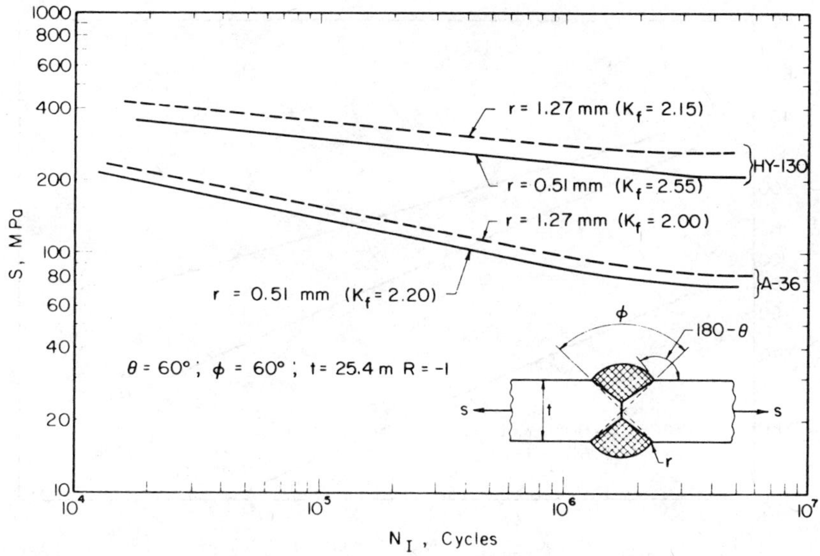

FIG. 9—*Predicted effect of altered weld toe radius* (r) *on the fatigue crack initiation life* (N_I) *of butt welds having ASTM Specification A 36 and HY-130 steel base metal properties at the weld toe.*

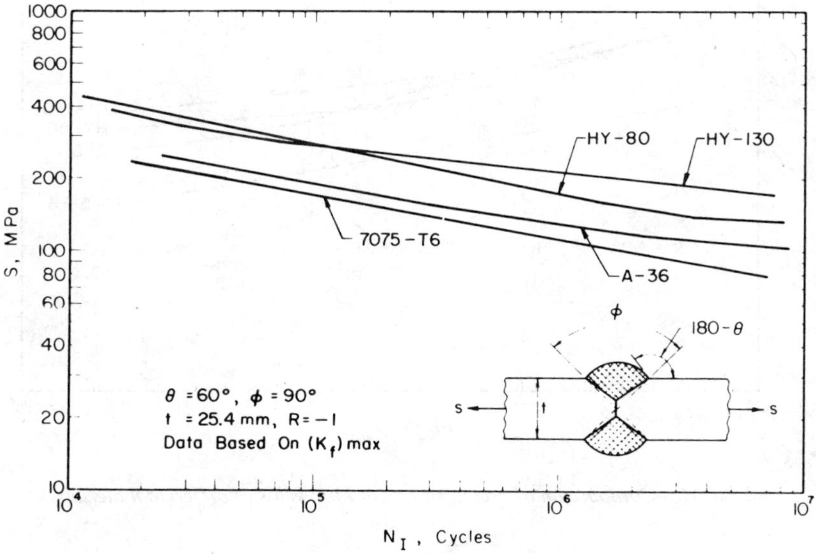

FIG. 10—*Comparison of predicted* N_I *in butt welds of different materials.* $(K_f)_{max}$ *conditions and base metal fatigue properties are assumed.*

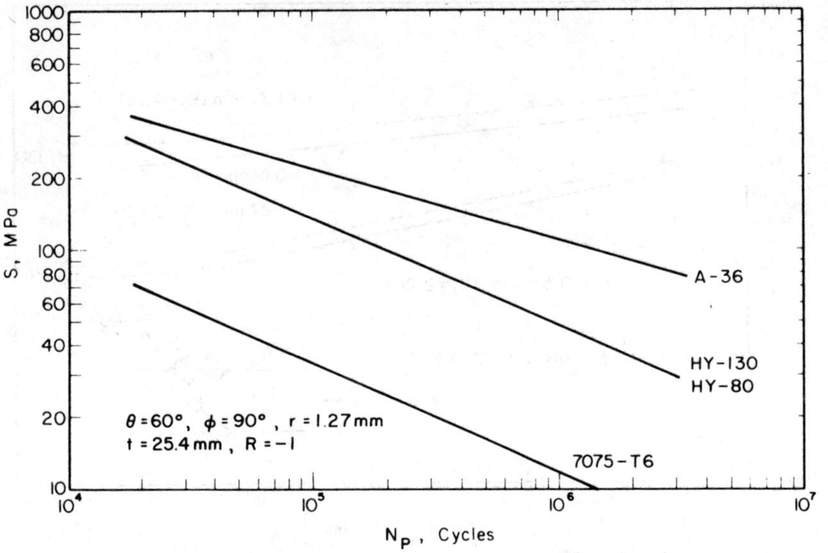

FIG. 11—*Comparison of predicted* N_P *for butt welds of different materials.* ΔK *was truncated at* $K = 0$ *to approximate* $R = -1$ *conditions.*

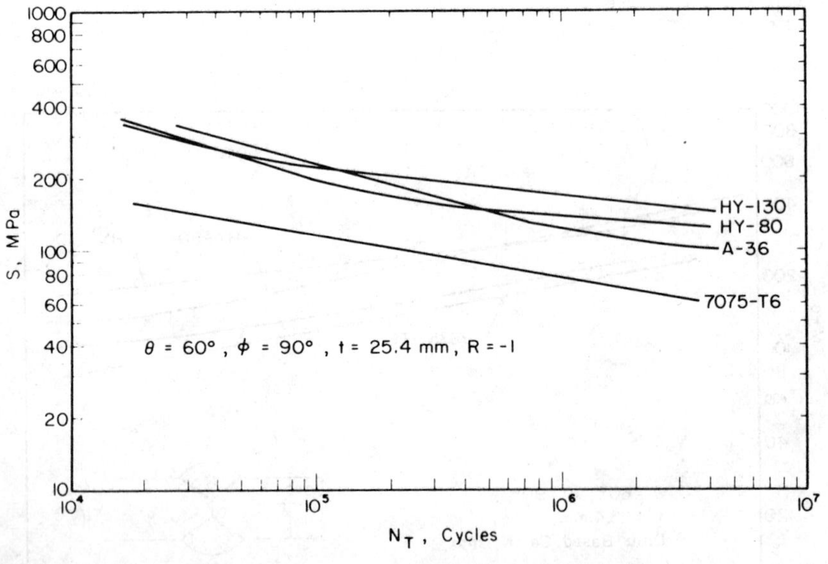

FIG. 12—*Comparison of predicted* N_T *for butt welds of different materials.*

The percentage of the total fatigue life of the four materials which is spent in crack initiation is plotted in Fig. 13 as a function of total life. Although these results are approximate, the following observations can be

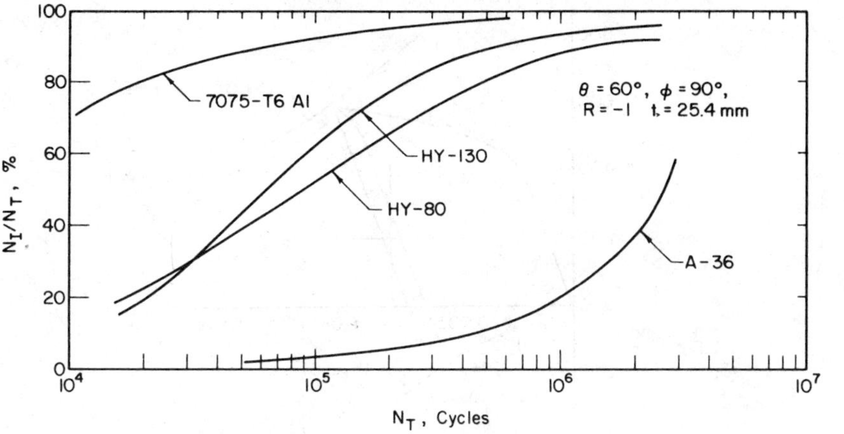

FIG. 13—*Percentage of total life devoted to fatigue crack initiation as a function of total life for butt welds of different materials.*

made. Mild steel materials such as ASTM Specification A 36 steel should be considered "propagation materials" (although initiation rapidly becomes important at lives greater than 10^6 cycles). Initiation is relatively more important in the HY materials and particularly important in the aluminum alloy investigated. Crudely speaking, it appears that some materials are "initiation materials" while others are propagation materials. The initiation materials are characterized by short transition fatigue lives (N_{tr}, the life at which the elastic strain amplitude equals the plastic) whereas, the propagation materials, by long N_{tr}. The relative importance of initiation and propagation for a given material should vary with R or the level of residual stress at the notch root, or both.

Influence of Residual Stress

The thermal strains which accompany welding can induce yield-point residual stresses (σ_r) at the critical notch in a weld. These residual stresses will result in a mean stress (σ_0) which may, however, diminish with cycling. As suggested by the Basquin relationship

$$\Delta S K_f = (\sigma'_f - \sigma_0)\,[2N_f]^b \tag{13}$$

a tensile mean stress can greatly reduce the fatigue crack initiation life. The influence of residual stresses has been studied using the simulation methods discussed [*22*]. The initial, notch-root residual stress is considered to be limited by the yield point of the base metal. The computer simulation is used to reproduce the ensuing cyclic history until the hysteresis loops become reasonably stable (100 to 1000 reversals), see Fig. 14. The stabilized value of mean stress (σ_{0s}) is then allowed to relax according to Eq 8.

Assuming that the notch root strains are essentially elastic ($2N_f > 2N_{tr}$)

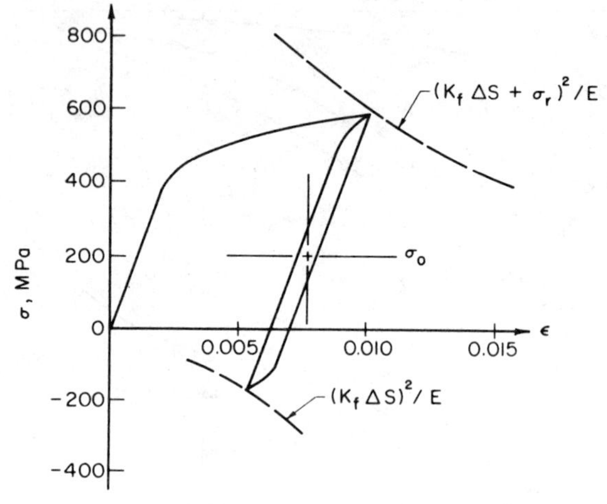

FIG. 14—*Simulation of base metal yield point (241 MPa) tensile residual stresses at the toe of an ASTM Specification A 36 steel butt weld* (K_f = *2.0,* ΔS = *413 MPa notch in HAZ).*

$$\left(\frac{1}{2N_{f_e}}\right)=\left(\frac{\sigma_a}{\sigma'_f}\right)^{\frac{-1}{b}}\left(1-\frac{\sigma_0}{\sigma'_f}\right)^{\frac{-1}{b}} \tag{14}$$

Using Eqs 5, 8, and 14 and passing to the integral, crack initiation should result when

$$\int_1^{2N_f}\left[\left(\frac{\sigma_a}{\sigma'_f}\right)\right]^{\frac{-1}{b}}\left[\left(1-\frac{\sigma_{0s}(2N_i)^k}{\sigma'_f}\right)\right]^{\frac{-1}{b}} dN = 1 \tag{15}$$

The upper limit of integration $2N_f$ can be found by integrating Eq 15, and solving for $2N_f$.

The mean stress relaxation exponent (k) was determined for several materials, and the variation of k with the total strain amplitude ($\Delta\varepsilon/2$) is shown in Fig. 15. The materials with the longest N_{tr} give the highest value of k and, thus, exhibit the most rapid cyclic relaxation of mean stress at a given level of $\Delta\varepsilon/2$. The degree to which residual stresses will influence the N_I of a notch of material will depend upon two main factors

1. The value of the stabilized, notch-root mean stress after the initial reversals (σ_{0s}) and
2. The rate at which σ_{0s} will subsequently decay.

Little notch-root yielding may occur with strong materials such as high strength steel welds subjected to high cycle fatigue, so that σ_{0s} will be large. The k for a high strength steel is relatively low; consequently, the σ_{0s} will not relax rapidly, and the influence of residual stress on N_I for such a condition would be large. The k for 5083 aluminum is also low, but there may be essentially no σ_{0s} after several cycles because of extensive plastic deformation

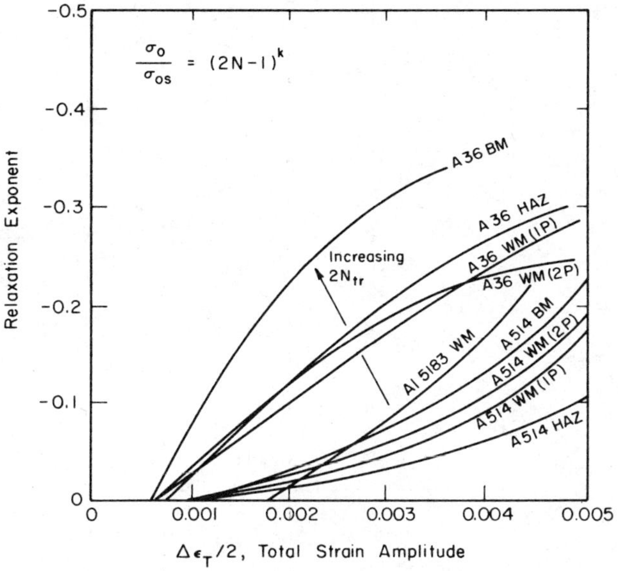

FIG 15—*Dependence of the mean stress relaxation exponent* (k) *upon total strain amplitude* ($\Delta\epsilon_t/2$) *for the microstructures of several welds (ASTM Specification A 36, ASTM Specification A 514, and 5183 Al WM).*

at the notch root during the initial loading (set-up cycle); therefore, in the case of the 5083 aluminum, welding residuals should be of little consequence (in the life range 10^5 to 10^6 cycles). Mild steel such as ASTM Specification A 36 steel is an intermediate case. There is usually substantial σ_{0s} after the first few reversals, but depending upon the total strain amplitude and the consequent value of k, σ_{0s} may or may not relax quickly.

The fatigue crack initiation life of ASTM Specification A 36 steel butt welds was predicted using the integral of Eq 15, and the results are plotted in Fig. 16. The relaxation of the externally imposed mean stress at the notch root ($\sigma_r = 0$, $R = 0$) imparts a peculiar inflection to the *S-N* curve which is further accentuated by the superposition of full, tensile residuals at the weld toe ($\sigma_r = +240$ MPa, $R = 0$). The measured N_I fall between the $\sigma_r = +240$ MPa predictions and possibly exhibit the same inflection.

Similar predictions for ASTM Specification A 514 steel welds are shown in Fig. 17. The predictions and experimental results are for butt welds with either tensile or compressive (HAZ) yield-point residual stresses at the weld toe. Since mean stresses do not relax rapidly in this material (for the life range 10^5 to 10^6 cycles), the effect of welding residual stresses is potentially very large.

Summary

The independent estimation of initiation life and propagation life leads to

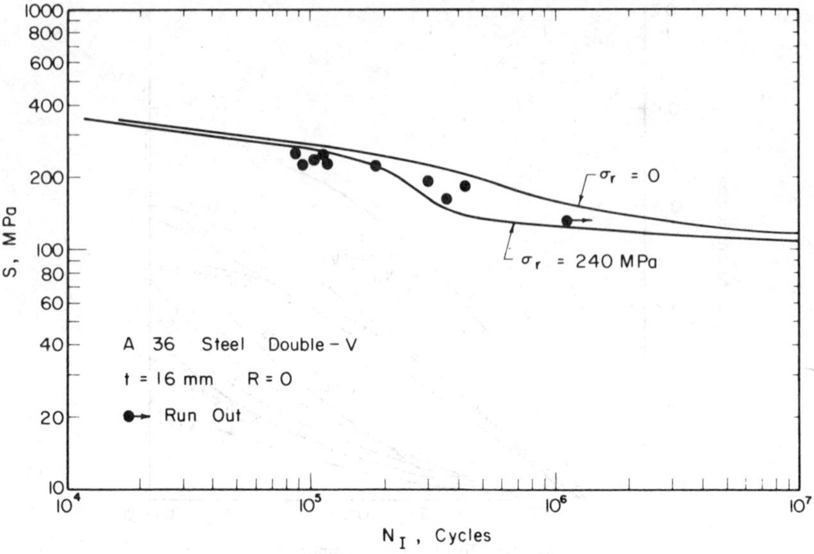

FIG. 16—*Predicted effects of mean stress and residual stress relaxation on* N_I *and in ASTM Specification A 36 steel welds compared with* R = *0 test data.*

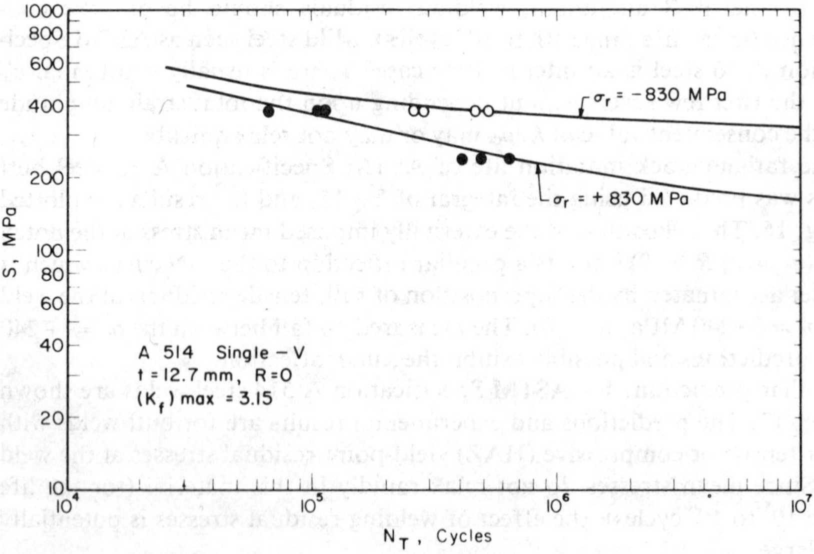

FIG. 17—*Predicted and measured effects of residual stresses on ASTM Specification A 514 steel welds. Solid data points indicate as-welded specimens. Open data points are specimens given a pretest tensile overload.*

surprisingly accurate predictions of total life, even though one is forced to adopt arbitrary boundaries between crack initiation and propagation. Key developments in obtaining this good agreement were the assumption of $(K_f)_{max}$ conditions (the largest value of K_f possible for a given weld macrogeometry), the use of computer simulation methods which modeled cyclic hardening and softening as well as mean stress relaxation effects, and the use of the strain-controlled fatigue properties of the weld zone in which the critical notch was located. Using these methods of estimating N_I, it was demonstrated that

1. N_I will be increased by any alteration of weld geometry which will reduce K_f below $(K_f)_{max}$ (or to a lower value of $(K_f)_{max}$).
2. Certain materials such as ASTM Specification A 36 steels appear to be propagation-dominated materials, whereas, others such as the higher strength steels and aluminums appear to be initiation dominated (in the 10^5 to 10^6 cycle life range).
3. The importance of welding residual stresses varies with the material and can be predicted accurately for constant amplitude loading using the damage integral method proposed.

Acknowledgments

The studies which this paper summarizes were supported by the University of Illinois Fracture Control Program which is funded by a consortium of midwest industries.

The authors are indebted to JoDean Morrow, Department of Theoretical and Applied Mechanics, University of Illinois, for his invaluable advice and help, including the use of his testing equipment for some of the experiments reported. The authors wish to thank W. H. Munse, Department of Civil Engineering, University of Illinois, for many illuminating discussions and useful suggestions.

References

[*1*] Gurney, T. A., *Fatigue of Welded Structures*, Cambridge University Press, Cambridge, United Kingdom, 1968.

[*2*] Topper, T. H., Wetzel, R. M., and Morrow, JoDean, *Journal of Materials*, Vol. 4, 1969, p. 200.

[*3*] Paris, P. C. and Erdogan, F., *Journal of Basic Engineering, ASME Transactions*, Vol. 85, Series D, No. 4, 1963, p. 528.

[*4*] Maddox, S. J., *Metal Construction*, Vol. 2, No. 8, 1970, p. 327.

[*5*] Maddox, S. J., "A Fracture Mechanics Analysis of The Fatigue Behavior of a Fillet Welded Joint," Report No. E/50/72, The Welding Institute, Jan. 1973.

[*6*] Maddox, S. J., "An Analysis of Fatigue Cracks in Fillet Welded Joints," Report No. E/49/72, The Welding Institute, Jan. 1973.

[*7*] Maddox, S. J., *Welding Research Supplement*, Vol. 53, No. 9, 1974, p. 401S.

[*8*] Lawrence, F. V., *Welding Research Supplement*, Vol. 52, No. 5, 1973, p. 212S.

[*9*] Burk, J. D. and Lawrence, F. V., *Welding Journal*, Vol. 56, No. 2, Feb. 1977, p. 61-S-66-S.

[*10*] Topper, T. H. and Morrow, J., Eds., "Simulation of the Fatigue Behavior at the Notch Root in Spectrum Loaded Notched Members (U)," TAM Report No. 333, Department

of Theoretical and Applied Mechanics, University of Illinois at Urbana-Champaign, Urbana, Illinois, Jan. 1970.

[11] Neuber, H., *Journal of Applied Mechanics*, Vol. 8, Dec. 1961, pp. 544-550.

[12] Peterson, R. E. in *Metal Fatigue*, Sines and Waisman, Eds., McGraw-Hill, New York, Chapter 13, 1959.

[13] Morrow, J. in *Internal Friction, Damping, and Cyclic Plasticity, ASTM STP 378*, American Society for Testing and Materials, 1965, pp. 45-87.

[14] Martin, J. F., "Fatigue Damage Analysis For Irregular Shaped Structures Subjected to Representative Loads," Fracture Control Program Report No. 10, College of Engineering, University of Illinois, Urbana, Ill., Dec. 1973.

[15] Martin, J. F., Topper, T. H., and Sinclair, G. M., *Materials Research and Standards*, Vol. 11, No. 2, p. 23.

[16] Jhansale, H. R. and Topper, T. H. in *Cyclic Stress-Strain Behavior—Analysis, Experimentation, and Fatigue Prediction, ASTM STP 519*, American Society for Testing and Materials, 1973, p. 246.

[17] Plummer, F. B., Jr., "Cyclic Plasticity and Structural Energy Dissipation," Ph.D. thesis, Theoretical and Applied Mechanics, University of Illinois, Urbana, Ill., 1973.

[18] Miner, M. A., *Journal of Applied Mechanics*, Vol. 12, 1945, p. A159.

[19] Peterson, R. E., "Analytical Approach to Stress Concentration Effect in Fatigue of Aircraft Structures," WADS Symposium, Wright Air Development Center, Aug. 1959.

[20] Mattos, R. J., "Estimation of the Fatigue Crack Initiation Life in Welds Using Low Cycle Fatigue Concepts," Ph.D. thesis, Department of Metallurgy and Mining, University of Illinois, Oct. 1975.

[21] Tobe, Y. and Lawrence, F. V., Jr., "Effect of Lack of Penetration on Fatigue Resistance of High-Strength Structural Steel Welds," Report to Construction Engineering Research Laboratory, Champaign, Ill. Sept. 1976.

[22] Burk, J. D. and Lawrence, F. V., Jr., "Effects of Lack of Penetration and Lack of Fusion on the Fatigue Properties of 5083 Aluminum Alloy Welds," Report to Naval Sea System Command, July 1976.

[23] Higashida, Y. "Strain Controlled Fatigue Behavior of Weld Metal and Heat-Affected Base Metal in A 36 and A 514 Steel Welds," Ph.D. thesis, Department of Metallurgy and Mining, University of Illinois, June 1976.

[24] Landgraf, R. W. in *Achievement of High Fatigue Resistance in Metals and Alloys, ASTM STP 467*, American Society for Testing and Materials, 1970, p. 3.

S. J. Maddox[1] *and D. Webber*[2]

Fatigue Crack Propagation in Aluminum-Zinc-Magnesium Alloy Fillet-Welded Joints

REFERENCE: Maddox, S. J. and Webber, D., "**Fatigue Crack Propagation in Aluminum-Zinc-Magnesium Alloy Fillet-Welded Joints,**" *Fatigue Testing of Weldments, ASTM STP 648*, D. W. Hoeppner, Ed., American Society for Testing and Materials, 1978, pp. 159–184.

ABSTRACT: The fatigue behavior of aluminum-zinc-magnesium (Al-Zn-Mg) alloy fillet-welded joints was analyzed in fracture mechanics terms. Basic crack propagation data were obtained with $-2 \leqslant R \leqslant +0.5$ and correlated using formulas in the literature and, more successfully, in terms of ΔK_{eff}, based on the results of crack closure experiments. The form of the *da/dN* versus ΔK relationship was influenced by the specimen geometry.

A fracture mechanics analysis of the fatigue life of Al-Zn-Mg alloy fillet welds based on the *da/dN* versus ΔK_{eff} relationship indicated that the weld toe was less severe from the fatigue viewpoint than the same region in a steel fillet weld. This was compatible with the fact that metallurgical examination of Al-Zn-Mg alloy fillet welds has failed to reveal toe defects similar in magnitude to those which act as fatigue crack initiators at the toes of steel fillet welds. The analysis showed that the fatigue life obtained from the Al-Zn-Mg alloy weld could be predicted on the basis that defects only one tenth the size of those observed in steel were present.

Fatigue failure from the weld root in a cruciform joint was also analyzed and the optimum weld design, which gives an equal chance of failure from the root and toe, was determined. The analysis was supported by fatigue test results. Comparison with results obtained for steel added confirmation to the finding that if toe defects are present in Al-Zn-Mg alloy welds, they are smaller than those in steel.

KEY WORDS: fatigue tests, weldments, aluminum alloys, crack propagation, fatigue (materials), fracture mechanics, fillet welds, defects, cracking, initiation, durability, design

Nomenclature

a Length of edge, depth of surface crack, half-length of enclosed crack (mm)

[1] Head of Fatigue Laboratory, The Welding Institute, Cambridge CB1 6AL, England.

[2] Principal scientific officer, Military Vehicles and Engineering Establishment, Christchurch, Dorset BH 23 5AN, England.

$2b$	Plate width (mm)
B	Plate thickness (mm)
$2c$	Width of surface crack (mm)
C, C_1, C_2, etc.	Constants in crack propagation and *S-N* relationships
da/dN	Rate of fatigue crack propagation (mm/cycle)
$\Delta\sigma$	Stress range (N/mm^2)
$\Delta\sigma_{eff}$	Effective stress range (N/mm^2)
$\Delta\sigma_p$	Stress range in plate (N/mm^2)
$\Delta\sigma^*$	Generalized stress parameter
ΔK	Stress intensity factor range (Nmm$^{-3/2}$)
ΔK_{eff}	Effective value of ΔK for which crack is open (Nmm$^{-3/2}$)
ϕ	Proportion of applied stress at which crack closes as a percentage of range
ϕ_o	Complete elliptic integral
H	Weld leg length (mm)
I	Crack propagation integral
K_c	Fracture toughness of test specimen (Nmm$^{-3/2}$)
m	Index in crack propagation relationship
M_s	Correction term dependent on crack front shape in solution for K
M_t	Correction term dependent on crack front shape and crack depth in solution for K
M_k	Correction term dependent on weld profile stress concentration in solution for K
N	Fatigue endurance (cycles)
R	Stress ratio defined as $\dfrac{\text{arithmetical minimum stress}}{\text{arithmetical maximum stress}}$
U	Proportion of stress range over which crack tip is open
W	$H + B/2$ (mm)
Y	Product of terms required to correct infinite plate solution for K for geometry.

Factors that affect crack initiation and crack propagation were studied as part of a general investigation of the fatigue behavior of aluminum-zinc-magnesium (Al-Zn-Mg) alloy fillet-welded joints, with particular reference to methods of improving their fatigue strength and their tolerance to fatigue damage.

In general, the fatigue process consists of crack initiation and crack propagation. As far as as-welded joints are concerned, there is some doubt whether the crack initiation phase really exists. In the case of steel it has been shown *[1]*[3] that weld toe fatigue cracks initiate at small sharp nonmetallic intrusions which can be regarded as preexisting cracks and that

[3] The italic numbers in brackets refer to the list of references appended to this paper.

the fatigue strength of fillet-welded joints which fail from the toe can be calculated using fracture mechanics concepts [*2–4*]. Comparable work to that described in Ref *1* on welded Al-Zn-Mg alloy [*5*] failed to locate any sharp defects forming sites for fatigue cracking, suggesting that crack initiation might occupy a significant proportion of the fatigue life.

In an attempt to determine the relative magnitudes of the crack initiation and propagation phases in the fatigue lives of fillet-welded joints failing from the weld toe, the relationship between the crack propagation rates and fatigue test data for the joint was explored. In particular, the crack propagation relationship was used to calculate the average size of the defect that would propagate to failure during the known life of the joint using fracture mechanics. To do this basic crack propagation data for the material were required, and they were obtained using notched specimens. To enable the effects of mean stress and residual stresses to be included in the analysis, results were obtained at various stress ratios. They were correlated in fracture mechanics terms using the method proposed by Forman [*6*] and in terms of a crack closure stress [*7*]. In this context, experiments were carried out to determine the proportion of stress range during which the crack remains open.

Finally, the crack propagation relationship was used in an analysis of fillet-weld failure from the weld root. Transverse load-carrying fillet welds may fail from the weld toe or root depending on the geometry, the former failure mode giving the highest fatigue strength. A practical design problem is the determination of the optimum size of weld and depth of weld penetration such that there is an equal chance of failure from the weld toe and root. Such conditions have been determined for welded joints in steel using fracture mechanics [*8*]. Test results were obtained from specimens with varying degrees of weld penetration to check the analysis.

Material

The material was an Al-4Zn-2Mg alloy prepared to Director General of Fighting Vehicles and Engineer Equipment Specification (DGFVE) 232 in the solution treated and precipitation hardened (TF) condition. The specified chemical composition and tensile properties are given in Table 1. The alloy is similar to United States 7005 alloy. The material was supplied in the form of 9.5-mm-thick plates.

Fatigue Crack Propagation Studies

Experimental Details

The specimens consisted of 162-mm-wide plates with 6-mm-long central notches. The notches were saw-cuts made on each side of a 1-mm-diameter drilled hole.

The specimens were tested axially in air under constant amplitude cyclic loading in hydraulic fatigue machines using frequencies in the range 5 to 11 Hz. In the laboratory, the temperature varied between 13 and 22°C while the humidity was typically in the range 55 to 70 percent. Stress ratios (R) between -2 and $+0.5$ were used and, in general, for each stress ratio a number of tests under different applied stress ranges were carried out. In most tests, the progress of the fatigue crack was monitored using an automatic photographic technique described elsewhere [*9*], but in some a computerized system using the electrical resistance method [*10*] was used.

In addition, the center-notched specimens were used to determine the crack closure stress for $R = 0$ and -1. This stress was deduced from electrical resistance strain gage measurements on statically loaded specimens. First, a crack was propagated to a suitable crack length at the stress ratio of interest. The testing machine was stopped then and 3-mm gage length strain gages were attached to the surface of the specimen close to the crack and parallel to the direction of stressing. The specimen was loaded and unloaded in increments, and strain gage readings were taken at each load level. The procedure was repeated several times. From the readings, plots were made of applied stress against corresponding strain, and an abrupt change of slope of the stress-strain line was assumed to coincide with the load at which crack closure occurred.

Analysis of Results

The crack propagation readings were used to plot smooth curves of crack length (a) against number of cycles (N). The rates of crack propagation, da/dN, at selected crack lengths were determined by drawing tangents to the curves and calculating their slopes. The corresponding value of the stress intensity factor range (ΔK) was calculated using the formula for a center-cracked plate of width $2b$ containing a crack of total length $2a$ and subjected to a cyclic stress range $\Delta\sigma$ [*11*]

$$\Delta K = \Delta\sigma \left[\pi a \sec \frac{\pi a}{2b} \right]^{1/2} \tag{1}$$

The test results are presented in terms of da/dN and ΔK or, in the case of the data obtained with $R < 0$, the maximum stress intensity factor, K_{max}, in other words the value of K which corresponded to the tensile part of the stress cycle on logarithmic scales in Fig. 1.

Referring to the crack closure measurements, typical results, plotted as the percentage ϕ of applied stress at which closure occurred, are given in Fig. 2. In Fig. 2(a), by extrapolation to the crack tips, an average crack closure occurred at approximately 25 percent of the tensile stress. Thus, the proportion of stress range for which the crack was open, U, was 0.75. Similarly from Fig. 2(b), crack closure occurred at approximately 18 percent of the tensile stress, so that the crack was open for 41 percent of the range, giving $U = 0.41$. Values of U determined for a number of crack lengths ranged from 0.64 to 1.0 for $R = 0$ while for $R = -1$ values were

TABLE 1—*Properties of Al-Zn-Mg alloy to DGFVE Specification 232.*

Chemical composition (nominal)

Element	Al	Zn	Mg	Mn	Zr	Cr	Si	Fe	Ti	Cu
%	93.5	4.0	2.0	0.35	0.15	<0.05	<0.2	<0.3	<0.05	<0.1

Mechanical properties

	0.2% Proof Strength, N/mm^2	Ultimate Tensile Strength, N/mm^2	Elongation on $5.65\sqrt{S_o}$
From tension tests	326 to 380 / 355	384 to 437 / 411	...
From specification (minimum)	332	386	8%

less scattered, ranging from 0.375 to 0.45. The average values were 0.73 and 0.41, respectively.

Interpretation of Fatigue Crack Propagation Results

Referring to Fig. 1, it will be seen that da/dN increases with an increase in positive R value, and for $R < 0$ the compressive portion of the cycle contributes some fatigue damage. Nevertheless, the scatter is not great and for some applications its upper limit would provide an adequate crack propagation relationship for $-2 \leqslant R \leqslant +0.5$.

It may be possible to reduce the scatter in Fig. 1 by using one of the correlation methods which allow for stress ratio [*12*]. Forman et al [*6*] suggested the relationship

$$\frac{da}{dN} = \frac{C_1(\Delta K)^m}{(1 - R)K_c - \Delta K} \tag{2}$$

which tries to take account of the acceleration of da/dN as K_{max} approaches K_c. They demonstrated its value by correlating data obtained from 7075–T6 and 2024–T3 aluminum alloys. However, Pearson [*13*] found that for low toughness alloys, a modified form of Eq 2, using the square root of the denominator, was required.

The present results, reanalyzed in terms of Eq 2, are given in Fig. 3. The value of K_c was estimated on the basis of the crack length at final fracture of the specimens as 2300 $Nmm^{-3/2}$ and for data obtained with $R < 0$, K_{max} was used instead of ΔK. It will be seen that reasonable correlation is achieved. A similar degree of correlation was obtained using Pearson's equation which is reasonable since the present material had relatively high toughness.

A second method of correlation makes use of the effective stress intensity factor range, ΔK_{eff}, which is the proportion of the range over which the crack is open, that is, $U\Delta K$, where U is determined from crack closure experiments [*7,12,14*]. Values of U for the present material, obtained

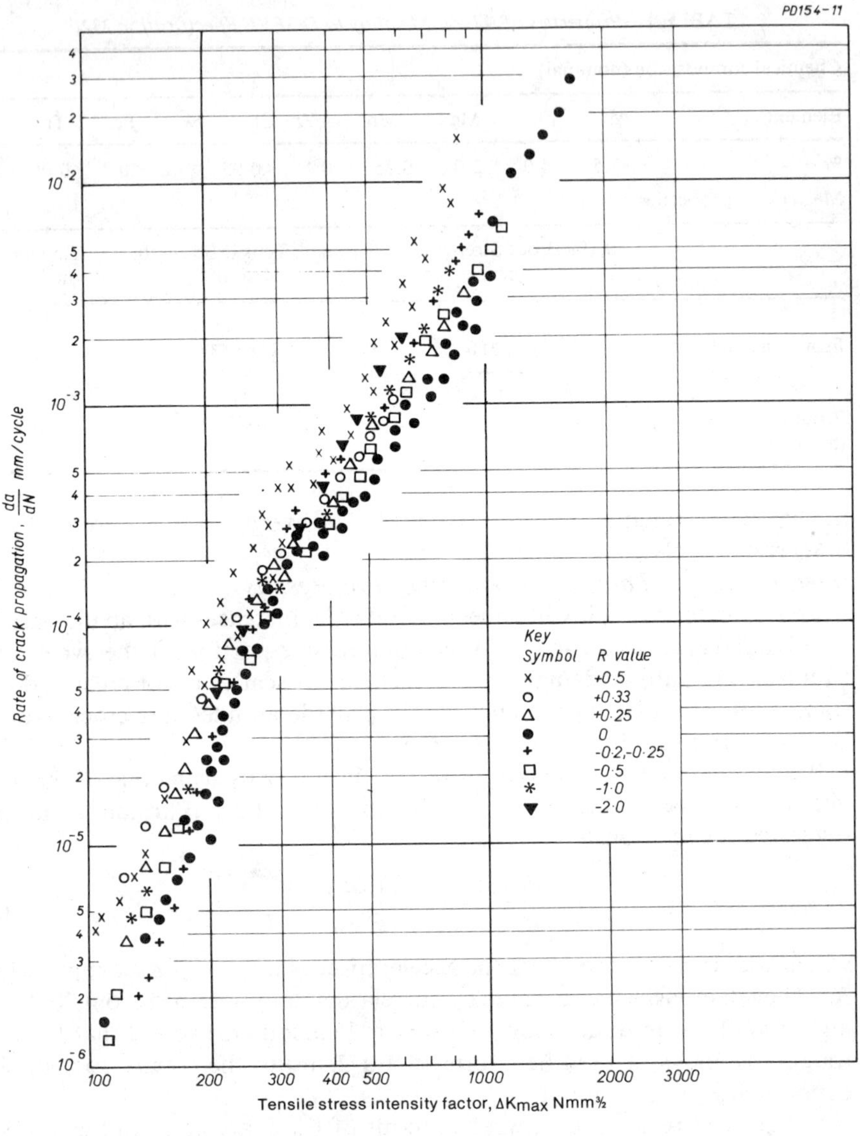

FIG. 1—*Fatigue crack propagation results plotted in terms of the tensile stress intensity factor range.*

earlier, have been used to replot the data for $R = 0$ and -1, as shown in Fig. 4. It will be seen that the correlation is good, suggesting that the simple concept of an effective ΔK is useful. It can be pursued by assuming that all the data can be correlated and then predicting the U values needed to achieve such correlation. The procedure used is to assume that for a given

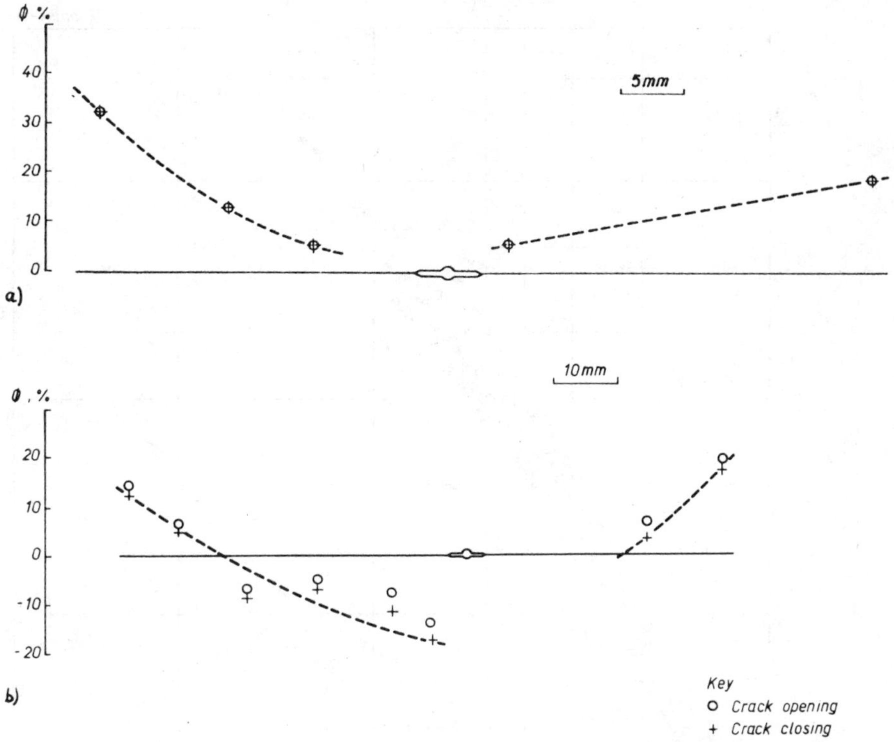

FIG. 2—*Typical crack closure results for* (a) R = *0 and* (b) R = *−1 in terms of the proportion of applied stress (+ve values for tensile and −ve for compressive part of cycle) and position along the crack.*

value of da/dN, ΔK_{eff} is the same for all R values. First, the most reliable crack closure result, $U = 0.41$ for $R = -1$, is used to obtain ΔK_{eff}. Then

$$\Delta K_{eff} = 0.41 \times \Delta K_{R=-1} = U_{R=i} \times \Delta K_{R=i} \tag{3}$$

Equation 3 was applied for each R value at three values of da/dN, 10^{-5}, 10^{-4}, and 10^{-3} mm/cycle, and the resulting average values of U are plotted with the measured values in Fig. 5. The relationship obtained by Elber [*14*], which gives lower U values, is also shown. A line parallel to Elber's provides a reasonable fit to the results, giving

$$U = 0.4R + 0.72 \tag{4}$$

while for $R < -0.5$, the results fell on the curve shown. Based on U values from Fig. 5, the crack propagation data, replotted in terms of ΔK_{eff}, are included in Fig. 4.

Comparing Figs. 1, 3, and 4, it appears that all the methods of correlation are reasonable. Considering only the results obtained with $R = 0$ and -1, the use of ΔK_{eff} results in the least scatter and on that basis would appear to be the best method. Scatter is increased in the other methods because the

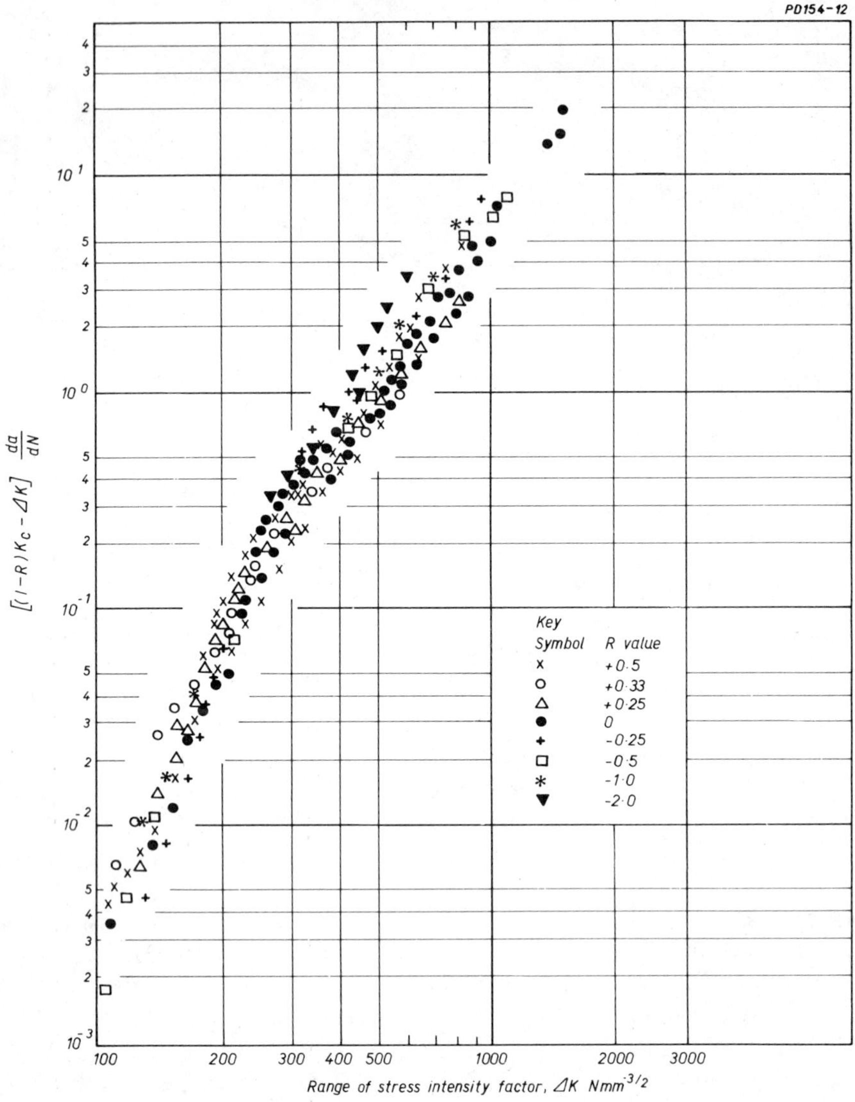

FIG. 3—*Fatigue crack propagation results correlated using Forman's equation* [6].

compressive part of the cycle for results obtained with $R < 0$ is ignored. Clearly, there is scope for combining the method proposed by Forman with the use of ΔK_{eff} by using ΔK_{eff} instead of ΔK in Eq 2 [*12*].

Crack Propagation Relationship

These crack propagation data were obtained in order to determine a re-

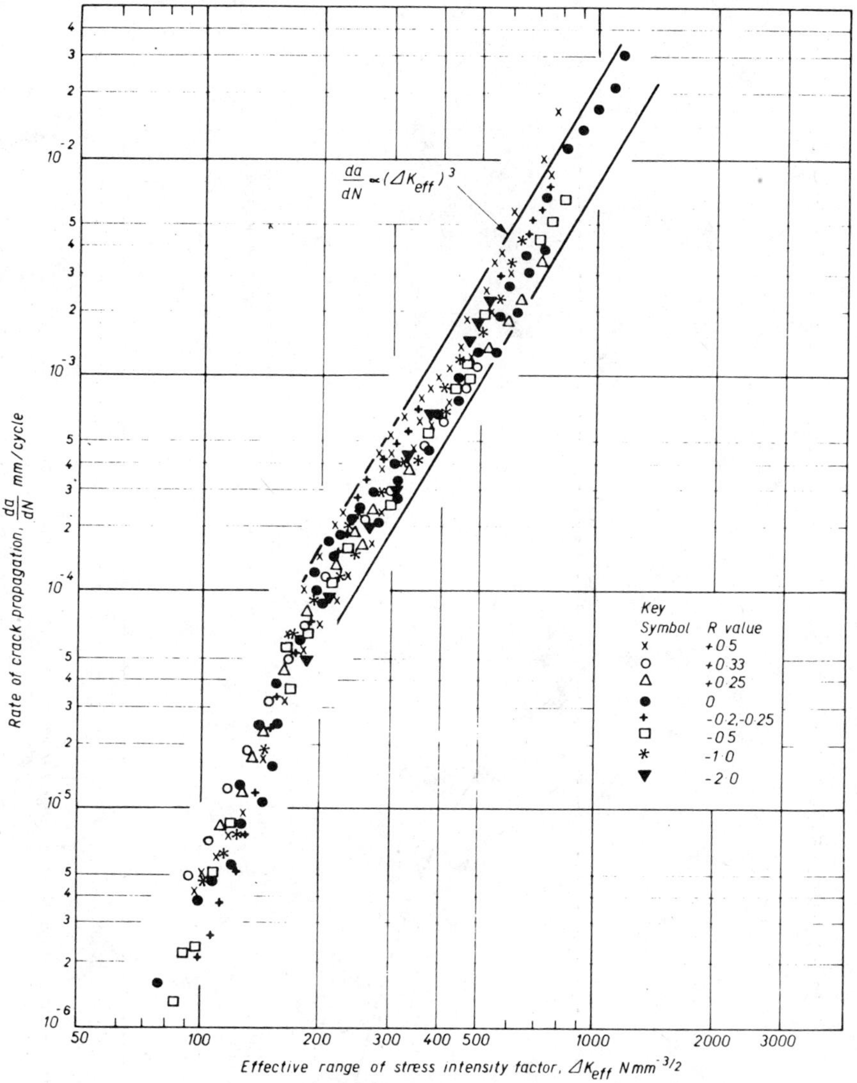

FIG. 4—*Fatigue crack propagation results correlated using the effective stress intensity factor range.*

lationship which could be used to describe fatigue crack propagation in welded joints. This means that the relationship must be independent of geometry. The crack propagation relationship based on ΔK_{eff} is the simplest of those which gave good correlation of results but it is not known whether the value of U depends on geometry. However, as will be noted, it is encouraging to find that fatigue test results obtained from fillet-welded joints

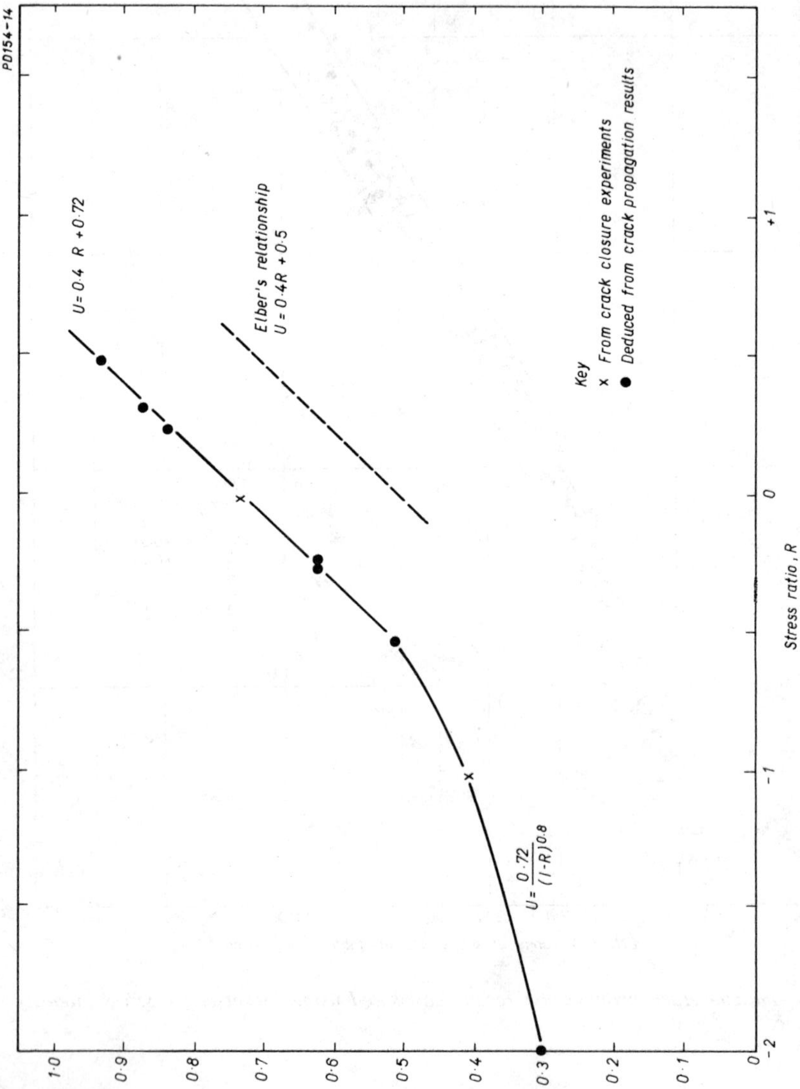

FIG. 5—Proportion of stress range for which crack tip is open (U) versus stress ratio (R).

under various stress ratios have been correlated on the basis of $\Delta\sigma_{eff} = U\,\Delta\sigma$ [*15*], suggesting that the crack propagation relationship based on ΔK_{eff} would be appropriate in the context of welded joints.

The form of crack propagation relationship depends on the shape of the *da/dN* versus ΔK curve. Referring to Figs. 1, 3, and 4, it will be seen that the present data had the form shown in Fig. 6.

In a study of fatigue crack propagation in center-notched steel plate [*3,8*], it was found that the shape of the curve was related to certain features on the fracture and that these features depended on specimen geometry. The data which were found to be relevant in the context of fatigue of welded joints [*4,8*] were those which fell on a straight line, so that, for a given stress ratio

$$\frac{da}{dN} = C(\Delta K)^m \tag{5}$$

and for which the corresponding fracture was predominantly plane strain,

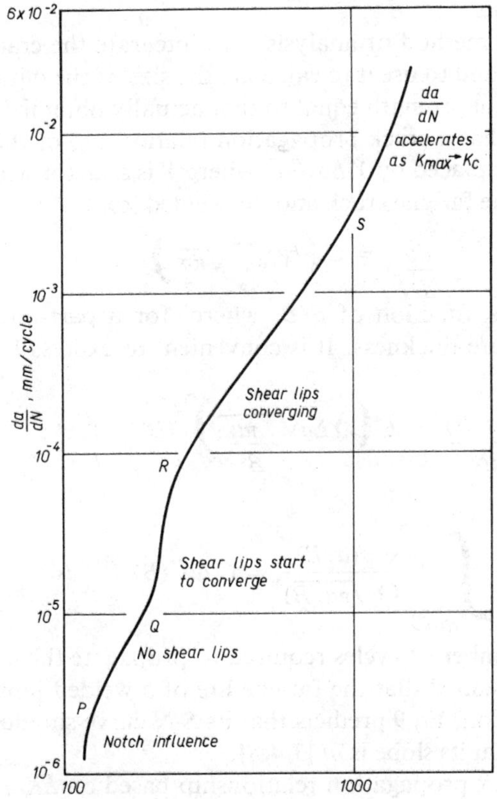

FIG. 6—*General shape of crack propagation curves and corresponding features related to specimen geometry.*

so that the fracture surface was mainly normal to the plate surface and shear lips were converging from the plate surfaces. In the present case, the corresponding region is that above S in Fig. 6. It is clear from the data that this is also the only region in which a stable relationship between da/dN and ΔK is obtained. In view of these factors, the relationship for use in the analysis of welded joints is based on the data which fell above R. In all cases (Figs. 1, 3, and 4), these data fell on a straight line and for a given stress ratio may be expressed as Eq 5 with $m = 3$ approximately. In the following analyses, where possible a relationship based on ΔK_{eff} is used, that is

$$\frac{da}{dN} = C_2 (\Delta K_{eff})^3 \tag{6}$$

where the values of C_2 on the scatterband limits are 1.7×10^{-11} and 6.9×10^{-12} mm/cycle (Nmm$^{-3/2}$)$^{-3}$, while the average (geometric mean) is 1.08×10^{-11} mm/cycle (Nmm$^{-3/2}$)$^{-3}$.

Analysis of Fatigue Failure in a Fillet Welded Joint

Method of Analysis

Essentially, the method of analysis is to integrate the crack propagation relationship [4,8] and to use it to calculate the size of the initial defect which would give a fatigue strength equal to that actually obtained for the welded joint. Assuming that a crack propagation relationship of the form of Eq 5 applies, with ΔK replaced by $Y \Delta\sigma\sqrt{\pi a}$, where Y is a factor which depends on the geometry of the fatigue crack and the welded joint

$$\frac{da}{dN} = C\left(Y\Delta\sigma \sqrt{\pi a}\right)^m \tag{7}$$

In general, Y is a function of a/B, where, for a part-through thickness crack, B is the plate thickness. It is convenient to express Eq 7 in terms of a/B. Then

$$\frac{d(a/B)}{dN} = C\left(Y\Delta\sigma\sqrt{\frac{\pi a}{B}}\right)^m (B)^{m/2-1} \tag{8}$$

Integrating

$$\int_{a_i/B}^{a_f/B} \frac{d(a/B)}{(Y\sqrt{\pi a/B})^m} = C\,\Delta\sigma^m (B)^{m/2-1} N \tag{9}$$

where N is the number of cycles required to propagate the crack from a_i/B to a_f/B. If it is assumed that the fatigue life of a welded joint consists only of crack propagation, Eq 9 predicts that its S-N curve should be linear on a log-log plot and that its slope is m [3,4,8].

Adopting a crack propagation relationship based on ΔK_{eff} (Eq 6) so that the influence of R is taken into account, Eq 9 becomes

$$\int_{a_i/B}^{a_f/B} \frac{d(a/B)}{(Y\sqrt{\pi a/B})^3} = I = C_2 B^{1/2} (\Delta\sigma_{eff})^3 N \tag{10}$$

where $\Delta\sigma_{eff} = U\,\Delta\sigma$, and C_2 has the values noted earlier.

Fatigue Test Results for Fillet-Welded Joint

The welded detail considered was a plate with longitudinal attachments fillet welded to one or both surfaces, as illustrated in Fig. 7. The welds were made by the gas metal arc (MIG) process. The plate was the same as that used for the crack propagation tests. The specimens generally failed as a result of fatigue crack propagation through the plate thickness from the weld toe at the end of an attachment. A typical fracture surface is shown in Fig. 7(*c*).

Test results obtained from these types of specimens were obtained under a number of stress ratios and are presented in Ref *15*. Reasonable correlation of the results obtained with $R > -1$ was achieved on the basis of $\Delta\sigma_{eff} = U\Delta\sigma$, where U was as given in Fig. 5, as indicated in Fig. 8. It will be noted that a slope of 3 was reasonable, that is, the slope that would have been predicted from Eq 10 if the whole fatigue life consisted of crack propagation. In other words, the test data suggest that this is the case.

Analysis of Fatigue Behavior of Joint

The fatigue behavior of welded specimens of the type shown in Fig. 7(*a*), but made from 12.7-mm-thick steel, has been analyzed successfully using fracture mechanics [*4*]. The analysis included detailed consideration of the fatigue crack shape and weld profile and their effects on the stress intensity factor [*16*]. Since the present joint is geometrically similar to the steel joint considered, it is reasonable to assume that the geometric correction term (Y) in the stress intensity factor for the fatigue crack should be the same. In detail, Y is made up of the factors indicated in Fig. 9. Curves relating the correction terms with the relevant dimensions were presented in Ref *16*.

A complication in the fracture mechanics analysis of weld toe cracks is that the crack front shape ($a/2c$), which affects M_s, M_t, and ϕ_o, may change as the crack propagates. In the study of the fillet-welded joint in steel, a simple relationship between a and $2c$ was found [*16*]. Examination of fracture surface markings on aluminum alloy specimens tested under programmed loading (for example, Fig. 7(*c*)) confirmed that the same relationship was applicable in the present analysis. This was $2c = 6.71 + 2.58a$ [*16*].

The weld profiles in the present specimens were reasonably uniform and, in general, θ (see Fig. 9) was between 40 and 50 deg. Therefore, the appropriate relationship between M_k and a/B [*16*] is that for $\theta = 45$ deg.

The integral in Eq 10, called I, was evaluated for $a_f/B = 1$ and Y corre-

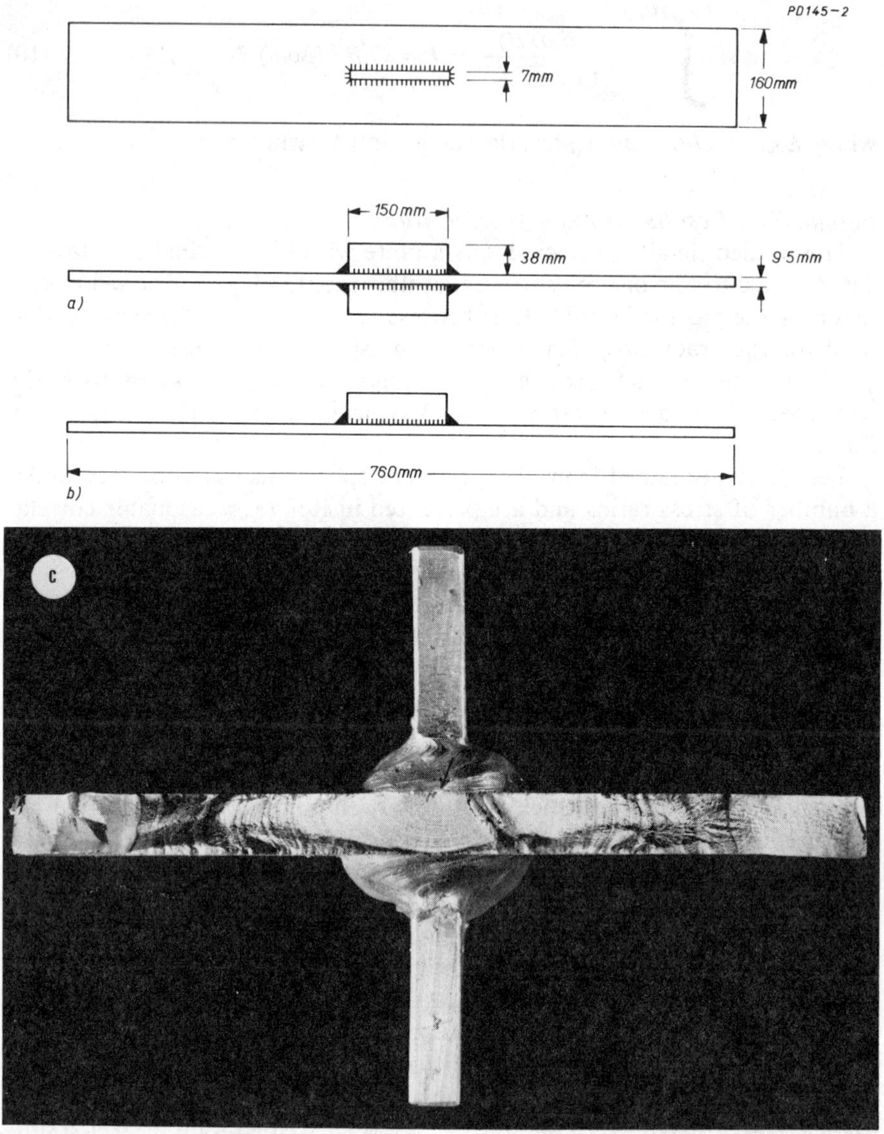

FIG. 7—*Geometry of welded joints considered in fracture mechanics analysis:* (a) *Type 1 specimen,* (b) *Type 2 specimen, and* (c) *fracture surface obtained from both types of specimens. Crack propagated from arrowed region.*

sponding to the assumptions just discussed. The resulting curve relating I and a_i/B is given in Fig. 10. It should be noted that the curve in Fig. 10 has been extended to crack sizes below those for which $a/2c$ was known. To do this, for $a_i/B < 0.01$, I was calculated for the extreme $a/2c$ values for an el-

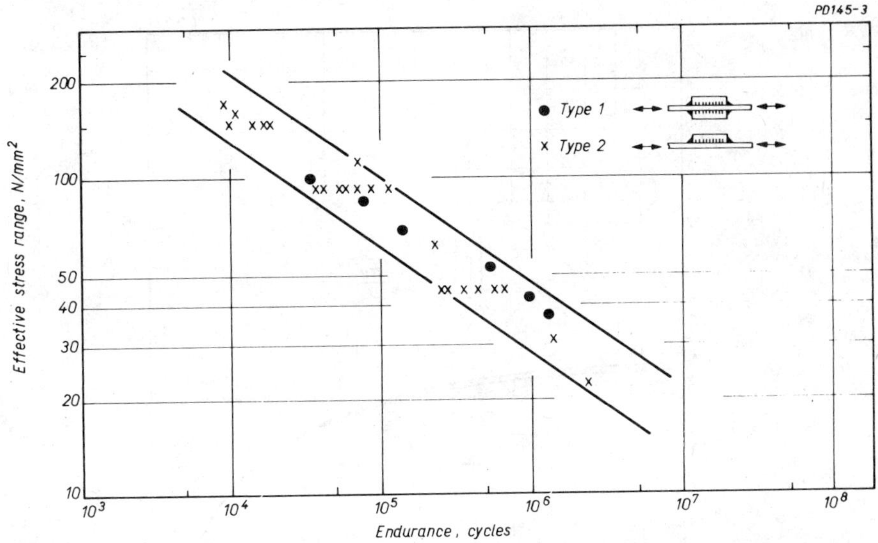

FIG. 8—*Fatigue test results for specimen Types 1 and 2 plotted in terms of* ΔK_{eff} [15].

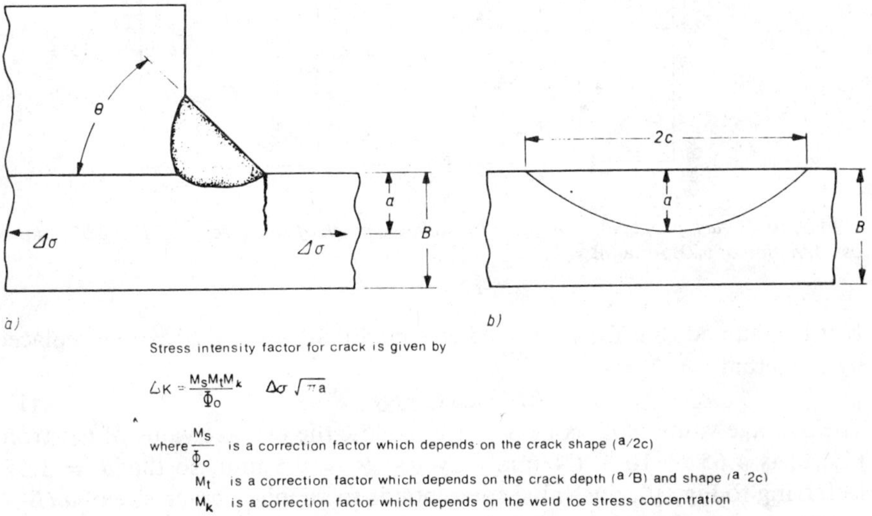

FIG. 9—*Details of weld toe crack:* (a) *section, and* (b) *plane of fracture.*

liptical crack front of 0 and 0.5 and the resulting curves form the boundaries of the thick curve plotted in Fig. 10, the upper curve relating to $a/2c = 0.5$.

The equation of the $\Delta\sigma_{eff}$ versus N curve for the joint (Fig. 8) is

$$(\Delta\sigma_{eff})^3 N = C_3 \tag{11}$$

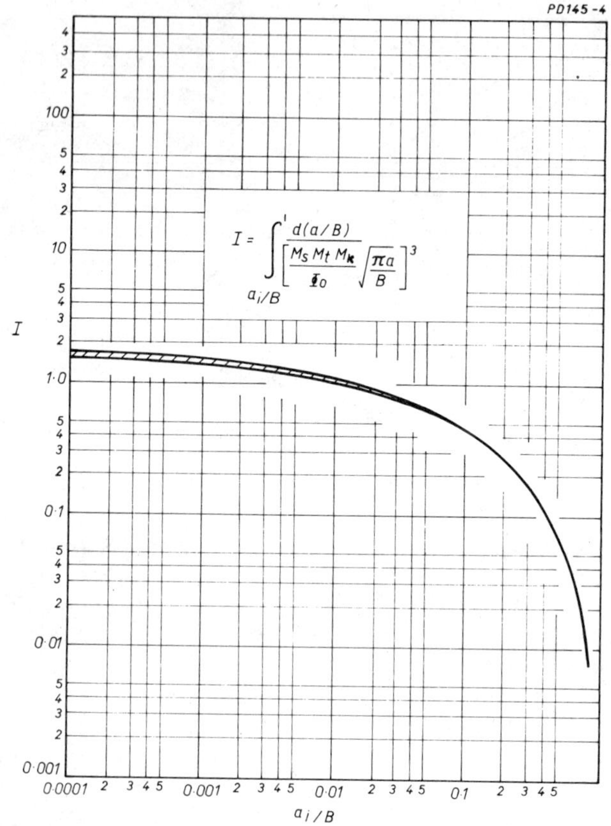

FIG. 10—*Crack propagation integral for weld toe crack of shape* 2c = *6.71 + 2.58*a *plotted as a function of* a_i/B *for* $a_f/B = 1$.

If it is assumed that Eq 11 is predicted by Eq 10, $(\Delta\sigma_{eff})^3 N$ can be replaced by C_3 so that

$$I = C_2 C_3 \sqrt{B} \tag{12}$$

The average value of C_2 is 1.08×10^{-11} while the average value of C_3, from Fig. 8, is 4.65×10^{-10} $(\text{N/mm}^2)^3$ cycles. $B = 9.5$ mm, so that $I = 1.55$. Referring to Fig. 10, this value corresponds to an initial crack size of $a_i/B \simeq 0.0003$ or $a_i \simeq 0.003$ mm. This value is an approximation because the flatness of the I versus a_i/B curve in this region leads to difficulties in interpretation. For example, for the lower limit curve, crack depths corresponding to $I = 1.4$ and 1.5 differ by an order of magnitude. Thus, small variations in the values substituted into Eq 12 can lead to large differences in estimates of crack depth. Another factor which is particularly important when a/B is small is M_k, which depends on the weld profile [*16*]. Small differences in the stress concentration at the weld toe have a significant effect on the value of I [*3,4*]. In general, it is reasonable to conclude that the analysis indicates that a fracture mechanics analysis predicts the fatigue life of the joint if it is assumed that initial cracks of the order of 0.01 mm are present.

Discussion of Results of Analysis

Metallographic examination of fillet-weld toes in Al-Zn-Mg alloy failed to reveal defects of the order of magnitude of those present in steel, only undercut and acute angles between the weld plate [*5*]. This suggests that if fatigue crack propagation occurs early on in the life of fillet welds in aluminum alloy, the cracks initiate at even smaller defects. The present analysis indicates that the whole life of the weld would consist of crack propagation if defects of the order of 0.01 mm in depth were present at the toe. This is an order of magnitude smaller than the size of defect found in steel. The accuracy of the method of analysis is supported by the fact that similar work predicted an initial defect depth of approximately 0.2 mm in steel [*3*], which is within the range measured [*1*]. In practice, surface discontinuities with depths of the order of 0.01 mm or more are certain to occur near weld toes, particularly on the weld itself. In fact, fatigue cracks have been observed to propagate even from weld ripples near the toe. Furthermore, as was just noted, if the estimated defect depths are realistic, the local stress concentration due to the weld profile has a strong influence on the rate of propagation of the small crack and hence on the total fatigue life. This expectation was to some extent realized in the results obtained from specimens with varying weld profiles in Ref *17*. However, in the absence of any direct evidence that fatigue cracks propagate from crack-like defects, the possibility that there is a crack initiation period in the life of Al-Zn-Mg fillet welds cannot be ruled out.

From the practical viewpoint, the analysis indicates that a fatigue crack at the toe of a fillet weld is very small for a large part of its life. I is directly proportional to fatigue life (Eq 10) so that for $a_i/B = 0.002$ the endurance is proportional to 1.35. When the crack reaches $a/B = 0.1$, that is, approximately 1 mm in a 9.5-mm-thick specimen, the remaining life is proportional to 0.45. Thus, two-thirds of the endurance has been spent producing a 1-mm-deep crack. In a real case [*18*], a crack depth of 0.75 mm was reached after 45 percent of the life. Under these circumstances, it is not surprising that the presence of a fatigue crack cannot be detected easily early in the life of aluminum welded joints [*18*].

Analysis of Fatigue Failure in Transverse Load-Carrying Fillet Welds

The application of fracture mechanics in the determination of the fatigue strength of transverse load-carrying fillet welds is useful for two reasons. First, although the conventional *S-N* curve is suitable for representing the fatigue strength for toe failures, in view of the range of geometries, it is not suitable for root failures [*8*]. Fracture mechanics analysis of this mode of failure leads to a method of describing fatigue strength. Second, a comparison between the optimum weld designs for aluminum alloy and steel welds throws further light on the magnitude of weld toe defects in aluminum alloy welds.

Method of Analysis

The method of analysis is identical with that used to determine the optimum design of steel welds [*8*]. Essentially, Eq 9 is used to calculate the fatigue strength of a transverse load-carrying fillet-welded joint in the form of a cruciform, as illustrated in Fig. 11. Assuming that as in the case of steel [*19*] the rate of crack propagation is the same in aluminum alloy weld metal and plate, from Eq 9 with $m = 3$, the appropriate relationship for weld failure in the cruciform joint becomes

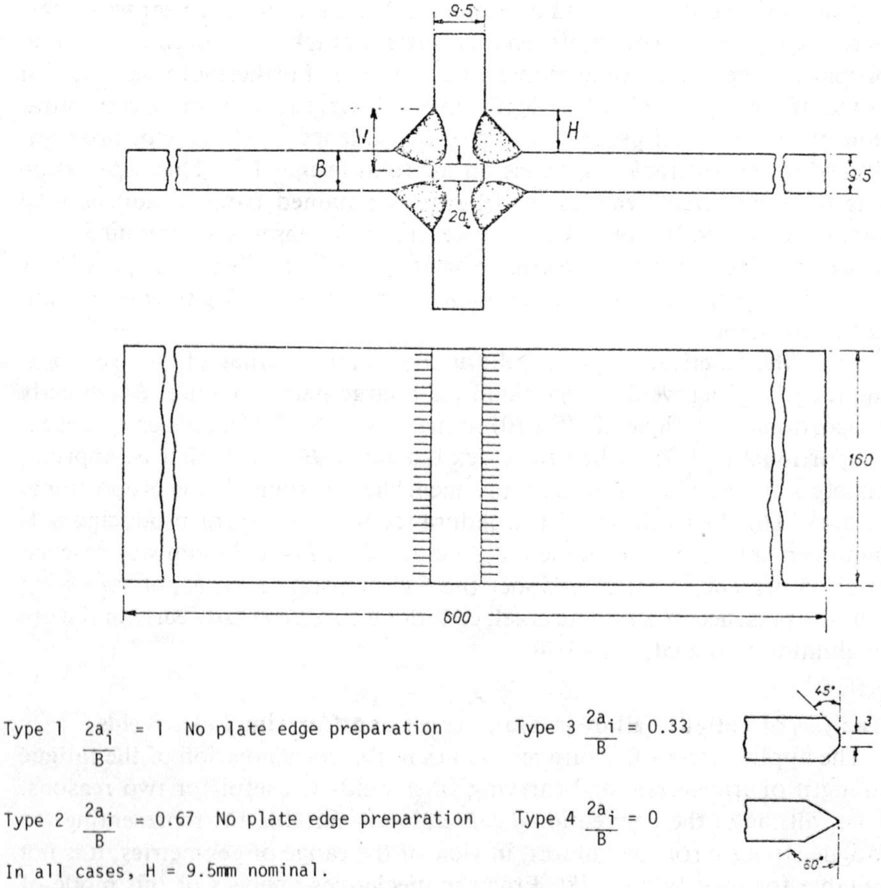

FIG. 11—*Cruciform specimen designs.*

$$\int_{a_i/W}^{a_f/W} \frac{d(a/W)}{\left(Y\sqrt{\frac{\pi a}{W}}\right)^3} = I = C\,(\Delta\sigma_p)^3\, W^{1/2} N \tag{13}$$

where $\Delta\sigma_p$ is the stress range in the plate. This is reexpressed in the form

$$(\Delta\sigma^*)^3 N = \frac{1}{C} \tag{14}$$

where $\Delta\sigma^*$, called the generalized stress parameter [*4,8*], depends on applied stress, initial crack size, crack size at failure, geometry of weld, and the crack propagation relationship for the material, and is given by

$$\Delta\sigma^* = \Phi\sigma_p \left(\frac{W^{1/2}}{I}\right)^{1/3} \tag{15}$$

Note that Eq 14 actually predicts the $\Delta\sigma^*$ versus N relationship since C is known from crack propagation results. Thus, if $\Delta\sigma^*$ can be calculated for any geometry of cracked component in a given material, its fatigue life is predicted.

It is useful to express the fatigue strength of the cruciform joint in the form of Eq 14 because it can be compared easily with the *S-N* relationship which describes the fatigue strength for toe failure. This should have the form

$$(\Delta\sigma)^3 N = C_4 \tag{16}$$

Optimum design conditions occur when the fatigue strengths for the two failure modes are the same, that is, $(\Delta\sigma)^3 N$ from Eq 16 can be substituted into Eq 14 to give

$$I = CC_4\sqrt{W} \tag{17}$$

I depends on a_i and W so that knowing C and C_4, the geometry of the load-carrying weld can be specified.

Fatigue Testing of Cruciform Joints

In order to check the basis of the analysis, Eq 14, a number of cruciform joint specimens with varying geometry were fatigue tested. The results were analyzed in terms of $\Delta\sigma^*$ and compared with the $\Delta\sigma^*$–N scatterband predicted on the basis of the crack propagation results.

Specimens—The specimens were designed to give both weld and plate failures by varying the degree of weld penetration. Four types of specimens were used, as shown in Fig. 11. The plate thickness and weld leg length were kept constant.

Each series of specimens was made in a single piece 710 mm wide which was then cut into specimen widths of 160 mm. This method of construction was chosen to enable the welds to be made using a mechanized system. The object was to produce welds of uniform profile, to reduce scatter in results for toe failures, and constant root penetration to simplify the fracture mechanics analysis.

The welds were made automatically using a mechanized MIG system. The filler material was 1.6-mm-diameter NG61 wire and the shielding gas was argon.

Test Method—The specimens were tested axially under cyclic tension loading with zero minimum stress ($R = 0$). Failure occurred soon after through-section cracking.

Test Results—The results are given in Table 2, and Fig. 12 shows typical failures.

Fracture Mechanics Analysis of Fatigue Test Results

A solution for the stress intensity factor of a crack at the root of a cruciform joint (Y in Eq 13) was determined by Frank [*20*] and used in the analysis of steel welds [*8*]. Full details of the solution and its use in conjunction with Eq 13 to determine the crack propagation integral as a function of a_i/W and H/B (see Fig. 11) were given in Ref *8*. That information was used to calculate I for each of the present specimens for which a_i/W and H/B were known, and $\Delta\sigma^*$ was calculated using Eq 15. The values obtained are given in Table 2.

Although the specimens were tested with $R = 0$, the presence of tensile residual stresses will have effectively increased the tensile mean stress [*15*]. Thus, ideally, the crack propagation relationship used to determine C in Eq 13 should take account of mean stress. However, the magnitude of residual stresses in the present specimens is not known, and so the assumption will be made that they were present and could have produced effective values of R within the range $R = 0$ to $R = +0.5$. Then, values of C based on the lower limit of crack propagation data obtained with $R = 0$ and the upper limit for data obtained with $R = +0.5$ (Fig. 1) will be used in the analysis to give a predicted $\Delta\sigma^*$–N scatterband.

The weld failure results are plotted in terms of $\Delta\sigma^*$ in Fig. 13 together with the scatterband predicted. It will be seen that all the data lie above the lower limit and all but one result fell within the scatterband. There is a tendency for the data to follow a slope slightly greater than 3. This may indicate that the fatigue crack propagation relationship for the weld metal is slightly different from that for the parent plate. However, in general, the agreement between predicted and actual results is good and supports the method of analysis.

Determination of Optimum Weld Design

The optimum weld design is determined using Eq 17. The resulting design is critically dependent on the choice of the two constants, C and C_4. In the present case, a value of C based on the upper limit of the crack propagation

TABLE 2—*Fatigue test results for cruciform joints.*

Specimen Type	Specimen	$\Delta\sigma_p$, N/mm^2	$\frac{H}{B}$	$2a_i$, mm	$2W$, mm	$\frac{a_i}{W}$	I	$\Delta\sigma^*$	Endurance Cycles	Location of Failure
	9	77	1.03	10.5	29.0	0.36	0.70	135	51 500	in weld from root
	10	54	1.05	10.0	29.5	0.34	0.81	91	287 000	in plate from toe; crack in weld almost through
	11	108	1.05	11.0	29.5	0.39	0.60	200	12 000	in weld from root
1	12	39	1.05	10.5	29.5	0.37	0.70	69	1 591 500	in weld from root
	17	62	1.05	10.0	29.5	0.34	0.81	104	85 000	in weld from root
	18	46	1.05	10.0	29.5	0.34	0.81	77	319 000	in weld from root
	20	46	1.05	9.0	29.5	0.31	0.98	75	342 500	in weld from root
	1	77	0.97	8.0	28.0	0.29	0.95	122	144 000	in weld from root
2	3	39							455 500	in plate from toe
	4	108	0.92	8.0	27.0	0.30	0.80	179	33 500	in weld from root
	5	39	0.92	8.0	27.0	0.30	0.80	65	1 185 000	in weld from root
3	6	93	0.92	8.25	27.0	0.31	0.76	159	30 500	in weld from root
	7	54							172 500	in plate from toe
	13	77							520 500	in plate from toe
	14	54							230 500	in plate from toe
4	15	39							770 500	in plate from toe
	16	108							54 000	in plate from toe

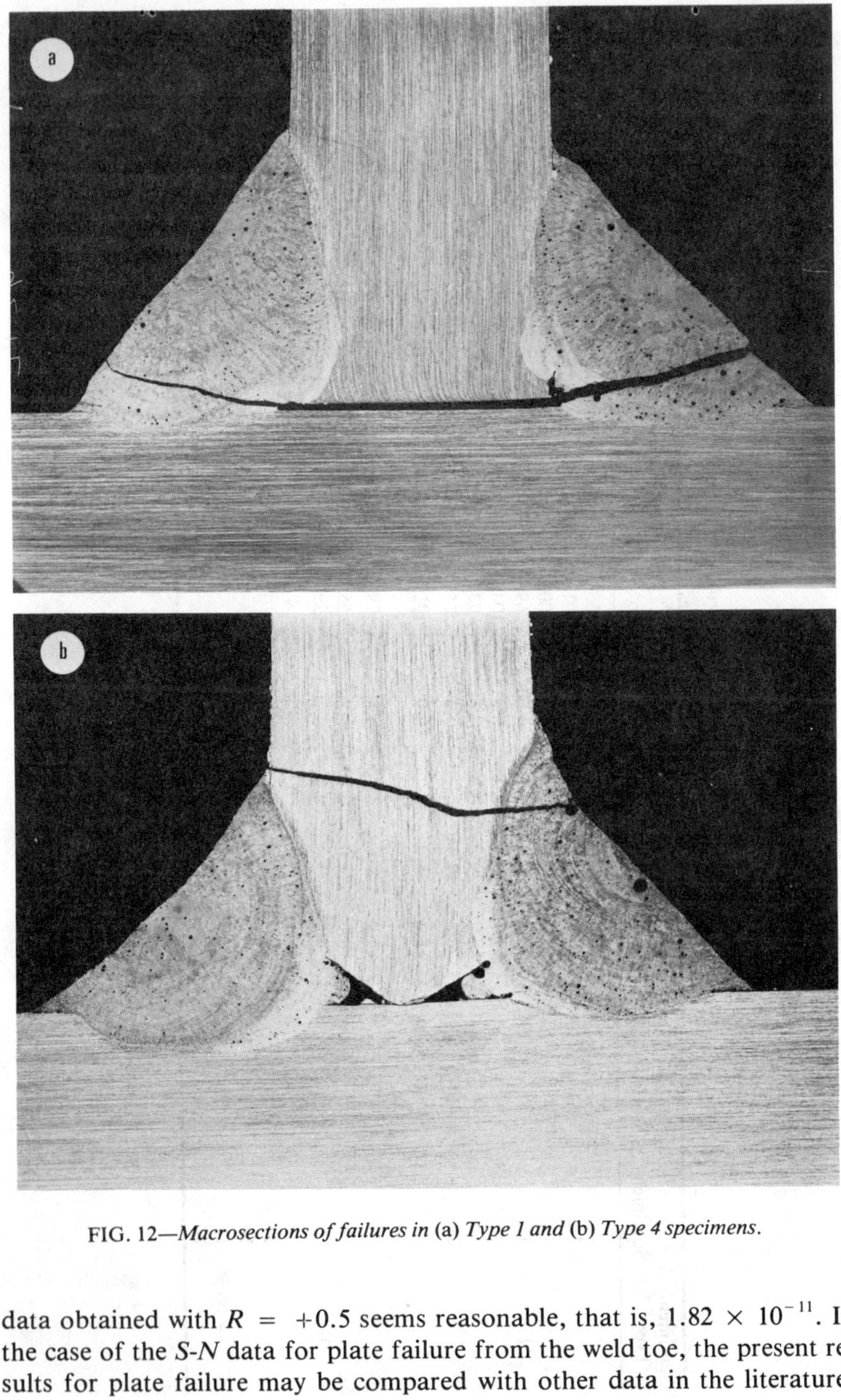

FIG. 12—*Macrosections of failures in* (a) *Type 1 and* (b) *Type 4 specimens.*

data obtained with $R = +0.5$ seems reasonable, that is, 1.82×10^{-11}. In the case of the *S-N* data for plate failure from the weld toe, the present results for plate failure may be compared with other data in the literature, including data for nonload-carrying transverse welds since the weld toe geometry is the same, to establish a lower limit. Such data are given in Fig. 14, which also indicates the scatterband for the present results drawn at a slope of $m = 3$, which appears reasonable. The value of C_4 corresponding to the lower limit of the scatterband is 2.7×10^{10}.

Thus, substituting for C and C_4 in Eq 17 gives $I/\sqrt{W} = 0.49$. Presenting this information in the way used in Ref *8*, that is, in terms of the corresponding values of weld size/plate thickness (H/B) and initial defect size/plate thickness ($2a_i/B$) gives the curves shown in Fig. 15. In that figure, coordinates to the right of the curves refer to weld failure while those to the left refer to plate failure.

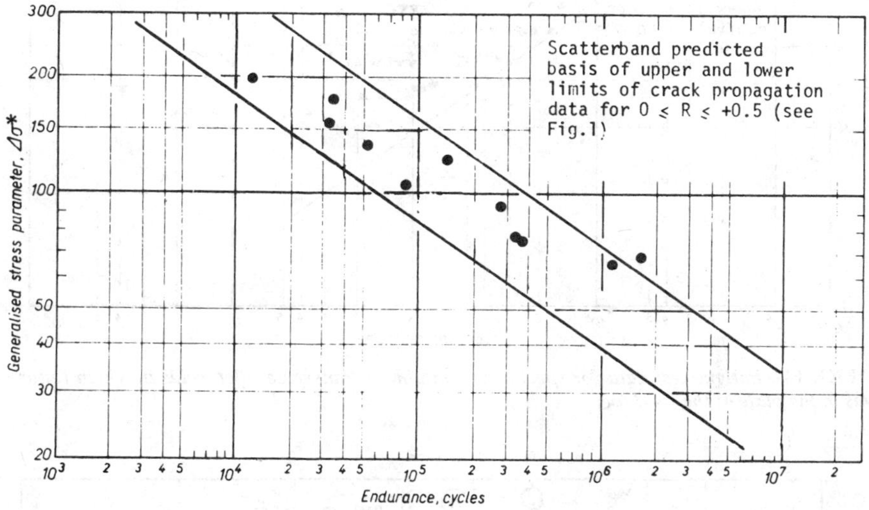

FIG. 13—*Fatigue test results for weld failure in cruciform specimens plotted in terms of* $\Delta\sigma^*$.

Discussion of Results

It is of interest to compare the optimum weld design for aluminum with that for steel [*8*], as shown in Fig. 15. On the assumption that the geometries of similar size fillet welds in steel and aluminum alloy are the same, it was anticipated that the optimum weld designs also would be similar. However, it will be seen that they are significantly different, the curves for steel predicting much larger initial defect sizes at the transition from toe to weld failure. For instance, for $B = 12.5$ mm and $H/B = 1$, the transition in aluminum occurs at $2a_i/B = 0.47$ while in steel it occurs at 0.97. Thus, a steel joint with $2a_i/B = 0.9$ would fail in the plate while an aluminum joint would fail in the weld. The implication of this difference in behavior is that the weld toe in an aluminum joint is less severe from the fatigue viewpoint than the weld toe in a steel joint. Since the geometries are similar, this suggests that the weld toe defects are less severe (smaller) in aluminum than in steel. This is the same conclusion which was drawn earlier, thus providing further evidence to support it.

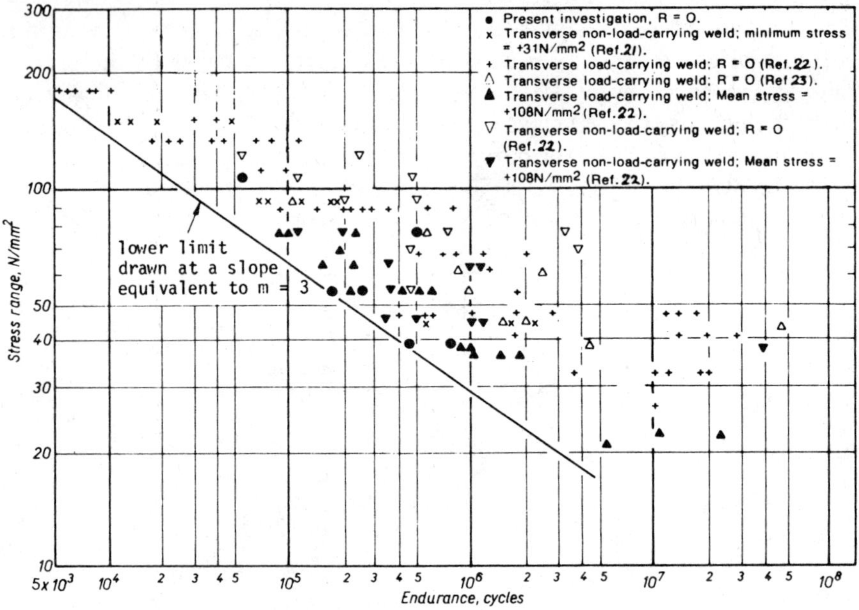

FIG. 14—*Fatigue test data for specimens containing transverse fillet welds in which failure was in the plate at the weld toe.*

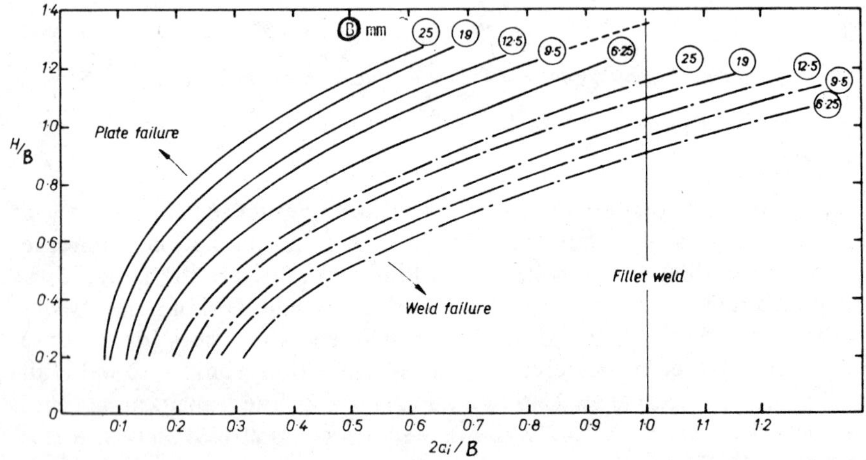

FIG. 15—*Geometric conditions for optimum design of transverse load-carrying fillet welds. Solid lines refer to Al-Zn-Mg alloy and broken lines refer to steel* [8].

The information given in Fig. 15 is a basis for the design of transverse load-carrying fillet welds. However, as noted earlier, it depends on the choice of values for C and C_4 in Eq 17. Further test results are needed for weld failure to ensure that the value of C used, which from Fig. 13 could be regarded as an overestimate, is reasonable. Clearly, the problem is complicated by the presence of residual stresses due to welding, and further work is needed to determine their magnitude in the weld root region.

Conclusions

Analysis of the fatigue behavior of Al-Zn-Mg fillet-welded joints made by the MIG process which fail from the weld toe and the root indicates that in each case most of the fatigue life can be considered to be spent propagating a crack. The weld toe region appeared to be less severe than the same region in steel welds in that preexisting defects may be an order of magnitude smaller or that a significant proportion of the life is spent producing a small fatigue crack. Consequences of this feature are that cracks are very small for a large proportion of their lives making early detection in service more difficult than for steel welds where fatigue cracks propagate from relatively large preexisting defects. Also, the optimum design of transverse load-carrying welds required a greater weld or smaller initial root defect than was the case for steel joints.

The analysis made use of a crack propagation relationship based on test results obtained from notched specimens. There was some doubt about the interpretation of these data, particularly the value of m, and although the results of the analysis indicated that the choice of $m = 3$ was reasonable, further clarification of this is required. Also, the validity of the assumption that crack propagation rate is the same in plate and weld metal needs checking. In view of the apparent success of the crack closure stress for determining ΔK_{eff}, further experiments to determine U could prove to be profitable. The resulting crack propagation relationship has the advantage of being simple.

Finally, the crack propagation results showed that the alloy used was sensitive to mean stress. Since as-welded joints contain residual stresses due to welding, their magnitude must be taken into account when the effective mean stress is determined.

Acknowledgments

This work was carried out at The Welding Institute and sponsored by the Ministry of Defence Procurement Executive. The authors are grateful to their colleagues for their assistance, particularly T. R. Gurney at The Welding Institute.

References

[1] Signes, E.G., Baker, R.G., Harrison, J.D., and Burdekin, F.M., *British Welding Journal,* Vol. 14, No. 3, March 1967.

[2] Maddox, S.J., "Fracture Mechanics Applied to Fatigue in Welded Structures," Proceeding of the Conference on Fatigue of Welded Structures, The Welding Institute, 1971.

[3] Maddox, S.J., "A Fracture Mechanics Analysis of the Fatigue Strength of Welded Joints," Ph.D. thesis, University of London, 1972.

[4] Maddox, S.J., *Welding Research,* Vol. 6, No. 5, 1976, p.1.

[5] Watkinson, F.W., "Examination of Fatigue Tested Al–Zn–Mg Welds," Contract Report No. 3318/2/72, The Welding Institute, Aug. 1972.

[6] Forman, R.G., Kearney, V.E., and Engle, R.M., *Journal of Basic Engineering,* Vol. 89, No. 3, Sept. 1967.

[7] Elber, W. *Engineering Fracture Mechanics,* Vol. 2, 1970, p.37.

[8] Maddox, S.J., *Welding Journal,* Vol. 53, No. 9, Sept. 1974, pp.401s-409s.

[9] Gurney, T.R., *Metal Construction,* Vol. 1, No. 2, Feb. 1969, pp.89-96.

[10] Boulton, C.F. and Maddox, S.J., "On the Determination of Fatigue Crack Propagation Data under Complex Loading," Proceedings of the Conference on Fatigue Testing and Design, Society of Environmental Engineers, 1976.

[11] Fedderson, G. in *Plane Strain Fracture Toughness Testing of High Strength Metallic Materials, ASTM STP 410,* W.F. Brown, Jr. and J.E. Srawley, American Society for Testing and Materials, 1967, p.77.

[12] Maddox, S.J., *International Journal of Fracture,* Vol. 11, No. 3, June 1975.

[13] Pearson, S., "The Effect of Mean Stress on Fatigue Crack Propagation in Half Inch Thick Specimens of Aluminum Alloys of High and Low Fracture Toughness," Technical Report 68297, Royal Aircraft Establishment, Dec. 1968.

[14] Elber, W. in *Damage Tolerance in Aircraft Structures, ASTM STP 486,* American Society for Testing and Materials, 1971.

[15] Maddox, S.J., "The Influence of Residual Stresses on the Fatigue Behaviour of Al–Zn–Mg Alloy Fillet Welds," Proceedings of the Conference on Fatigue Testing and Design, Society of Environmental Engineers, 1976.

[16] Maddox, S.J., *International Journal of Fracture,* Vol. 11, No. 2, April, 1975.

[17] Webber, D., "Screening Tests on Possible Life Improvement Methods for Al–Zn–Mg Alloy Fillet Welds," presented at this symposium.

[18] Webber, D. and Maddox, S.J. in *Detection and Measurement of Cracks,* The Welding Institute, 1976, Chapter 4.

[19] Maddox, S.J., *Welding Research,* Vol. 4, No. 1, 1974.

[20] Frank, K.H., "The Fatigue Strength of Fillet Welded Connections," Ph.D. thesis, Lehigh University, 1971.

[21] Maddox, S.J., "Fatigue Strength of Helium Shielded MIG Welds in Al–Zn–Mg Alloy," Contract Report No. 3318/7/74, The Welding Institute.

[22] Gunn, K.W. and McLester, R., *British Welding Journal,* Dec. 1962.

[23. Person, N.C., *Welding Journal,* Vol. 50, No. 2, Feb. 1971.

J. P. Sandifer[1] *and G. E. Bowie*[1]

Fatigue Crack Propagation in A537M Steel

REFERENCE: Sandifer, J. P. and Bowie, G. E., "**Fatigue Crack Propagation in A537M Steel,**" *Fatigue Testing of Weldments, ASTM STP 648*, D. W. Hoeppner, Ed., American Society for Testing and Materials, 1978, pp. 185–196.

ABSTRACT: Fatigue crack propagation rates are presented for weldments in A537M steel at room temperature. Rates are determined for cracks propagating along the fusion line, in the melt zone of welds made with two shielding gas ratios, and in the base metal. Static tension and Charpy impact tests conducted on specimens from the base material and melt zone also are presented. Results indicate that crack growth resistance of A537M steel is not degraded by proper welding procedures, and the weld zone may impede growth rates. Shielding gas ratios, which affect tensile properties, are shown not to affect fatigue crack growth rates significantly. An interim *da/dN* versus ΔK design curve is computed for the data sets, and its application to flawed plate life prediction is discussed.

KEY WORDS: fatigue tests, steels, crack propagation, weldments, melting points, fusion line, weld defects, Charpy impact properties, tensile data, life prediction

Increased service requirements of advanced naval vehicles and petroleum exploration equipment necessitate the use of material with higher strength and toughness yet available at a reasonable cost. The A537M steel was developed to satisfy these present requirements. It is a quenched and tempered low carbon-manganese-silicon steel used in sections up to 50.8 mm (2 in.) thick. The low carbon and increased manganese content results in a material with very low nil ductility transition temperature thus making the alloy particularly well suited for use in the extremes of the arctic environment. For wide usage, however, such an alloy must have good weldability from the fabrication and service standpoints. The latter standpoint, service, is the prime concern of this investigation.

Higher performance of man-rated equipment or items where catastrophic

[1] Senior scientist and senior research scientist, respectively, Fatigue and Fracture Mechanics Laboratory, Lockheed-California Company, Burbank, Calif. 91520.

failure would be prohibitively expensive necessitates the application of fracture control techniques to assure that life requirements of the structure are met. Since all materials, particularly welds, contain defects, it is essential to know if these defects will propagate to a critical size under service loads and environmental conditions where failure would occur. It is necessary, therefore, to establish the crack growth rates under cyclic fatigue loading for cracks located in the melt zone, fusion line, and base metal away from the heat affected zone (HAZ).

Experimental Procedure

This investigation used a quenched and tempered steel prepared according to ASTM Specification for Pressure Vessel Plates, Heat-Treated, Carbon-Manganese-Silicon Steel (A 537-76) and modified. The chemical composition of this steel is given in Table 1, and the mechanical properties are given in Table 2.

TABLE 1—*Chemical composition of A537M 4.45-cm (1 ¾-in.) thick plate and weld metal.*

Element	Base Material, weight %	Weld Metal, weight %	
		98Ar-2O	95Ar-5O
C	0.12	0.06	0.065
Si	0.30	0.91	0.83
Mn	1.47	1.47	1.42
P	0.018	0.009	0.011
S	0.011	0.011	0.010
Ni	0.24	...	...
Mo	0.07	...	...
		...[a]	...[a]

[a] Hydrogen 1.4 ppm.

TABLE 2—*Mechanical properties of A537M base metal.*

Tensile	
Yield strength, MPa (psi)	428 (62 200)
Tensile strength, MPa (psi)	549 (79 650)
Elongation, %	35
Reduction of area, %	78.7
Modulus, MPa (psi)	2.03×10^5 (29.5×10^6)
Charpy V-notch impact	
Temperature	−51°C (−60°F)
J (ft·lb)	111 (82)

Two plates, 4.445 cm (1¾ in.) thick, were welded by a semiautomatic inert gas- shielded metal arc-spray transfer weld process with two different shielding gas ratios: 98Ar-2O and 95Ar-5O. Electrodes were dual shield 9000–C1 ⅙ in. Heat input was 863.6 J/m (34 000 J/in.) average, with a maximum of 1 397 J/m (55 000 J/in.) per weld pass.

Charpy, tension, and compact tension specimens were fabricated from each plate as shown in Fig. 1. For crack propagation measurements, the

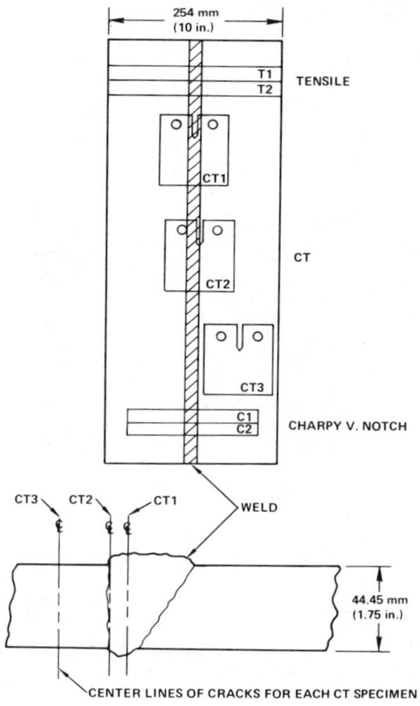

FIG. 1—*Specimen location and orientation in welded plate.*

compact tension specimens were fabricated per ASTM Test for Plane-Strain Fracture Toughness of Metallic Materials (E 399–74); the dimensions are shown in Fig. 2. The compact tension specimens were installed in a MTS servocontrolled electrohydraulic fatigue machine and precracked under constant amplitude sinusoidal loading. During precrack, loads were stepped down so that the final 0.5 mm (0.020 in.) of growth occurred at a maximum stress intensity equal to that at which subsequent testing was to be started. The precrack load shedding procedure was conducted in a manner such that the last crack growth increment ensured that the crack tip was beyond any previous plastic zones. Crack length was measured on both sides of the specimens using a traveling microscope. After precracking, specimens were then monotonically loaded to failure or fatigue cycled to failure.

As shown in Fig. 1, the compact tension specimens were machined from the plates so that the notches were located in three areas: (*a*) center of the melt zone, (*b*) on the fusion line, and (*c*) in the base material beyond the HAZ.

Charpy V-notched specimens were machined from each plate as shown in Fig. 1 and tested at $-51\,°C$ ($-60\,°F$) per ASTM Notched Bar Impact Testing of Metallic Materials (E 23–72).

Tension test specimens taken from each plate were cut perpendicular to the weld as shown in Fig. 1 and longitudinally in the melt zone for weld

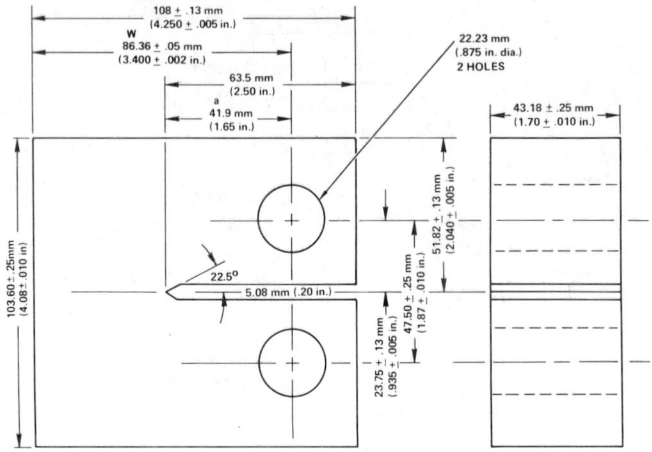

FIG. 2—*Compact tension (CT) specimen geometry.*

metal as well as base metal properties and tested per ASTM Tension Testing of Metallic Materials (E 8–17a).

Fractographic examinations were made on each failed specimen. Scanning electron microscopy and X-ray analyses were conducted in areas of apparent abnormality. Chemical analyses were made to verify constituent elements, and quantitative hot extraction analysis was performed on sections of weld metal to determine hydrogen content.

Presentation and Discussion of Results

Fracture Toughness

Charpy impact tests conducted in the base metal at −51°C (−60°F) gave average values of 111 J (82 ft·lb), while tests conducted in the melt zones of the two gas mixtures gave values of approximately 33.9 J (25 ft·lb). These data are shown in Tables 2 and 3.

Static fracture toughness tests conducted on the compact tension crack growth specimens did not yield K_{Ic} values since plain strain conditions were not met. ASTM Test E 399–74 requires that both thickness, B, and the uncracked ligament must exceed $2.5\ (K_Q/\sigma_{ys})^2$. Tests conducted in the base metal and on the fusion line gave average K_Q values of 145.2 MPa $\sqrt{\text{m}}$ (132 ksi $\sqrt{\text{in.}}$). No significant difference was noted in values for the two locations. Observations of the failures indicated possibly higher toughness in the melt and HAZs.

Subcritical Fatigue Crack Growth Rates

During fatigue crack growth testing, crack lengths were measured at least every 0.5 mm (0.02 in.). From these data, the points shown in Fig. 3 were

TABLE 3—*Mechanical properties of weld metal using two shielding gas ratios from A537M plates.*

Plate 95-5-03 (95Ar-5O)	
Tensile	
yield strength, MPa (psi)	660 (95 740)
tensile strength, MPa (psi)	715 (103 885)
elongation, %	22.25
reduction of area, %	57.7
modulus, MPa (psi)	2.23×10^5 (32.3×10^6)
Charpy V-notch impact	
temperature,	−51°C (−60°F)
J (ft·lb)	34.7 (25.6)
lateral expansion	0.0538 mm (0.0212 in.)
Plate 98-2-03 (98Ar-2O)	
Tensile	
yield strength, MPa (psi)	724 (105 140)
tensile strength, MPa (psi)	747 (108 415)
elongation, %	8.0
reduction of area, %	17.0
modulus, MPa (psi)	1.07×10^5 (30.0×10^6)
Charpy V-notch impact	
temperature,	−51°C (−60°F)
J (ft·lb)	33.9 (25)
lateral expansion	0.0513 mm (0.0202 in.)

calculated by the following equation

$$\Delta K = (1 - R)\,K_Q \qquad (1)$$

where K_Q is defined in ASTM Test E 399–74.

These data points in Fig. 3 represent pairs of measured crack growth rate, da/dN, and stress intensity range, ΔK, for the five crack growth specimens tested. The test environment was laboratory ambient air (temperature 21 ± 3°C (70 ± 5°F), relative humidity, 40 ± 10 percent). The applied load waveform was sinusoidal with a frequency of 5 Hz and ratio of minimum to maximum load $R = 0.1$. For crack growth in the base metal, as shown in Fig. 4, crack extension were relatively uniform. Data reduction for this case included the effect of crack front bowing. For crack growth in the weld metal and fusion zone, nonuniform growth was evidently caused by effects of weld inclusions, as shown in Figs. 5 and 6. Occasional retardation of crack growth was particularly evident for fusion zone specimen CT 2–98–2–03, Fig. 6. Data reduction for weld metal and fusion zone specimens was performed with the premise that the average of crack extensions measured optically on opposite specimen faces represented crack front positions.

Fatigue crack growth for the fusion zone specimens, Fig. 6, was observed to occur without deviation of the crack front from the fusion zone. During subsequent application of monotonically increasing tensile loading of these specimens fast fractures propagated away from the weld, through the HAZ, and into the base material.

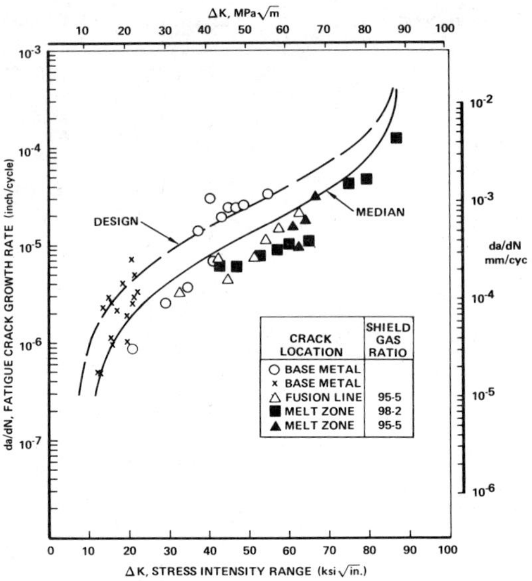

FIG. 3—*Comparison of crack growth rates for A537M steel weldments in 21°C (70°F), 40 ± 10 percent relative humidity air at 5 Hz and at a stress ratio of* R = *0.1. Median and design curves for weldment and base metal are shown.*

The total number of pairs of (da/dN, ΔK) measurements given in Fig. 3 is equal to 47. There are insufficient numbers of data points for any one specimen or condition to warrant separate crack growth rate curve fitting. A decision was made to derive a single median curve fit for all 47 points, using data reduction procedures developed by the second author, described elsewhere [*1*].[2] The curve titled "Median" in Fig. 3 was plotted by means of the empirical relation

$$da/dN = -1 + \exp\left[e + (v - e)(-\ln(1 - \Delta K/K_b))^{1/k}\right] \qquad (2)$$

where

$K_b =$ 95.7 MPa $\sqrt{m}$ (87 ksi $\sqrt{in.}$),
$e =$ −9.7377,
$v =$ 2.9044, and
$k =$ 8.2946.

Quantities e, v, and k are referred to as the threshold parameter, characteristic value, and shape parameter, respectively. Parameters e and v depend upon calculated values of linear regression coefficients as described in Ref *2*. The curve fitting parameter, K_b, is the stress intensity range value where da/dN becomes indefinitely large. In computations, K_b is selected by a trial and error procedure. For example, K_b is never less than or equal to the largest ΔK value in the data set. The shape parameter, k, and the stress intensity upper limit, K_b, are optimized iteratively with the aid of correlation

[2] The italic numbers in brackets refer to the list of references appended to this paper.

FIG. 4—*Fracture surface of base metal specimen.*

coefficient comparisons. The fitting relation, Eq 2, for the set of 47 pairs of (da/dN, ΔK) measurements was obtained with a sample correlation coefficient equal to 0.87 and a sample standard deviation in the plane of linear regression equal to 0.05. By examination of Fig. 3, it is seen that base metal data points tend to lie above the median curve, and the fusion line and melt zone data points lie below the curve. Evidently crack growth is slower in the weld areas than in the base metal. This is partly attributable to the pinning of the crack by inclusions and defects. While for a propagating crack these may be beneficial in this regard, they are not desirable since they can also act as stress concentrations and crack initiation sites. The effects of segregation during multiple passes and on strength are clearly shown in Fig. 5 where each pass is visible in the fracture face.

It is also noted by examination of Fig. 3 that fatigue crack growth rates in the melt zones for 98Ar-2O_2 and 95Ar-5O_2 gas ratios are essentially the same, even though tensile properties listed in Table 3 indicate apparent gas ratio effects. These crack growth rates for A537M are very close to those obtained by Socie and Antolovich [3] for unmodified A537 weld and bare metal.

FIG. 5—*Melt zone specimens and fracture surface showing multiple pass weld.*

In view of the scatter in the data sets, a decision was made to suggest for interim design purposes a baseline crack growth rate curve for ambient air conditions arbitrarily displaced one standard deviation toward the faster growth rate side of the median curve. The equation of the interim design curve retains the values of K_b and k in the aforementioned expression, but has modified values of e and v as follows: $e = -9.1390$ and $v = 3.5030$.

The interim design curve is sigmoidal when plotted in the format of Fig. 3 and has the property of predicting rapidly increasing crack growth rates as ΔK approaches the value $K_b = 95.7$ MPa $\sqrt{\text{m}}$ (87 ksi $\sqrt{\text{in.}}$). The da/dN versus ΔK curve-fitting equation permits estimation of the design threshold stress intensity for subcritical fatigue crack growth, as follows

$$\begin{aligned} K_{th} &= K_b \left\{ 1 - \exp\left(-\left(\frac{-e}{v-e} \right)^k \right) \right\} \qquad (3) \\ &= 6.27 \text{ MPa} \sqrt{\text{m}} \ (5.7 \text{ ksi} \sqrt{\text{in.}}) \end{aligned}$$

Application of Design Curve to Flawed Plate Life Prediction

As discussed by Lindh and Peshak [*4*] initial flaws at the surface of a plate tend to be more harmful from a stress concentration standpoint and therefore with respect to fatigue crack growth than imbedded flaws of

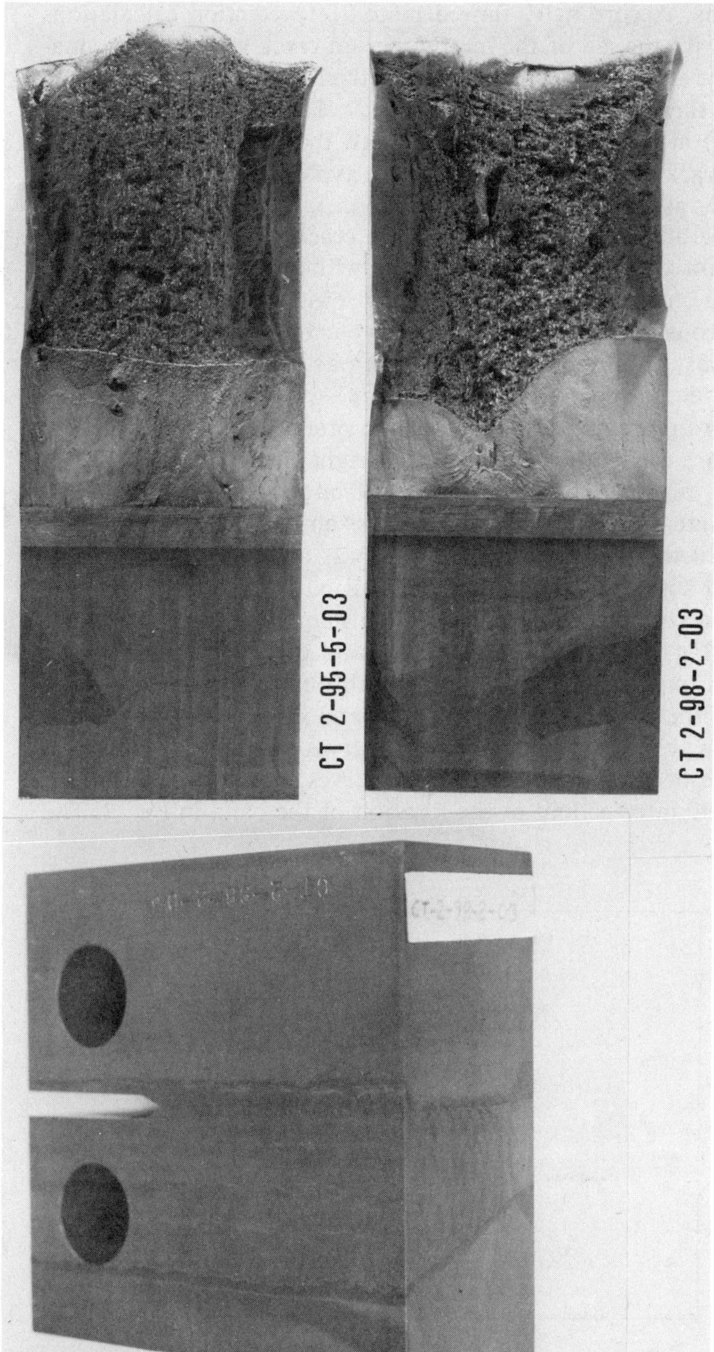

FIG. 6—*Fusion line specimens and fracture surfaces.*

similar dimensions. Accordingly, flawed plate life prediction calculations were performed with the aid of the interim design crack growth rate equation for the case of an idealized surface flaw in the base metal. The shape of the initial surface flaw assumed is shown in Fig. 7. The ratio of the depth (a) to the length ($2c$) at the plate surface for both the initial flaw and subsequent fatigue crack was assumed to have the typical value $a/2c = 0.45$. Plate thickness was assumed to be 4.45 cm (1.75 in.).

Numerical integrations of the interim design crack growth rate equation were performed for two initial flaw sizes: one with a depth equal to 0.127 cm (0.05 in.), and one with depth of 2.54 cm (1.0 in.). The 0.127-cm (0.05-in.)-deep flaw is considered representative of typical weldment defects such as pores, inclusions, grind marks, or corrosion pits. The 2.54-cm (1.0-in.)-deep flaw is considered representative of damage which could conceivably occur in service or in rare multiple-pass welding practice. The numerical integration procedure was performed at each of eight equally spaced applied stress amplitudes, σ_0, with the starting point of integration at an assumed initial flaw size. Integration at a particular stress amplitude was performed incrementally until the predicted critical flaw was achieved. For practical purposes, applied stress amplitudes are normalized by means of the base metal tensile yield strength, σ_{ty}. Results of the numerical integrations are demonstrated in Fig. 7. Note the horizontal band of uncertainty between the upper part-through crack (PTC) and the through-the-thickness crack (TTC) regimes of the graph. This uncertainty band was drawn on the basis of a comparison of separate prediction curves developed during a preliminary phase of analysis. The separate curves, not shown in Fig. 7, were faired to yield intersections at the location of the band in Fig. 7. The

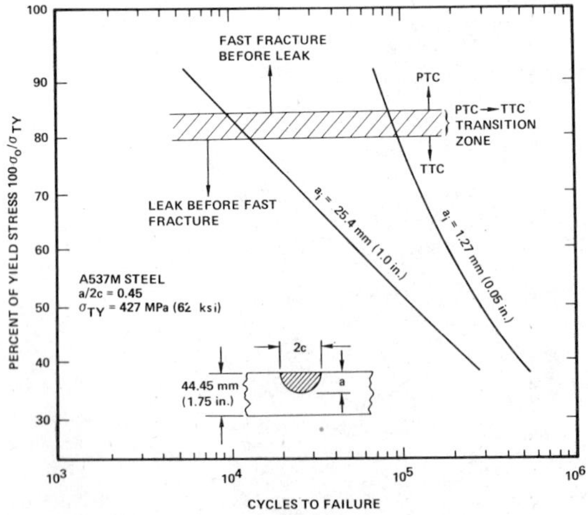

FIG. 7—*Tentative design curves for assumed initial flaw in base metal.*

height of the band at the intersection was established as a matter of practical judgment on the part of the authors.

The ordinate in Fig. 7 is ratio of applied stress amplitude to yield stress, 100 σ_0/σ_{ty}, and the abscissa is the number of cycles to failure, N. There are two curves on the graph, one labeled a_i = 0.127 cm (0.05 in.), and the other, a_i = 2.54 cm (1.0 in.). At any one ratio, 100 σ_0/σ_{ty} on the ordinate, the difference between the first and second curve read on the abscissa predicts the number of cycles for growth from a_i = 0.127 cm (0.05 in.) to a depth of 2.54 cm (1.0 in.). To illustrate the use of Fig. 7, consider a stress ratio above the transition band in Fig. 7, such as 100 σ_0/σ_{ty} = 90 percent. At this relatively high stress condition, it is predicted that fast fracture will occur prior to penetration of a PTC through the plate thickness. At some lower stress ratios, such as 100 σ_0/σ_{ty} = 50 percent, the fatigue crack will propagate through the plate thickness and continue to grow as a TTC prior to fast fracture. The PTC–TTC transition zone was identified on the graph to reflect uncertainties in crack front behavior which can occur when a deep fatigue crack approaches the opposite surface of a plate from the surface where initiation occurred.

For discussion, suppose that a plate structure is thought to have surface flaws with depths not deeper than 0.127 cm (0.05 in.). In this event, exceptional cases can occur when a deeper flaw or fatigue crack with a depth of 2.54 cm (1.0 in.), for example, can be present. It can be seen from examination of Fig. 7 that selection of inspection periods and methods under the conditions of the discussion are indeed more critical for relatively high than for low stress ratios 100 σ_0/σ_{ty}.

Conclusions

1. Crack growth resistance in A537M steel is not degraded by proper welding procedures.
2. No difference in crack growth rates were noted for cracks propagating in the melt zone compared to ones in the fusion line.
3. Shielding gas ratios, while they affect tensile properties, do not significantly affect crack growth rates.
4. Operational design stresses may be selected from a life prediction curve so that a crack will leak prior to fast fracture.

Recommendations

All tests presented in this paper were conducted in 21°C ± 3° (70 + 5°F) air with a relative humidity of 40 ± 10 percent and at a test frequency of 5 Hz. Use of these data for design applications should be done only for those test conditions since a substantially different environment or loading waveform could drastically alter these predictive curves. Particularly significant changes would be anticipated in a marine environment with very low load frequencies (1→0.01 Hz), hold times of minutes, or a sustained load [5].

Additional tests must be conducted under the planned usage environmental and load conditions to obtain the required design reliability.

Acknowledgments

The authors wish to thank G. L. Goss for his assistance in the test programs.

References

[*1*] Bowie, G. E., Pettit, D. E., and Ryder, J. T., "Fracture of Engineering Materials," Lockheed-California Company Report LR 27012, 15 Jan. 1975.

[*2*] Bowie, G. E. and Hoeppner, D. W. in *Proceedings,* 1976 International Conference on Computer Simulations for Materials Applications, Gaithersburg, Md., 19-21 April 1976 pp. 1171-1178.

[*3*] Socie, D. F. and Antolovich, S. D., *Welding Research,* June 1974, pp. 267-271-S.

[*4*] Lindh, D. V. and Peshak, G. M., *Welding Research,* Feb. 1969, pp. 45-48.

[*5*] Pettit, D. E., Ryder, J. T., Krupp, W. E., and Hoeppner, D. W., "Investigation of the Effects of Stress and Chemical Environments on the Prediction of Fracture in Aircraft Structural Materials," AFML-TR-74-182, Air Force Materials Laboratory, Dec. 1974.

Pedro Albrecht[1]

A Study of Fatigue Striations in Weld Toe Cracks

REFERENCE: Albrecht, Pedro, "**A Study of Fatigue Striations in Weld Toe Cracks,**" *Fatigue Testing of Weldments, ASTM STP 648*, D. W. Hoeppner, Ed., American Society for Testing and Materials, 1978, pp. 197–217.

ABSTRACT: Striation spacing in weld toe cracks due to constant amplitude stress cycling was found to correlate with Bates' empirical equation. Adding closely spaced periodic overloads to the constant amplitude load history increases the striation spacing. Evidence of a decrease of striation spacing for widely spaced overloads, due to crack growth delay, could not be uncovered. About ten percent of the fatigue crack surface was striated, the remaining area being occupied by quasistriations and featureless regions. The lowest spacing detected in the direct surface scanning mode of electron microscopy was 3.3 x 10^{-8} m/cycle at a maximum magnification of X70 000. At that growth rate, about 80 percent of the crack propagation life had already expired. The microscopic striation spacings observed in this study, employing welded specimens fabricated from A588 structural steel, did not provide reliable indications of the average crack growth rate.

KEY WORDS: fatigue tests, weldments, striations, fatigue (materials), scanning electron microscopy, overloads, structural steels.

Typically, studies of fatigue striations and crack growth rates are done with specimens used in fracture mechanics testing such as those described in ASTM Test for Plane-Strain Fracture Toughness of Metallic Materials (E 399–74). In civil engineering steel structures, however, stress cycling usually causes initiation of surface cracks from points of stress concentration, of which weld toes are the most common, and crack propagation progresses through the weld material, the heat affected zone (HAZ), and the base metal. The microstructure changes between the various regions and, along with the thermal welding stresses, may significantly affect crack extension.

The purpose of this study was to examine fatigue striations in cracks emanating from the weld toe of a specimen whose configuration and fabri-

[1] Associate professor of Civil Engineering, University of Maryland, College Park, Md. 20742.

cation technique simulate conditions typical of a bridge weldment. In order to permit correlation of striation spacing with load, a simplified service stress history was selected.

Information was also sought on the extent of area striated, the maximum resolution of spacing, the portion of the propagation life that could be verified, and the reliability of estimating crack growth rates from striation spacing.

Experiment Design

Specimen Geometry

The fatigue specimens, shown in Fig. 1, consisted of a cruciform joint fabricated from ASTM A588 steel plates and connected by the automatic submerged-arc welding process. The main plate had a cross section of 10 by 26 mm and was 330 mm long. The two cross plates, 6.5 by 26 and 50 mm long, were attached to the main plate with 6.5-mm fillet welds.

Figure 2 shows the longitudinal section through the welded joint. The surface was polished and etched so that the extent of the HAZ, the weld material, and the base metal could be seen. A metallographic examination of the polished surface indicated an average grain size of 25 μm in the base metal and 4 μm in the fine-grained HAZ. The transition in grain size occurred very rapidly over a distance of about 50 μm. These same grain sizes were found on a metallographic section that coincided with the plane of crack propagation, Section A–A in Fig. 2.

Mechanical Properties

The following mechanical properties of the main plate were measured: 420-MPa yield strength, 565-MPa tensile strength, 22 percent elongation, and 75-Nm Charpy V-notch impact energy absorption at 21 °C. Additional impact specimens were machined from spare cruciform specimens such that the plane of the notch coincided with the plane of fatigue crack extension, that is, Section A–A through the weld toes in Fig. 2. The corresponding impact energy was 80 Nm. The slight increase in impact energy, over values for

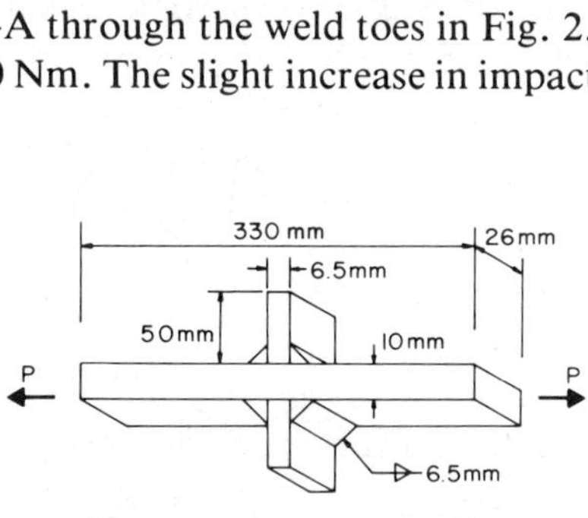

FIG. 1—*Specimen.*

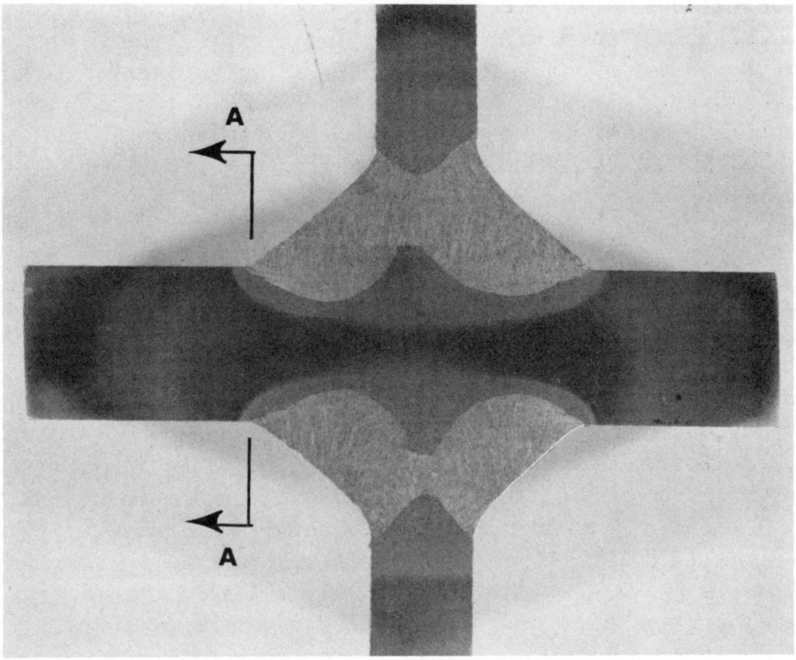

FIG. 2—*Polished and etched longitudinal section through weldment.*

specimens cut from areas remote from the weld, is attributed to residual welding compression stresses past the end of the HAZ, where the bottom of the 5-mm-deep notch was embedded.

Stress Variables

The experiment design consisted of a two-way factorial with three levels of constant amplitude stress range, six levels of overload spacing, and three replicates per cell. The stress ranges were 144, 176, and 229 MPa. The seven levels of overload spacing were 10, 10^2, 10^3, 10^4, 10^5, 10^6, and a single overload applied at the beginning, respectively. A constant ratio of 1.67 between overload stress range and constant amplitude stress range, σ_r, and a constant minimum stress of 3 MPa were maintained in all tests. A series of control specimens was tested under constant amplitude cycling, without intermittent overloads, to provide base line data for comparison.

Fatigue Test Data

Crack Initiation and Propagation

All cracks initiated at one or more points along the weld toe lines on the main plate. The cracks then propagated through the thickness of the main plate, in Section A–A of Fig. 2, until the net ligament eventually ruptured at a mean net section stress slightly larger than the uniaxial tensile strength. From calculations of crack propagation [*1*],[2] from an initial crack size corresponding to an assumed threshold value of $\Delta K = 5.5$ MPa$\sqrt{\text{m}}$ to failure, about one half of the total number of cycles to failure are accountable. Hence, the crack propagation life was preceeded by an equally long crack initiation life.

Fatigue Life

The fatigue test data are shown in Fig. 3, a plot of the overload spacing versus the fatigue life of the overloaded specimens as normalized by the fatigue life of the specimens tested without overloads. Each data point represents the log average of the three replicates in one cell of the experiment design. The arrows signify that testing of these specimens was stopped after 10 million load cycles and no visible cracking. To the left of the plot, the curves converge to unit value of normalized fatigue life. To the right, at $\Delta N = 0$, the normalized fatigue life was computed from the *S-N* curve for a stress range of 1.67 times the constant amplitude stress range.

The results show that overloads spaced every 10^3 cycles, or farther apart, increase the fatigue life. Conversely, overloads applied every 10^2 cycles, or less, decrease the fatigue life.

Delay Effects

It is well known that the rate of crack growth following a prior high load excursion decreases significantly. This slowdown may be attributed to an increase of the value of the load at which the crack opens near the leading edge [*2*]. It results from the additional clamping action of the residual compressive stresses within the oversized plastic zone of the prior load excursion. In fact, Von Euw [*3*] and Rice [*4*] have shown that the delay effect on the growth rate ends after the crack front has advanced through the prior plastic zone.

An approximate relationship between the delay length and the range of the stress intensity factor can be shown for various overload spacings with the aid of Paris' crack growth law

$$\frac{da}{dN} = C\,\Delta K^n \tag{1}$$

and the expression for the overload plastic zone size

$$2r_Y = \frac{1}{\pi}\left(\frac{K_{OL}}{\sigma_Y}\right)^2 \tag{2}$$

Assuming average growth rates for ferrite-pearlite steels [*5*] in metres per cycle, the coefficients in Eq 1 are $C = 4.8 \times 10^{-12}$ and $n = 3.0$ when ΔK is

[2] The italic numbers in brackets refer to the list of references appended to this paper.

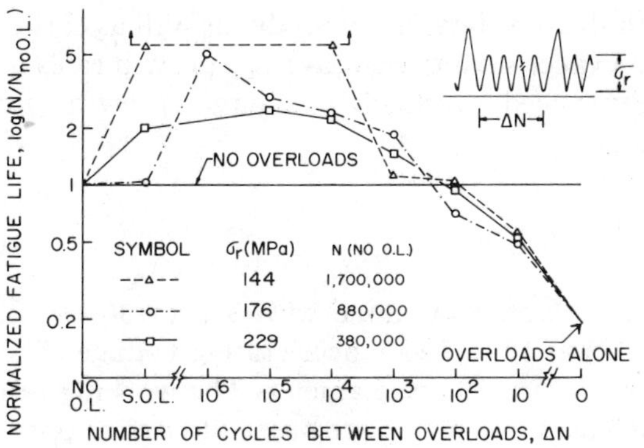

FIG. 3—*Summary of fatigue test data.*

substituted in units of MPa$\sqrt{m}$. The plane strain yield strength, σ_Y, was elevated by a factor of $\sqrt{3}$ over the uniaxial yield strength. Since in this study $K_{OL} \cong 1.67\ \Delta K$, the ratio of undelayed crack growth rate between two overloads, Δa, and the plastic zone size is approximately proportional to the number of cycles between overloads and the range of the stress intensity factor, that is,

$$\frac{\Delta a}{2r_Y} \cong 1.67\,\pi\,\sigma_Y^2 C\,\Delta K\,\Delta N \tag{3}$$

Equation 3 gives an upper bound for Δa. Since crack growth rates through the overload plastic zone are delayed, the true value of crack extension between overloads would be smaller.

Because of the exponential nature of crack growth, most of the propagation life is spent while cycling occurs at small values of ΔK. For example, when cycling at 176 MPa constant amplitude stress range, 80 percent of the crack propagation life, N, computed by integration of Eq 1 to yield

$$N = \frac{1}{C} \int_{a_i}^{a_f} \frac{1}{\Delta K^n}\, da \tag{4}$$

is consumed between the threshold value ΔK_{th} = 5.5 MPa $\sqrt{m}$ and ΔK = 20 MPa$\sqrt{m}$ [*1*]. The integration limits, initial crack size, a_i, and final crack size, a_f, corresponding to the two ΔK values just specified, were obtained from Eq 5. Keeping this result in mind, and substituting the various values of ΔN into Eq 3, one finds that overloads spaced every 10^4 cycles are reapplied after the crack has grown deeply into the delay region, $0 < \Delta a/2r_Y < 1$. When $\Delta N \geq 10^5$, the crack front extends beyond the delay region. Conversely, closely spaced overloads, $\Delta N < 10^3$, are reapplied after a small penetration of the overload plastic zone size. In this case, the delay mechanism was not activated and the closely spaced overloads shortened the life. In accordance with the test data in this study, as well as earlier findings [*3, 4*], one would then expect smaller than average growth rates in specimens subjected to widely spaced overloads and larger growth rates under closely spaced overloads.

Procedure

Equipment

The microscopic work was done with a 0 to 30-kV scanning electron microscope (SEM) at 15 to 20-kV accelerating voltage. The instrument has an electron beam of 5 x 10^{-9}m diameter. The working resolution is about 10^{-8}m. Since at least three points are needed to define one fatigue striation, the smallest crack growth rate (striation spacing) one could expect to see at the microscopic level would be 3 x 10^{-8}m. At a magnification of X70 000, the highest used in this study, the striations would be spaced every 2.1 mm on the cathode ray tube (CRT) display.

Specimen Preparation

The specimens were cut with a saw to a width of about 10 mm from the fatigue crack surfaces of the samples to a size that allowed SEM observation over the full thickness of the cross section, measured in the direction of crack propagation. The specimens were cleaned with methanol in an ultrasonic bath, mounted on an aluminum stub with silver conducting paint, sputter etched, and coated with a gold-palladium alloy to prevent charging. Since the work was done with a SEM, and not with a scanning transmission electron microscope, no surface replicas had to be made. This greatly simplified the specimen preparation.

Stress Intensity Factor

In order to correlate striation spacing with average crack growth rates, one needs an expression for the range of the stress intensity factor, ΔK. For a part-through crack growing from the weld toe, see Section A–A in Fig. 2, ΔK is given by [*6*]

$$\Delta K = \frac{1.12}{E_k} \quad F_G\, \sigma_r \sqrt{\pi a} \sqrt{\frac{2t}{\pi a} \quad \tan \frac{\pi a}{2t}} \tag{5}$$

where

a = crack depth measured normal to the toe line,
σ_r = nominal stress range remote from the crack,
t = plate thickness, and
E_k = complete elliptical integral of the second kind.

Equation 4 consists of the well known K estimate for a part-through crack in a flat plate [*7*] multiplied by a geometry correction factor, F_G, which accounts for the stress concentration effect of the weldment [*6*].

Striation Spacing

The specimens were placed in the chamber of the SEM with a tilt of 30 deg towards the detector and were, in most cases, oriented such that a line in the direction of crack propagation remained horizontal. This was done to permit direct crack size measurement, from the weld toe to a desired point, with the scale on the x-direction control knob. The CRT display was automatically compensated for tilt.

Fatigue striations were usually sought at preselected ΔK values. After placing the electron beam at the crack depth corresponding to ΔK, as just described, the surface was searched in the y-direction for well defined striations. The microscopic crack growth rate (striation spacing) was computed by dividing a measured length on a one-to-one photographic reproduction of the CRT display by the number of striations in that length and also by the magnification factor. The length measurement was always done normal to the striations.

The striation spacing was compared with the average crack growth rate, defined by Eq 1, and with the semiempirical equation proposed by Bates and Clark [*8*]

$$S = 6 \left(\frac{\Delta K}{E} \right)^2 \tag{6}$$

where S is the striation spacing and E is the modulus of elasticity.

Macroscopic Examination

The term macroscopic is used in this study in connection with overload markings on the fatigue crack surface that were visible by the naked eye.

Overload Markings

Several specimens exhibited markings produced by the overload cycles. Typically, such clear traces were imprinted only after the crack depth had exceeded one half of the plate thickness, just prior to net section yielding.

Figure 4(*a*) shows the fatigue crack surface of a specimen tested at a 229-MPa stress range and an overload spaced every 10 000 cycles. The generally fine textured area of about quarter-elliptical shape is the fatigue crack surface while the rough textured surface is the shear type fracture of the net ligament. Within the fatigue crack surface, over a depth of about one-sixth of the plate thickness measured from the toe where the crack initiated, a very fine-grained strip is noticeable. This corresponds to the region of growth through the HAZ where the average grain size was about six times smaller than in the base metal. The arrows indicate the position of the crack front at the two last overload applications, 10 000 constant amplitude load cycles apart. The total life of this specimen was 718 000 cycles.

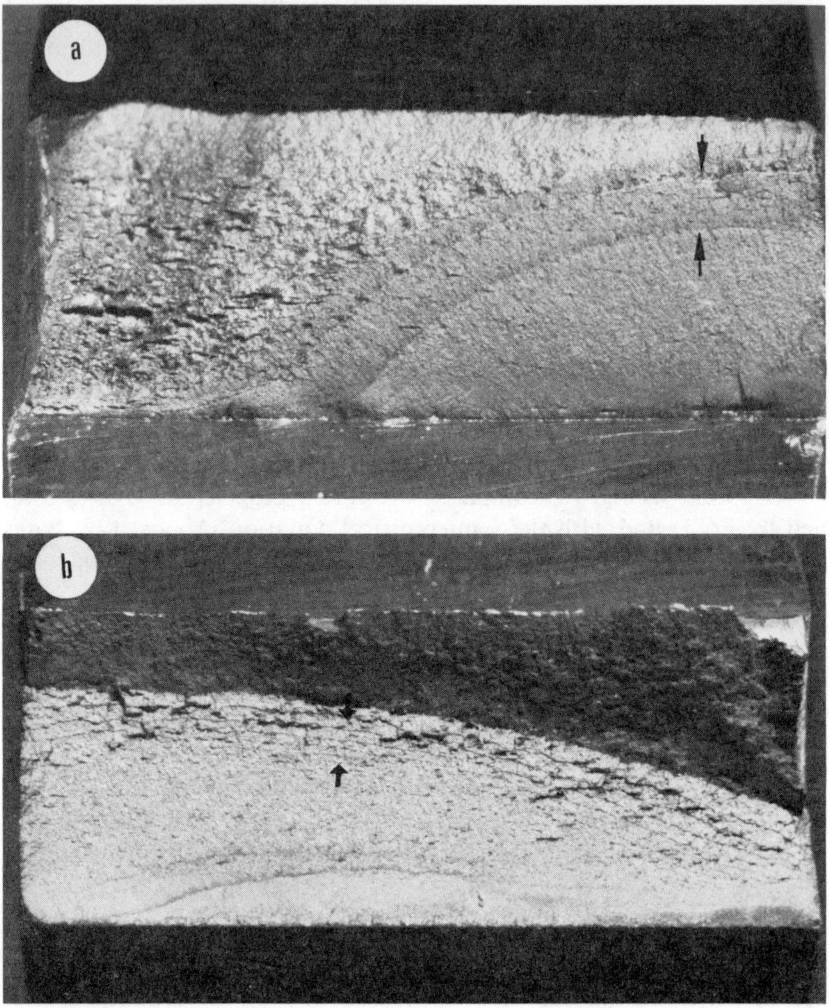

FIG. 4—*Overload markings on fatigue crack surfaces. See Section A–A of Fig. 2.* (a) σ_r = *229 MPa,* ΔN = *10 000, and* N = *718 000.* (b) σ_r = *176 MPa,* ΔN = *1 000, and* N = *1 741 000.*

The fatigue surface of a specimen with 1 000 cycles between overloads is shown in Fig. 4(*b*). It was tested at 176-MPa stress range and sustained 1 741 000 cycles prior to failure. Careful inspection of the fractograph revealed six overload markings in the band bound by the two arrows in Fig. 4(*b*).

Figure 5 is a low magnification micrograph of an overload marking taken at a crack depth at which the net section had already yielded. The marking was characteristically wide, in this case 1½ times the average grain size, and accompanied by slip bands. The surfaces on either side were offset by a step, created during overload crack extension, at an angle of about 45 deg. This was evident from a profile of the fatigue crack surface. The shadow cast by such steps, when exposed to low-angle incident light, makes the marking visible.

Markings from constant amplitude cycling cannot be seen without SEM magnification.

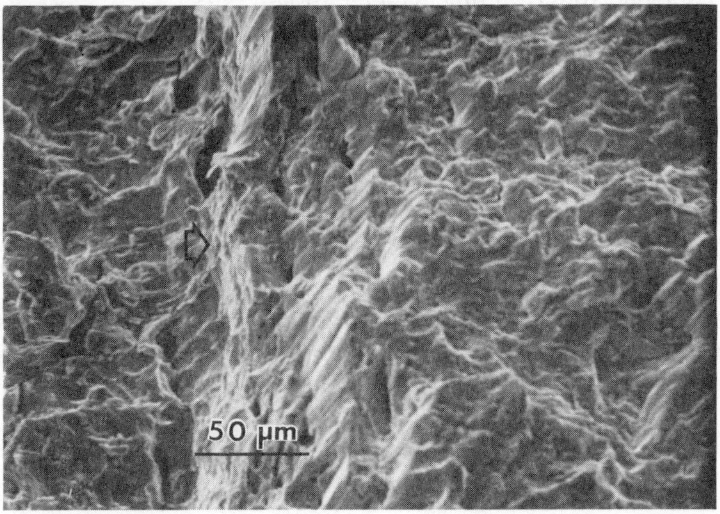

FIG. 5—*Overload marking viewed at low magnificantion (X290).* σ_r = *176 MPa,* ΔN = *10 000,* N = *1 333 000, and* a = *8.2 mm. Crack growth direction is to the right.*

Striation Spacing

Average values of macroscopic striation spacing were determined for four specimens with overload markings similar to those shown in Fig. 4. The striation spacing was computed as the ratio of measured distance and known number of cycles, $\Delta N + 1$, between two consecutive overloads. The results are plotted in Fig. 6 as a function of the range of the stress intensity factor for constant amplitude cycling taken at a point halfway between two consecutive overload markings. The open and solid circles indicate specimens tested under 176 and 229-MPa stress range, respectively. Net section yielding occurred where the lines connecting the data points for each specimen changed from solid to dashed.

Specimens subjected to one overload every ΔN = 1 000 cycles had been selected because markings from more closely spaced overloads could not be distinguished. On the other hand, overloads spaced every 10 000 cycles, or more, produced at the most two visible and widely spaced markings from which an average striation spacing for the interval could not be determined with reasonable accuracy.

The macroscopic striation spacings are compared in Fig. 6 with the band of growth rate data reported by Barsom [5] for four ferritic steels. The values measured in this study are about a factor of two higher. This could be expected because the average stress on the net ligament of the cruciform specimens was at general yielding. Under those conditions, plastic elongations across the net ligament would relieve the residual compressive stresses along the crack front and reduce the crack closure effect in the wake of the advancing front, thus resulting in larger growth rates.

The width of the marking in Fig. 5, excluding the slip band, is 27 μm. This is much larger than the overload crack extension, predicted by Eq 1 to be $\Delta a = 4.8 \times 10^{-12} (1.67 \times 176)^3 = 1.2\ \mu\text{m}$. Even so, the measured width is less than ten percent of the predicted crack extension, $\Delta a = 4.8 \times 10^{-12} \times 176^3 \times 1\,000 = 260\ \mu\text{m}$, during the next 1 000 constant amplitude cycles. A comparison of these values indicate that, although the overload cycle advanced the crack front by more than Eq 1 could predict, the additional crack increment was not nearly large enough to explain the increase of growth rate by a factor of two, as shown in Fig. 6.

As stated previously, plastic stress relief at general yielding remains then the plausible reason for the acceleration of the crack growth rate at high ΔK values.

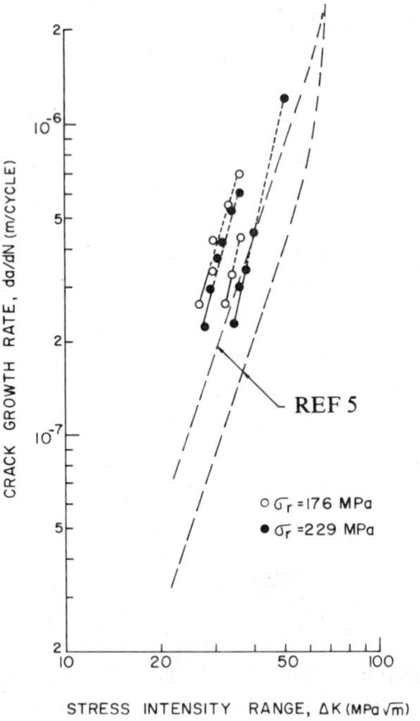

FIG. 6—*Comparison of macroscopic striation spacing and crack growth rates.*

Microscopic Examination

The term microscopic is used to characterize data obtained by SEM.

Overload Striations

When viewed with SEM, overload striations could be distinguished easily from those produced by constant amplitude cycling only when the overload was applied every 10 cycles. For example, seven overload striations were identified on the micrograph shown in Fig. 7. The fine lines between them represent the constant amplitude striations. Figure 8, a higher magnification micrograph of another area of the same specimen, shows the individual striations in more detail.

At the next higher overload spacing, ΔN = 100 cycles, overload striations could no longer be identified with certainty. This is apparent from the composite micrograph shown in Fig. 9. The band-like features, in line with the arrows, bear some resemblance to the overload striations of Figs. 7 and 8. Moreover, they are spaced at a distance which is nearly equal to the computed crack extension of 5.8 μm during 100 cycles of constant amplitude. Whether these features were indeed caused by overloads remains uncertain for two reasons. First, neither in Fig. 9 nor in other micrographs was an area found where a number of consecutive striations close to 100 could be counted. In fact, the highest number ever found in this study was in the order of 30. Second, where the crack front intersects grain boundaries or traverses pearlite colonies, the resulting marks appear, in some cases, as a band when viewed under SEM. They could, therefore, be mistaken as an overload striation.

Constant Amplitude Striations

Very high magnification micrographs of fatigue striations caused by constant amplitude load cycling are shown in Fig. 10. Such regular striation patterns provide evidence of the mechanism of progressive crack extension per load cycle and were reported by many authors for a variety of metals. They should not, however, mislead one to believe that large parts of the fatigue crack surface of specimens fabricated from structural steels are so marked. As is apparent from the scale markers, the striated areas shown in Fig. 10 cover only a fraction of an average grain size of 25 μm. As reported by Koterazawa et al [*9*] for a 0.38 percent carbon steel, and confirmed in this study on A588 steel, at most 20 percent of the total surface exhibited well defined striation patterns. The remainder is either covered with so called quasistriations, whose spacing for the A588 steel was an order of magnitude larger than the expected growth rate, or by rubbed and pressed areas. In contrast to steel, up to 70 percent of an aluminum alloy fracture surface is striated [*9*].

A typical example of the sparsity of striations and the roughness of the fatigue crack surface is shown in Fig. 11, a micrograph taken at the HAZ of

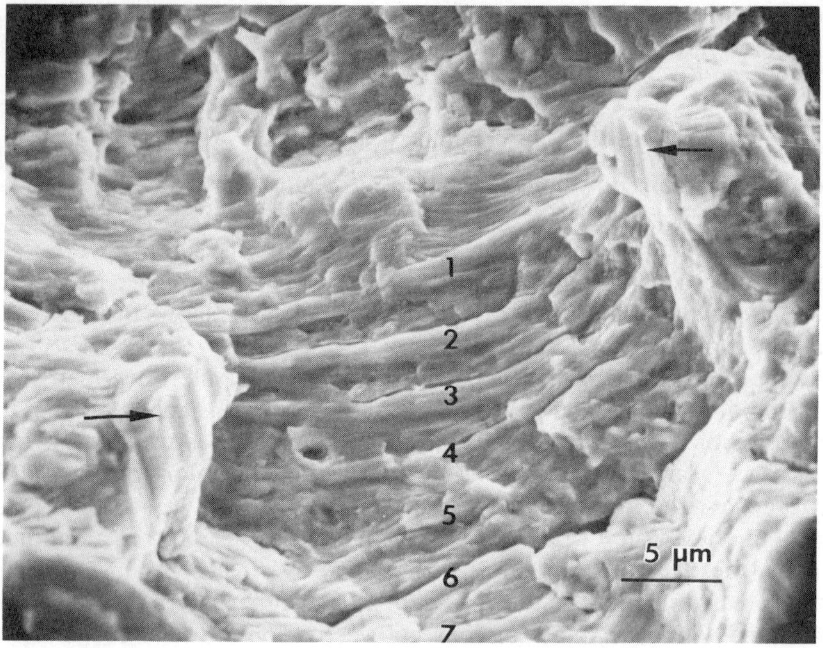

FIG. 7—*Overload and constant amplitude striations within area of about the same size as an average grain (X2700).* σ_r = *176 MPa,* ΔN = *10,* N = *422 000, and* a = *2.67 mm. Crack growth direction is down.*

a specimen subjected to constant amplitude load cycling. Large parts are featureless while others are occupied by quasistriations. Figure 12 is a 25 times higher magnification micrograph of an area of Fig. 11. It reveals an irregular striation pattern and an orientation of the local front at an angle of about 45 deg from the macroscopic crack front.

Striation Spacing

Figure 13 summarizes the microscopic striation data obtained from over 50 photomicrographs of 9 different specimens. Each micrograph was taken only after the particular area of a specimen had been carefully searched for clearly defined striations, a time consuming process. The dashed lines define a band of crack growth rates for ferritic steels [*5*], whereas the solid line is a plot of Eq 6.

The data points for the control specimens, shown with diamonds, fall along the trend line defined by Eq 6. At low ΔK values, the average microscopic striation spacing was about a factor of two larger than the average crack growth rate reported by Barsom [*5*] for ferritic steels. As ΔK increases, the data tend to cross over the crack growth rate band. The same trend was reported by other investigators, such as Von Euw [*3*] on 2024–T3

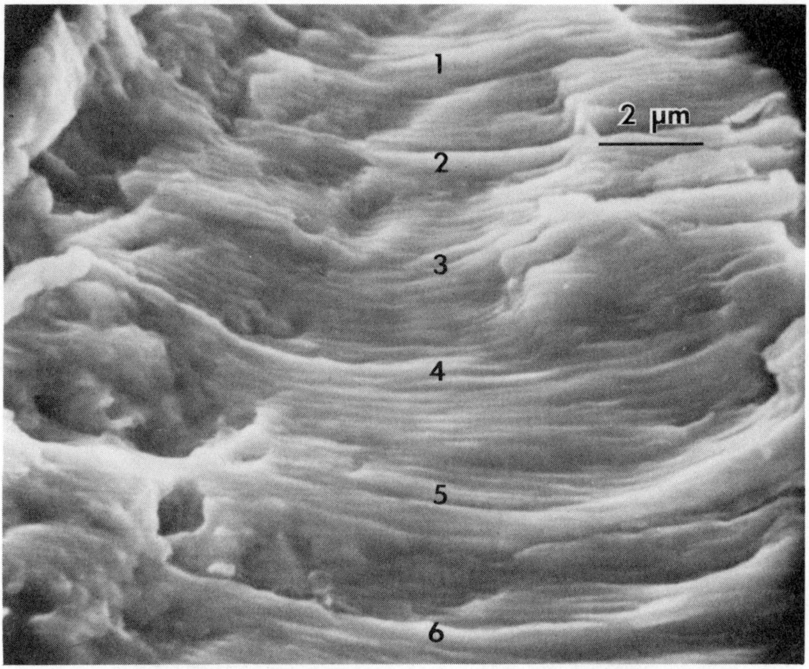

FIG. 8—*Overload and constant amplitude striations.* σ_r = *176 MPa,* ΔN = *10,* N = *422 000, and* a = *3.35 mm. Crack growth direction is down.*

aluminum and Koterazawa [*9*] on carbon steel and 5052–0 aluminum alloy.

The data points with the highest striation spacing, relative to the average growth rates, are plotted with asterisks. They come from specimens that were subjected to one overload every 10 cycles. This implies faster growth rates and, in fact, Fig. 3 shows a substantial decrease in fatigue life, for ΔN = 10 cycles. Mean striation spacings for this specimen were determined by dividing the measured distance between the first and the sixth overload striation by the number of cycles, see Fig. 8 for example, that is, 5 (10 + 1) = 55. Higher than average microscopic striation spacings were also observed when ΔN = 100 cycles.

With the exception of one point, the data for $\Delta N \geq$ 1 000 fall on or above the solid line, as shown in Fig. 13. This was surprising since widely spaced overloads delay growth rates as the crack front moves through the overload plastic zone size. Von Euw [*3*] measured a maximum slowdown of growth rate by a factor of 10. One would have expected to find more evidence of delayed growth, but the thorough scanning yielded only one point exhibiting a modest delay.

FIG. 9—*Composite micrograph. Arrows indicate suspected overload striations (X2700)* σ_T = *229 MPa*, ΔN = *100 cycles*, N = *503 000*, a = *2.26 mm. Crack growth direction is to the right.*

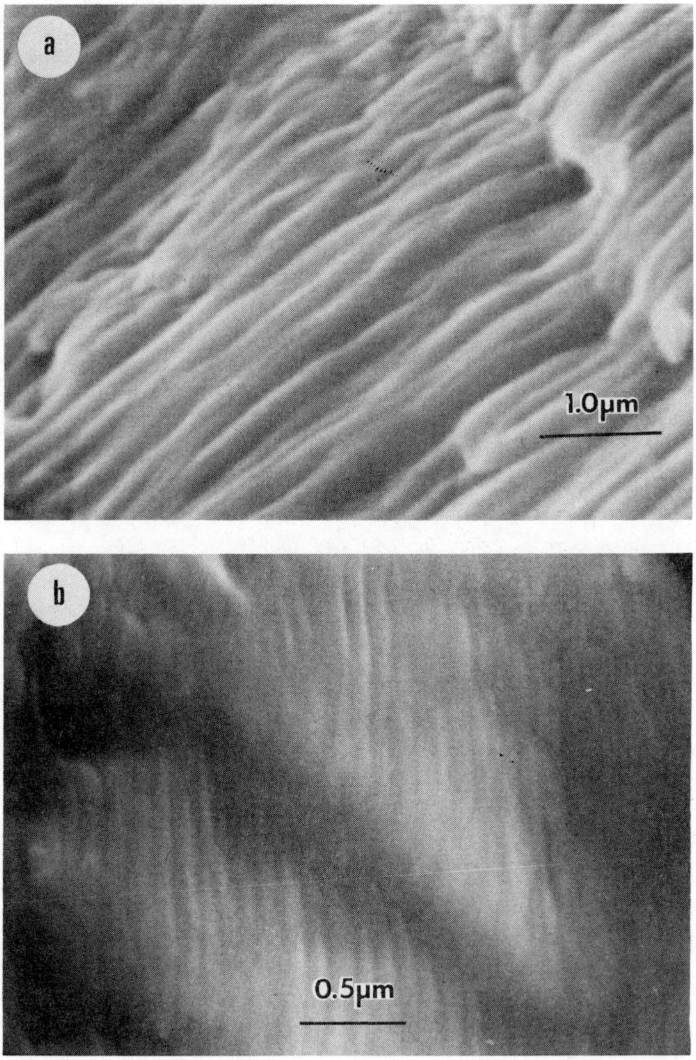

FIG. 10—*Fatigue striations.* (a) σ_r = *229 MPa,* ΔN = *10 000,* N = *718 000, and* a = *4.90. Crack growth direction is up (X14 100).* (b) σ_r = *176 MPa, no overloads,* N = *616 000, and* a = *3.20 mm. Crack growth direction is to the right (X 28 000).*

Discussion

It should be noted that the growth rate data of interest to fracture mechanists is of macroscopic nature and stem from observations of crack extension where the crack front intersects the side surfaces of the specimen. The material at these side surfaces is in a state of plane stress. Consequently, the crack tip plastic zone size at the edge is larger than that at the center of the

FIG. 11—*Fatigue crack surface in HAX (X2 800).* σ_r *= 176 MPa, no overloads,* N = 616 000, and a = 1.68 mm. Crack growth direction is to the right.

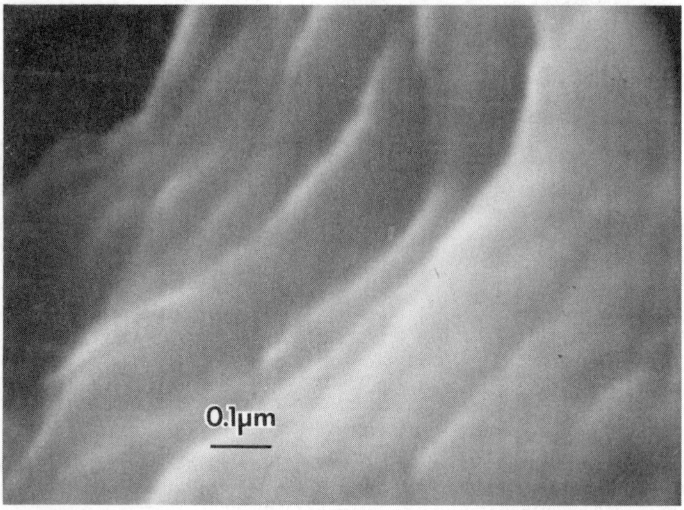

FIG. 12—*Very high magnification fractograph of area at center of Fig. 11 (X70 000). Crack growth direction is to the right.*

specimen. Because plane stress plastic zones are larger, the crack closure effect along the side surfaces is greater and this tends to restrain the growth rate along the crack front inside the specimen where the state of stress is predominantly one of plane strain. The amount of restraint depends on the de-

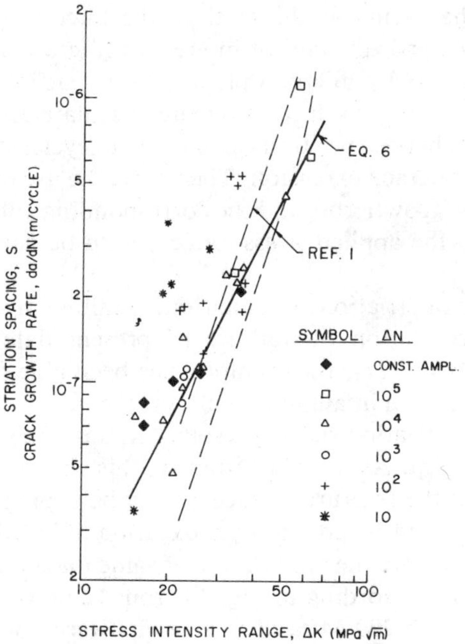

FIG. 13—*Comparison of microscopic striation spacing and crack growth rates.*

gree of plane strain provided by the specimen thickness. In fact, Hudson [*10*] found for a 7187–T6 aluminum alloy a tendency towards faster growth rates with increasing specimen thickness and degree of plane strain. One might therefore expect to observe along the deep front of a part-through crack, a growth rate larger than that measured on the side surface of a through-the-thickness crack. This explanation presumes the existence of a dominant process of kinematic crack extension controlled by the striation spacing.

One could also take an opposite viewpoint and argue that the main resistance to crack extension comes from the featureless sections which occupy most of the fatigue surface. Conceivably, ligaments of ductile ferritic material bridging between the crack surfaces could restraint the crack front. Only after their rupture would the kinematic process take over in surrounding areas. This would result in intermittent bursts of crack growth, or catch-up periods, and thus large striation spacings until newly formed ligaments again restrain the local crack front. Hudson [*10*] observed macroscopic pop-ins in fatigue tests of aluminum alloy specimens. Similar bursts at a microscopic scale would be a plausible reason why microscopic striation spacing is greater than average growth rates measured on the sides of the specimen.

To date, little concrete evidence was reported in the literature that satisfactorily links the events at the level of the microstructure to the mean macroscopic crack growth rates of interest to fracture analysts. Therefore, the explanations of why microscopic striation spacings at low ΔK values tend to be larger than growth rates remain speculative. Although a one-to-one relation exists between a striation and a load cycle, striations are not the only mechanism of crack extension. Therefore, the use of the striation spacing, by way of the growth rate and the corresponding value of ΔK, to derive the magnitude of the applied stress range should be done with appropriate caution.

The usefulness of striations in quantitative failure analysis is also greatly limited by the resolution capabilities of present date scanning electron microscopes. Pelloux [*11*], for example, has been able to resolve striations as small as 2.5×10^{-8}m in aluminum alloys, the metal that shows striations best. The smallest striation spacing resolved in this study on A588 structural steel was 3.3×10^{-8}m. As stated previously, this was accomplished through direct scanning of the fracture surface and without preparation of replicas. The portion of the computed crack propagation life of the specimen shown in Fig. 1 that falls above the growth rate of same magnitude, $da/dN = 3.3 \times 10^{-8}$m/cycle, varies according to Fig. 14 from 12 percent at $\sigma_r = 144$ MPa to 40 percent at $\sigma_r = 299$ MPa. These results were obtained by numerical integration of Eq 4 for the number of cycles needed to propagate a crack from a size corresponding to the threshold value, $\Delta K_{th} = 5.5$ MPa $\sqrt{m}$, or from Eq 1 a corresponding mean growth rate of 8.0×10^{-10}m/cycle, to failure. Given that the propagation model accounted only for about one half of the total fatigue life, with the balance presumably spent on crack initiation, striations became only visible after most of the fatigue life had expired.

Microstructural features such as grain size, segregations, and inclusions are known to affect crack front extension and striations. They cause the local crack front to advance in various planes and directions. The crack path through the ferrite matrix and the pearlite colonies was examined on SEMs of the fracture profile. The typical micrograph of the fracture profile in A588 steel, see Fig. 15, shows crack growth through two ferrite grains and a pearlite colony located between the two grains. No evidence was found to indicate that either one of the two constituents might provide a path of least resistance for fatigue crack growth, nor was the crack front observed to advance along grain boundaries.

The intersection of the crack plane and the fracture profile is not well defined in Fig. 15 because of the slight rounding of the edge during the polishing process and the bright reflections from the crack plane that rises behind the edge. Variations in crack plane levels at the microscopic scale may be attributed to the random spatial distribution of the constituents and the local joining of the crack front with advanced separations in the high-strain region ahead of the crack tip.

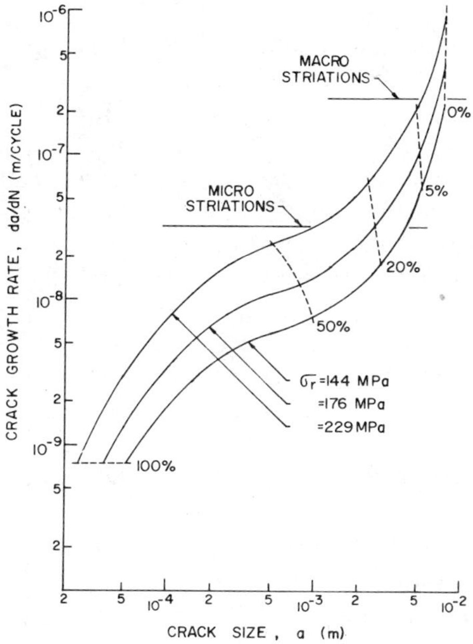

FIG. 14—*Portion of computed crack propagation life in which striations were found.*

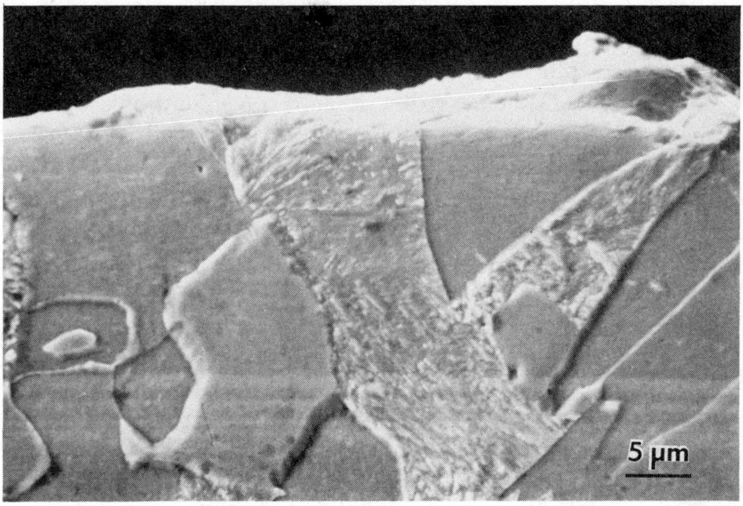

FIG. 15—*Profile of fatigue crack surface showing transgranular growth (X1 600).*

Summary

A SEM examination of fatigue cracks emanating from the weld toe of specimens fabricated from A588 steel, and subjected to constant amplitude cycling and periodic single load excursions, showed the following results.

1. Less than ten percent of the fatigue surface was striated.

2. The control specimens, those subjected to constant amplitude cycling alone, exhibited striation spacings that correlated in general with Eq 6 proposed by Bates and Clark [*8*]. For ΔK <30 MPa $\sqrt{m}$, striation spacing exceeded average crack growth rates.

3. When a periodic overload was applied after every ten cycles of constant amplitude, the microscopic overload striations could be seen. When overloads were spaced every 100 cycles, or more, microscopic overload striations could no longer be distinguished with certainty from other surface features.

4. Overloads spaced every 10 and 100 cycles increased average striation spacing.

5. Specimens with overloads spaced every 1 000 cycles, and more, exhibited striation spacings that were in good agreement with values predicted by Eq 6. Although delay effects of widely spaced overloads on macroscopic crack growth rates are known to exist, only one region of moderately reduced striation spacing following an overload was found in this study.

6. The microscopic striation spacings observed in this study, employing welded specimens fabricated from A588 structural steel, did not provide reliable estimates of the average crack growth rate. The same can be expected for other ferritic steels.

7. If it is desired to estimate from striation spacing, measured after failure, the value of ΔK and hence stress range, Eq 6 should be used in lieu of Eq 1.

8. The SEM work and supporting calculations indicate that a fatigue crack surface at a typical bridge weldment would exhibit striations, visible at a maximum magnification of X70 000, only during the last 20 percent of the propagation life.

References

[*1*] Albrecht, P., Abtahi, A., and Irwin, G. R., "Fatigue Strength of Overloaded Bridge Components," Civil Engineering Report, University of Maryland, College Park, Md., June 1976.

[*2*] Elber, W. in *Damage Tolerance in Aircraft Structures, ASTM STP 486,* American Society for Testing and Materials, 1971, pp. 230-242.

[*3*] Von Euw, E., Hertzberg, R., and Roberts, R. in *Stress Analysis and Growth of Cracks, ASTM STP 513,* American Society for Testing and Materials, pp. 230-259.

[*4*] Rice, R. C. and Stephens, R. I. in *Progress in Flaw Growth and Fracture Toughness Testing, ASTM STP 536*, American Society for Testing and Materials, 1973, pp. 95-114.

[*5*] Barsom, J. M., *Transactions*, American Society of Mechanical Engineers, Nov. 1971, pp. 1190-1196.

[*6*] Albrecht, P. and Yamada, K., *Journal of the Structural Division*, American Society of

Civil Engineers, Feb. 1977, pp. 377-389.

[*7*] Irwin, G. R., *Transactions*, American Society of Mechanical Engineers, Series E, Vol. 29, Dec. 1962.

[*8*] Bates, R. C. and Clark, W. G., Jr., *Transactions*, American Society for Metals, Vol. 62, No. 2, 1969, p. 380.

[*9*] Koterazawa, R. et al, *Transactions*, American Society of Mechanical Engineers, Oct. 1973, pp. 202-212.

[*10*] Hudson, C. M. and Newman, J. C., Jr., NASA Technical Note TN D-7173, National Aeronautics and Space Administration, April 1973.

[*11*] Pelloux, R. M., *Proceedings*, 1974 Symposium on Scanning Electron Microscopy, IIT Research Institute, Chiciago, Ill., April 1974, pp. 851-858.

C. R. Brinkman,[1] *J. P. Strizak,*[1] *and J. F. King*[1]

Elevated Temperature Fatigue Characterization of Transition Joint Weld Metal and Heat Affected Zone in Support of Breeder Steam Generator Development

REFERENCE: Brinkman, C. R., Strizak, J. P., and King, J. F., "**Elevated Temperature Fatigue Characterization of Transition Joint Weld Metal and Heat Affected Zone in Support of Breeder Stream Generator Development,**" *Fatigue Testing of Weldments, ASTM STP 648*, D. W. Hoeppner, Ed., American Society for Testing and Materials, 1978, pp. 218–234.

ABSTRACT: Piping systems in breeder reactor plants will require transition joint weldments between austenitic and ferritic materials. Specifically, annealed 2¼Cr-1Mo steel will be welded with ERNiCr-3 using the hot wire automatic gas tungsten-arc process. These weldments may see elevated temperature service for periods of up to 30 years, and, accordingly, prototypic weldments will require extensive mechanical property characterization. It was the objective of this effort to define the strain controlled low cycle fatigue behavior of as-deposited and stress-relieved ERNiCr-3 weld metal and to develop test methods for establishing the cyclic behavior of heat-affected-zone (HAZ) material. Low cycle fatigue and cyclic stress-strain response for ERNiCr-3 were established over the temperature range of 295 to 866 K. A number of low cycle fatigue test results are reported for hourglass-shaped specimens that demonstrated that the low cycle fatigue behavior of HAZ material adjacent to the fusion line could be characterized.

KEY WORDS: fatigue tests, weldments, fatigue (materials), high temperature fatigue, low alloy steels, Inconel, stress cycle, weld metals

Current plans call for construction of breeder reactor plant steam generator systems primarily out of annealed 2¼Cr-1Mo steel [*1*].[2] Many of the steam generator piping components will require transition weld joints to

[1] Group leader and mechanical engineer, Mechanical Properties Group, and metallurgist, Welding and Brazing Group, respectively, Metals and Ceramics Division, Oak Ridge National Laboratory, Oak Ridge, Tenn. 37830.

[2] The italic numbers in brackets refer to the list of references appended to this paper.

connect the ferritic material to austenitic piping leading to other components within the system. Pipe diameters will range from 50 mm to as large as 660 mm with design temperature as high as about 790 K [*2,3*]. A diagram showing the cross section of a typical transition joint between 2¼Cr-1Mo steel and Type 316 stainless steel is shown in Fig. 1. The weld filler metals shown are ERNiCr-3 (67 percent minimum Ni-18 to 22Cr-2.5 to 3.5Mn-3Fe-2.5Nb + Ta-0.75Ti-0.1C-0.5Si-0.5Cu, weight percent) and 16-8-2 (16Cr-8Ni-2Mo balance Fe, weight percent) stainless steel. One of the requirements for maintaining integrity of a transition joint subject to loading from static and dynamic thermal stresses is that there be a gradual transition in coefficient of thermal expansion across the weld joint as shown in Fig. 1. Hence, Alloy 800H is added as a spool piece.

Transition joints between 2¼Cr-1Mo ferritic steel and Type 316 stainless steel have been utilized successfully in commercial fossil-fired power plants subject to elevated temperatures for many years. However, incidences have been reported of failure resulting from primarily fatigue and creep in combination with surface oxidation occurring in the ferritic material adjacent to the fusion line, and metallurgical changes occurring within the heat-affected-zone (HAZ) of the ferritic material [*3,4*]. While it is believed that the operating temperatures of the piping transition joints within breeder systems will be low enough to preclude environmentally induced changes within the critical HAZ areas of the 2¼Cr-1Mo piping, the possibility of damaging mechanisms occurring over a design lifetime of 30 years cannot be totally discounted. Accordingly, systematic programs are underway at Oak Ridge National Laboratory (ORNL) and elsewhere to develop the mechanical properties of base, weld filler, and composite material. This program calls for the initial testing of weldment and base materials using essentially small specimens for tensile, creep, fatigue, and creep-fatigue evaluations. However, testing of full scale pipe transition joints is planned as a future effort. It is the objective of this paper to report continuous cycling strain controlled fatigue and cyclic stress-strain data and discuss results obtained from ERNiCr-3 all weld metal specimens taken from prototypic as-fabricated stress-relieved weldments. Further, interim results of continuous cycling fatigue tests are reported that were obtained using composite specimens taken from the HAZ in the 2¼Cr-1Mo steel side of the weldment, since this is the region currently considered to be most important in terms of resistance to environmental degradation.

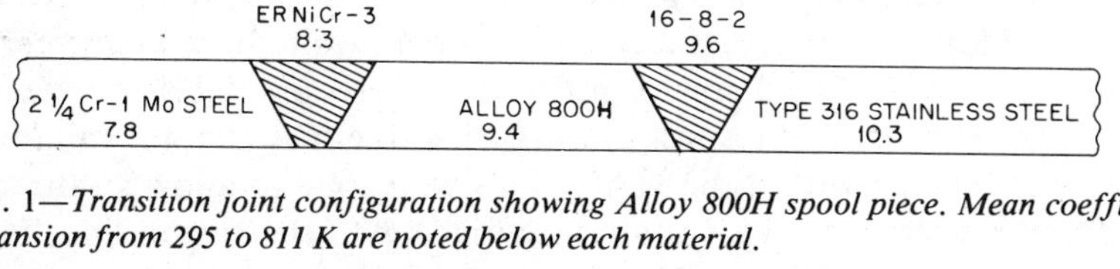

FIG. 1—*Transition joint configuration showing Alloy 800H spool piece. Mean coefficients of expansion from 295 to 811 K are noted below each material.*

Procedure

Material

Plates of 19-mm-thick annealed 2¼Cr-1Mo steel (ASME SA-387 Grade D) containing 0.11C were joined by multiple pass welding employing the automatic gas tungsten-arc process using either cold or hot-wire ERNiCr-3 filler additions. The weld seam was perpendicular to the plate rolling direction. The hot wire process is superior to cold wire welding for this application since it minimizes dilution and allows higher deposition rates. These weldments were prepared with a 30-deg-included-angle V-groove joint geometry with a 32-mm root opening and a backing strip. Typically, filling of the joint required 40 weld passes with the cold wire process and 16 to 18 with the hot wire process. Subsequent to welding, the plates were examined nondestructively by radiography and when found to be defect free, that is, linear indications were less than 0.25 mm wide and pores less than 0.25 mm in diameter, the plates were stress relieved for 1 h at 1005 K. A typical composition of the weld metal was found to be as follows.

Chemical Composition, weight %					
Cr	Fe	Mn	Nb	Ti	Ni
18	1.8	3.0	2.3	0.3	balance

Greater concentrations of iron were found in weld metal areas adjacent (<1 mm) to the fusion line indicating some dilution. Figure 2 shows the microstructure and associated hardness values of a typical multipass weldment fabricated in the aforementioned manner. As can be seen by the hardness profile, the ERNiCr-3 deposit is harder than the annealed 2¼Cr-1Mo steel. The HAZ consists of a tempered bainite structure due to the cooling rate following welding. The structure going across the weld metal through the HAZ into the annealed 2¼Cr-1Mo steel is quite complicated due to compositional differences, varying amounts of residual cold work induced during multipass welding, and the variation in response to heat treatment.

Following stress relief of the weldments, both fatigue and tension specimens were taken from essentially the central sections of the weld metal such that the uniform gage lengths consisted entirely of weld metal. The tension specimens (3.18-mm-diameter and 28.6-mm-long gage section) were fabricated with a longitudinal orientation, that is, major axis parallel to the fusion line, while the uniform gage fatigue specimens (6.35 mm diameter and 10.16 mm long) were fabricated with both a transverse and longitudinal orientation. In addition, hourglass-shaped specimens (minimum diameter was 5.08 mm with a radius-to-diameter ratio of 6) were also fabricated with a transverse orientation with respect to the fusion line such that the minimum diameter was located within the HAZ. This was done to allow strain controlled fatigue testing of HAZ material adjacent to the fusion line. Final

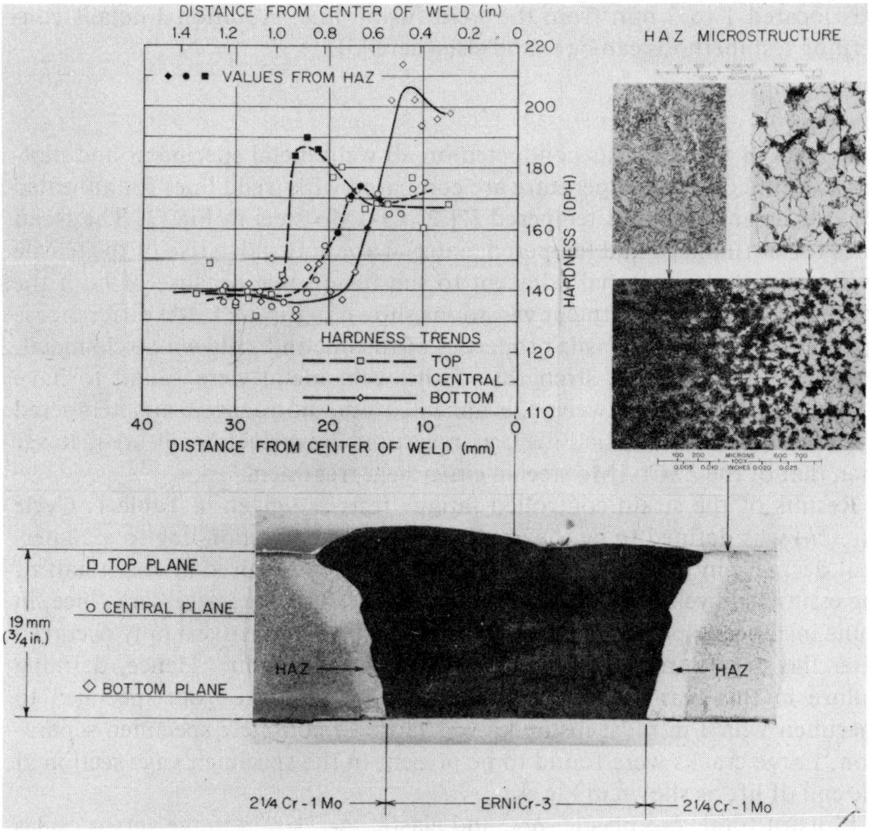

FIG. 2—*Hardness measurements on a weldment joining 2¼Cr-1Mo steel with ERNiCr-3 filler metal. The weldment was fabricated by the automatic gas tungsten-arc process with hot wire filler additions and then stress relieved at 1005 K for 1 h.*

buffing of all specimens was done longitudinally to remove all circumferential grinding marks and leave a 0.20 to 0.28 μm (8 to 11 μin.) surface finish. Additional information concerning specimen preparation can be found elsewhere [*3*]. Prior to testing, all specimens were radiographed and rejected if defects, that is, linear indications greater than 0.05 mm wide or pores greater than 0.15 mm in diameter, were found.

Testing Procedure

All of the fatigue tests were conducted at a strain rate of $4 \times 10^{-3} s^{-1}$ in axial strain control. Heating was accomplished using an induction coil with an air environment. Diametral extensometry was used for the hourglass HAZ specimens with an axial extensometer (9.53-mm gage length) used for the uniform gage specimens. The minimum diameter of the hourglass HAZ specimens, and hence the point of contact of the diametral extensometer, was located 1 to 2 mm from the weld fusion line. Additional details concerning test methods can be found elsewhere [5].

Results

Results of tension tests conducted on all weld metal specimens and plotted as a function of temperature are compared with trend lines for annealed [6] and normalized and tempered [7] 2¼Cr-1Mo steel in Fig. 3. The trend lines for normalized and tempered material are only indicative of the tensile properties of the material adjacent to the fusion line because of both the cold work and heat treatment variations shown in Fig. 2. Little difference is apparent between the tensile properties of the hot and cold wire weld metal. The yield and ultimate strengths of the weld metal were found to have values intermediate between the annealed and normalized and tempered 2¼Cr-1Mo steel with ductilities, as measured by reduction of area, lower than that of the 2¼Cr-1Mo steel in either heat treatment.

Results of the strain controlled fatigue tests are given in Table 1. Cycle life (N_f) was defined to be the number of cycles corresponding to a 20 percent decrease in the stress range; the stress range measured at about half of the estimated cycle life is taken as 100 percent. This was necessary since, in some instances, specimen failure (separation into two halves) only occurred after the stress range had decreased to a very low value. Hence, defining failure in this manner provided a uniform definition from specimen to specimen which in most instances was close to complete specimen separation. Large cracks were found to be present in the specimen gage section at the end of life as shown in Fig. 4.

Plots of total, $\Delta\varepsilon_t$, plastic, $\Delta\varepsilon_p$, and elastic, $\Delta\varepsilon_e$, strain range versus cycles to failure (N_f) were constructed from the results of all weld metal fatigue tests for each of the temperatures under construction. The Coffin-Manson-Basquin relationship

$$\Delta\varepsilon_t = \Delta\varepsilon_p + \Delta\varepsilon_e = AN_f^{-a} + BN_f^{-b} \qquad (1)$$

was generally found to be obeyed as shown in Figs. 5 and 6 for the 616 and 811 K tests, respectively. Similar plots at 291 and 866 were omitted for brevity; however, values of the constants and coefficients for Eq 1 as established by least squares analysis are summarized in Table 2 for all temperatures investigated. Best fit total strain range lines showing the influence of temperature on the low cycle strain controlled fatigue life of ERNiCr-3 are plotted in Fig. 7. The apparent crossover of the 811 and 616 K curves over that of the 295 K curve at strain ranges less than 0.75 percent is not well defined and may simply be due to scatter in the limited data.

Cyclic stress-strain curves were constructed from the hysteresis loops generated at various fractions of the cycle life, that is, ¼ cycle, cycle 10, and $N_f/2$. The data were fit by an equation of the form

$$\Delta\varepsilon_t/2 = \frac{\Delta\sigma}{2E} + \left(\frac{\Delta\sigma}{2A}\right)^{1/n} \qquad (2)$$

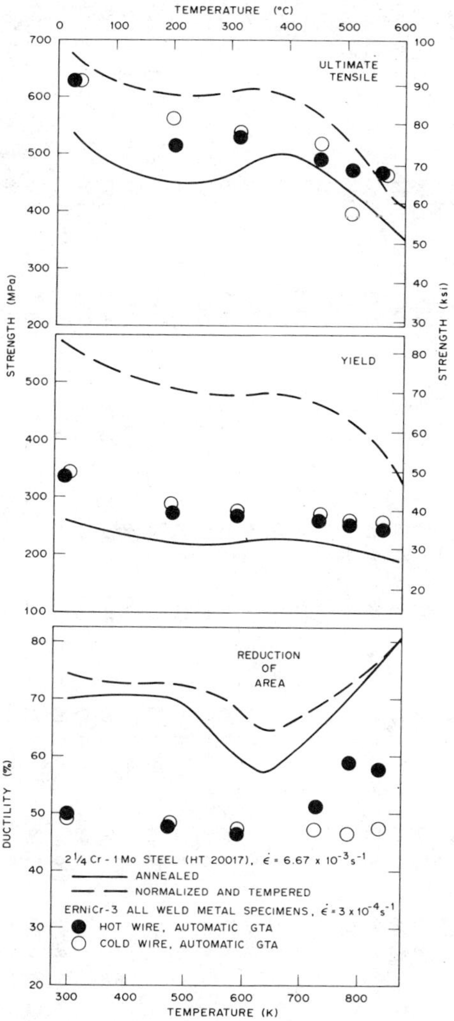

FIG. 3—*Comparison of tensile properties of base, HAZ, and ERNi-Cr-3 weld metal.*

where

$\Delta\sigma$ = stress range at indicated cycle,

E = Young's modulus taken as an average value from several first quarter hysteresis loops at a given temperature, and

n,A = cyclic strain hardening exponent and strength coefficient, respectively.

Examples comparing the 811 K cyclic stress-strain all weld metal data with the best fit curves are shown in Fig. 8. Cyclic hardening is apparent with little or no differences evident in the cyclic stress-strain behavior of hot

TABLE 1—*Results from strain-controlled fatigue tests for all weld metal and HAZ materials.*

Speci-men	Weld Process[a] and Specimen Orientation[b]	Temperature, K	Total Strain Range,[c] $\Delta\varepsilon_t$, %	Range at Tenth Cycle: Plastic Strain, $\Delta\varepsilon_t'$, %	Range at Tenth Cycle: Elastic Strain, $\Delta\varepsilon_e'$, %	Range at Tenth Cycle: Stress, $\Delta\sigma'$ MPa	Range at $N_f/2$: Plastic Strain, $\Delta\varepsilon_p$ %	Range at $N_f/2$: Elastic Strain, $\Delta\varepsilon_e$ %	Range at $N_f/2$: Stress, $\Delta\sigma$, MPa	Modulus of Elasticity,[d] E, GPa	Cycles to *Failure*,[e] N_f
8H511	H,T	295	2.99	2.29	0.70	987	2.35	0.64	1076	162	718
8H515	H,T	295	2.00	1.34	0.66	950	1.29	0.71	1090	158	1 066
8H501	H,T	295	0.96	0.53	0.43	811	0.44	0.52	901	171	3 547
8H514	H,T	295	0.70	0.27	0.43	780	0.19	0.51	933	182	12 152
8H219	H,T	295	0.57	0.27	0.30	696	0.16	0.42	811	205	13 831
8H205	H,T	616	2.99	2.33	0.66	809	2.17	0.82	1150	141	304
8H64A	H,L	616	1.95	1.40	0.55	793	1.26	0.69	1044	159	729
8H81A	H,L	616	1.46	0.95	0.51	694	0.80	0.66	936	143	877
8H211	H,T	616	0.96	0.58	0.38	603	0.38	0.58	931	166	2 944
8H63A	H,L	616	0.75	0.36	0.39	640	0.22	0.53	875	165	8 608
8H208	H,T	616	0.57	0.28	0.29	268	0.10	0.47	807	169	19 521
8W10	C,T	811	2.93	2.34	0.59	863	2.09	0.84	1212	166	154
8W23	C,T	811	2.00	1.44	0.56	843	1.30	0.70	1082	168	293
8W24	C,T	811	1.51	1.01	0.50	794	0.86	0.65	1039	172	520
8W22	C,T	811	1.49	1.00	0.49	737	0.86	0.63	961	150	389
8W02	C,T	811	1.00	0.58	0.42	637	0.39	0.61	864	150	2 411
8WO6	C,T	811	0.79	0.40	0.39	611	0.24	0.55	852	157	4 692
8W05	C,T	811	0.50	0.14	0.36	555	0.03	0.47	685	165	20 693
8W09	C,T	811	0.45	0.10	0.35	523	0.02	0.43	623	146	191 566
8W04	C,T	811	0.40	0.09	0.32	516	0.03	0.37	539	144	88 878
8W03	C,T	811	0.35	0.01	0.34	443	0.01	0.34	517	148	2 571 449
8W502	H,T	811	2.50	2.01	0.49	779	1.73	1.77	1141	162	299
8H507	H,T	811	2.00	1.48	0.52	723	1.27	0.73	1080	156	404
8H61A	H,A	811	1.68	1.68	0.45	762	1.01	0.67	1013	151	511
8H509	H,T	811	1.50	0.98	0.52	684	0.75	0.75	988	130	1 176
8H513	H,T	811	0.83	0.41	0.42	629	0.21	0.62	849	160	4 727
8H506	H,T	811	0.82	0.33	0.49	475	0.17	0.66	811	131	11 070
8H62A	H,A	811	0.77	0.37	0.40	622	0.24	0.53	783	155	5 159
8H508	H,T	811	0.57	0.17	0.40	560	0.05	0.52	722	140	15 530

8H201	H,T	811	0.45	0.14	0.31	461	0.06	0.39	599	153	65 229
8H512	H,T	811	0.41	—	—	—	0.03	0.38	557	149	150 798[f]
HC3	H,H	811	2.00	1.50	0.50	734	1.51	0.49	731	148	405
HC1	H,H	811	1.99	1.46	0.53	734	1.46	0.53	727	137	279
HC5	H,H	811	1.00	0.51	0.49	721	0.54	0.46	573	146	1 067
HC13	H,H	811	1.00	0.43	0.57	720	0.47	0.53	746	132	920
HC12	H,H	811	0.45	0.07	0.38	593	0.08	0.37	556	147	5 797
HC7	H,H	811	0.45	0.10	0.35	541	0.12	0.33	503	147	9 212
HC9	H,H	811	0.40	0.02	0.28	496	0.03	0.37	494	132	6 505
HC4	H,H	811	0.35	0.03	0.32	423	0.04	0.31	416	134	136 263[f]
8H209	H,T	866	2.90	2.33	0.57	699	2.11	0.79	984	133	125
8H206	H,T	866	1.96	1.45	0.51	713	1.25	0.71	1023	149	272
8H203	H,T	866	1.45	1.01	0.46	697	0.82	0.65	947	150	321
8H204	H,T	866	0.97	0.59	0.38	583	0.40	0.58	855	149	1 931
8H202	H,T	866	0.56	0.25	0.31	520	0.08	0.48	694	148	7 453
8H207	H,T	866	0.41[g]	0.09	0.32	457	0.01	0.34	485	141	442 600[f]

[a] C implies automatic gas tungsten-arc process with cold wire filler additions; H, hot wire.

[b] T implies all weld metal uniform gage specimen with longitudinal axis transverse to the weld direction; L, parallel; H, hourglass specimen with 2¼Cr-1Mo steel HAZ located at specimen minimum diameter.

[c] Strain rate $4 \times 10^{-3} s^{-1}$.

[d] Measured from hysteresis loop.

[e] Cycles to 20 percent decrease in stress range at $N_f/2$.

[f] Test discontinued with no cracks observed.

[g] Total strain range decreased early during the test to 0.35 percent.

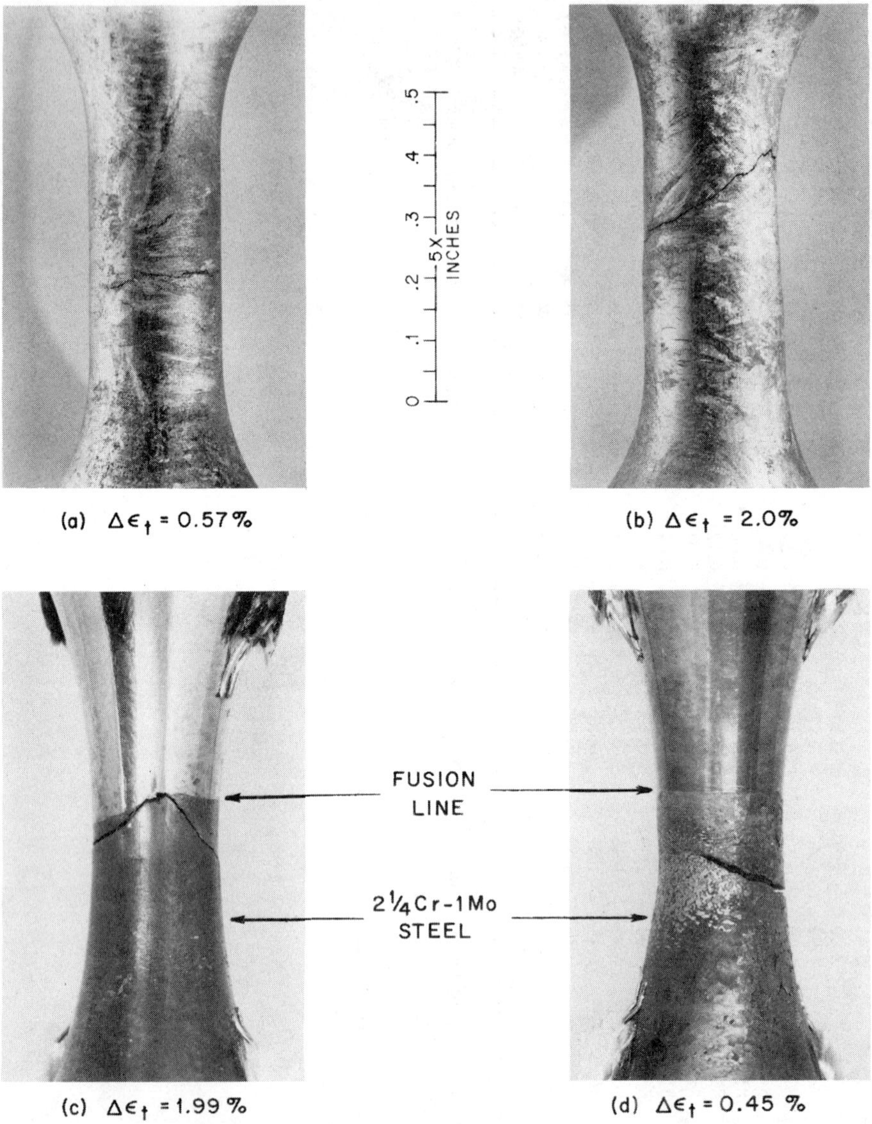

FIG. 4—*Photographs of post-fatigued specimen showing cracks, (a) Specimen 8H508 all weld metal, (b) Specimen 8H507 all weld metal, (c) Specimen HC1 composite, and (d) Specimen HC7 composite.*

or cold wire weld metal. Further, no dependence of specimen orientation (longitudinal versus transverse) is apparent. Similar plots for the other temperatures investigated were constructed (omitted for brevity), and the values of the coefficients and exponents are given in Table 3. Overall, the influence

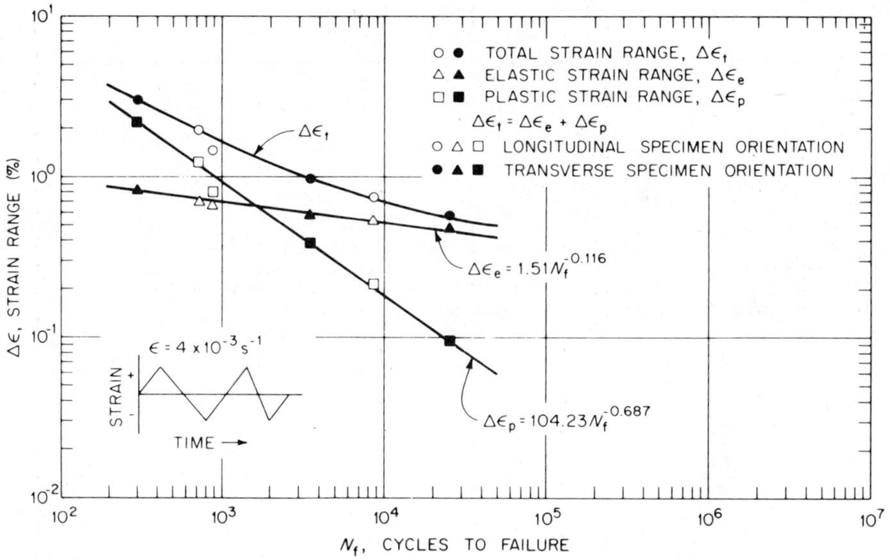

FIG. 5—*Strain-controlled fatigue behavior of ERNiCr-3 at 616 K. All weld metal specimens were prepared by gas tungsten-arc process with hot wire filler additions.*

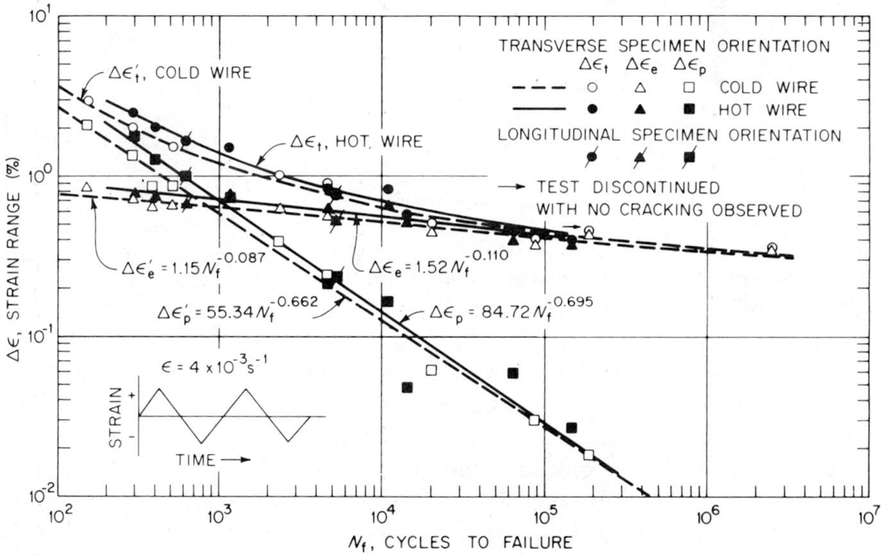

FIG. 6—*Comparison of the strain-controlled fatigue behavior of ERNiCr-3 at 811 K. All weld metal specimens were prepared by automatic gas tungsten-arc processes with hot and cold wire filler additions.*

TABLE 2—*Values for the elastic and plastic strain range constants for tests conducted at a strain rate of 4 × 10⁻³s⁻¹.*

Weld Process[a]	Temperature, K	$\Delta\varepsilon_t = \Delta\varepsilon_p + \Delta\varepsilon_e = AN_f^{-a} + BN_f^{-b}$			
		A	*a*	*B*	*b*
H	295	727.28	0.887	1.71	0.139
H	616	104.23	0.687	1.51	0.116
C	811	55.34	0.662	1.15	0.087
H	811	84.72	0.695	1.52	0.110
H	866	48.08	0.671	1.30	0.112

[a] C denotes automatic gas tungsten-arc welding process with cold-wire filler additions; H denotes automatic gas tungsten-arc welding process with hot wire additions.

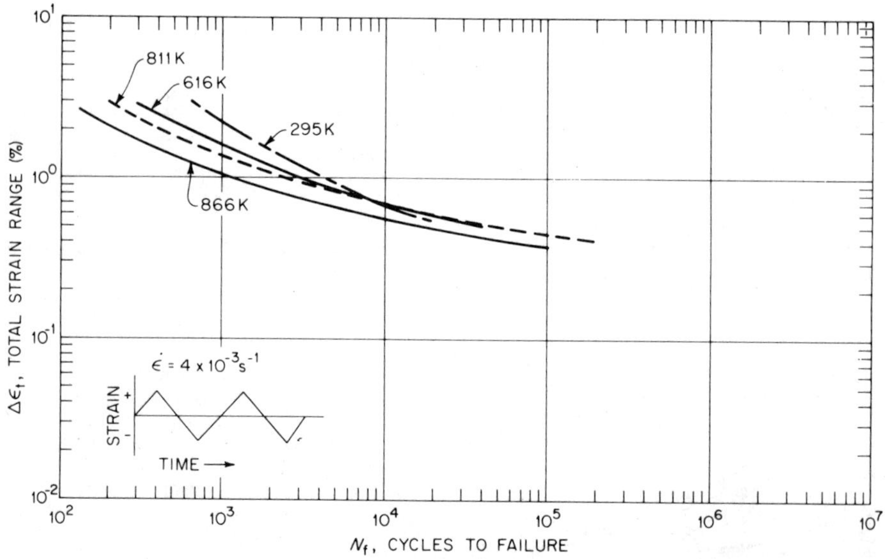

FIG. 7—*Strain controlled fatigue behavior of ERNiCr-3 as a function of temperature. All-weld-metal specimens were prepared by automatic gas tungsten-arc process with hot wire filler additions.*

of temperature on the cyclic stress-strain response, like the monotonic yield behavior shown in Fig. 3, was small over the strain amplitudes considered.

Cyclic life comparisons for results of strain controlled composite hourglass-shaped weldment specimens are made in Fig. 9, with best fit lines for annealed 2¼Cr-1Mo steel data [*1*] and the all-weld-metal data from Fig. 6. Reduced strain controlled fatigue life in comparisons to the other materials is apparent which was not surprising considering the variability in material properties and the presence of a metallurgical notch at the fusion line. The presence of the metallurgical notch was particularly important in influencing the fatigue life of the specimens tested at a strain range of 1 and 2 per-

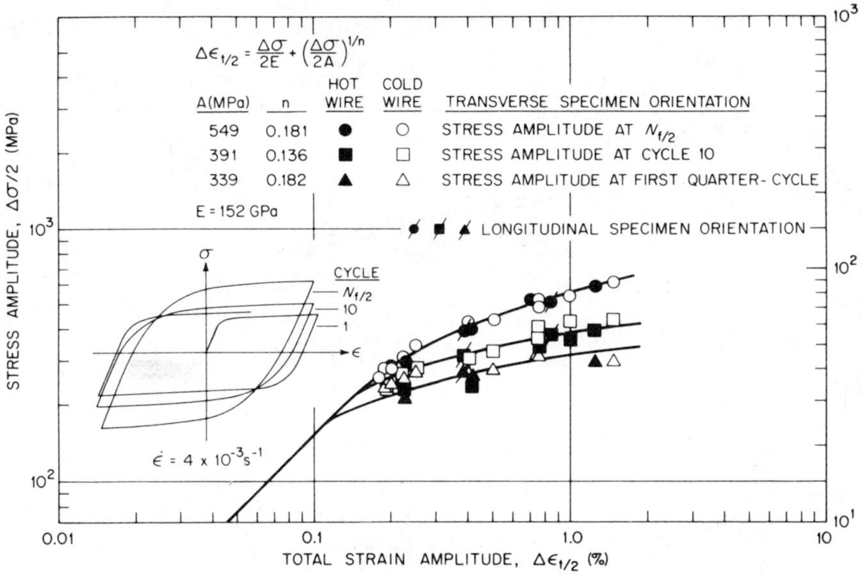

FIG. 8—*Cyclic stress-strain curves for ERNiCr-3 at 811 K.*

TABLE 3—*Cyclic hardening constants for weld and composite (HAZ) material.*

$$\Delta\varepsilon_t/2^a = \frac{\Delta\sigma}{2E} + \left(\frac{\Delta\sigma}{2A}\right)^{1/n}$$

Material	Temperature, K	Modulus of Elasticity, E, GPa	First Quarter Cycle A	First Quarter Cycle n	Tenth Cycle A	Tenth Cycle n	Cycle $N_f/2$ A	Cycle $N_f/2$ n
Hot wire	295	169	...	...	493	0.156	558	0.111
Hot and cold wire	616	160	...	...	417	0.211	557	0.113
Hot wire	811	152	339	0.182	391	0.136	549	0.181
Cold wire	866	149	244	0.066	352	0.143	520	0.126
Composite (HAZ)	811	141	393	0.223	407	0.115	412	0.154

[a] Cyclic stress-strain curves developed from strain-controlled fatigue tests at strain rate of $4 \times 10^{-3}s^{-1}$; $\Delta\varepsilon_t$ = total strain range, $\Delta\sigma$ = stress range.

cent in that in all cases crack nucleation occurred at the fusion line as shown in Fig. 4. The cracks then propagated away from the fusion line into the 2¼Cr-1Mo steel ($\Delta\varepsilon_t$ = 2 percent tests), Fig. 4, or followed the fusion line ($\Delta\varepsilon_t$ = 1 percent test). At the lower strain ranges, that is, $\Delta\varepsilon_t$ = 0.45 percent, crack nucleation always occurred in the base metal at distances of about 3 to 4 mm from the fusion line. The width of the HAZ in the composite specimens was also about 3 to 4 mm. Since the diametral extensometer was always situated such that the ceramic knife edge contact points were about 1 mm from the fusion line (minimum diameter of the specimen), it may be

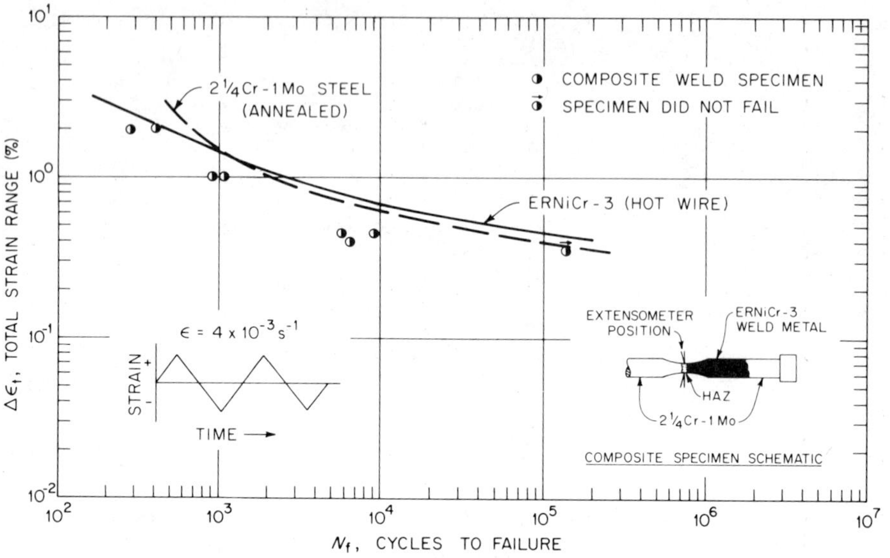

FIG. 9—*Comparison of fatigue data from composite weld specimens at 811 K with the fatigue behavior of the associated base materials, that is, 2¼Cr-1Mo steel and ERNiCr-3 weld metal.*

concluded that crack nucleation probably always occurred at a local stress or strain concentration point. For the test as $\Delta\varepsilon_t = 0.45$ percent, a localized region at the edge of the HAZ (3 to 5 mm from the fusion line) may have had increased ductility due to some thermal aging of the essentially annealed microstructure, that is, proeutectoid ferrite and pearlite. Indeed, there is some evidence of a softer region in this area as seen by the hardness profiles given in Fig. 2. Increased localized plastic strain in this area could have assisted nucleation of the resultant cracks.

A photomicrograph showing a cross section of the composite specimen after fatigue testing and local cracks in relation to the microstructure is given in Fig. 10.

Comparisons between the cyclic stress range induced at a constant strain range of 0.45 percent as a function of fraction of cycle life are given in Fig. 11. All of the data shown were generated at 811 K. Figure 11 shows that ERNiCR-3 weld metal cyclically hardened to higher stress levels than either the composite (HAZ) or annealed 2¼Cr-1Mo steel. At this strain range, the stress range stabilized after an initial period of cyclic hardening in the all weld metal specimens and showed evidence of cyclic softening in the composite (HAZ) and 2¼Cr-1Mo steel. At a higher strain range, for example, 2 percent, the differences between cyclic stress-strain behavior were more pronounced with the all weld metal specimens showing continuous hardening throughout most of the cyclic life to stress levels much higher than that

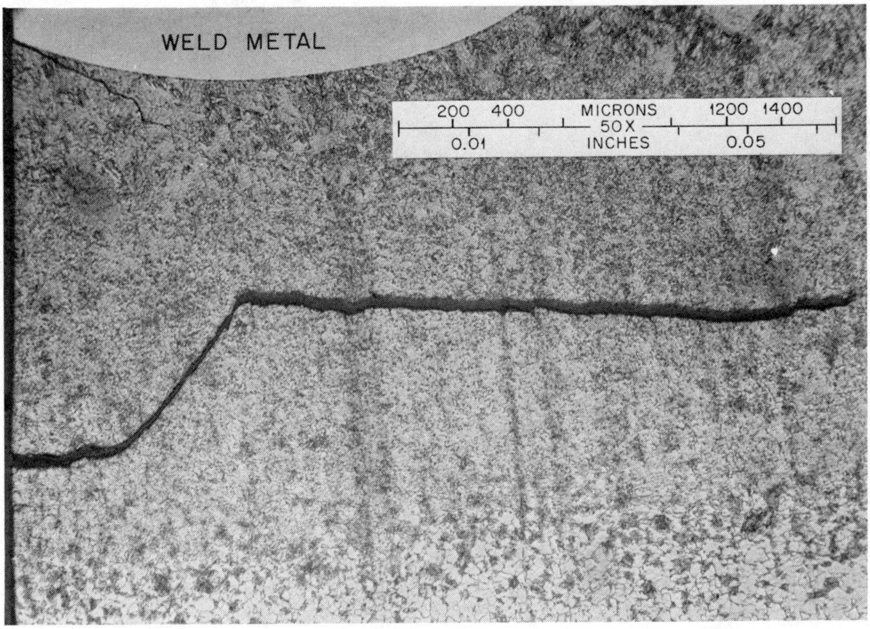

FIG. 10—*Photomicrograph showing fatigue cracking in HAZ.*

of either HAZ or annealed 2¼Cr-1Mo steel. A comparison at this strain range is shown in Fig. 12.

Summary Remarks

It has been observed that when fatigue induced failure of piping occurs, it usually is associated with cracking in the weld or HAZ material.

The causes of failure usually have been attributed to poor weldment design, presence of defects, high residual stress levels, or metallurgical changes occurring within the HAZ. Successful design of weldments to be exposed to elevated temperature for prolonged periods of time is complicated because of the number of factors, including the dissimilarity of mechanical properties of base, HAZ, and weld metal. Residual stresses induced by thermal-mechanical processing and metallurgical notches can cause high local discontinuity stresses. Further, weld-metal, like casting, is known to display anisotropy in mechanical properties.

The information reported herein represents a partial fulfillment of the mechanical property needs as specified by the design community for the successful design of the Clinch River Breeder Reactor Plant transition joints. Large differences in tensile properties were demonstrated between ERNiCr-3 weld metal and the HAZ and annealed 2¼Cr-1Mo steel. The low-cycle fatigue properties of as-deposited hot-and-cold-wire ERNiCr-3 weld

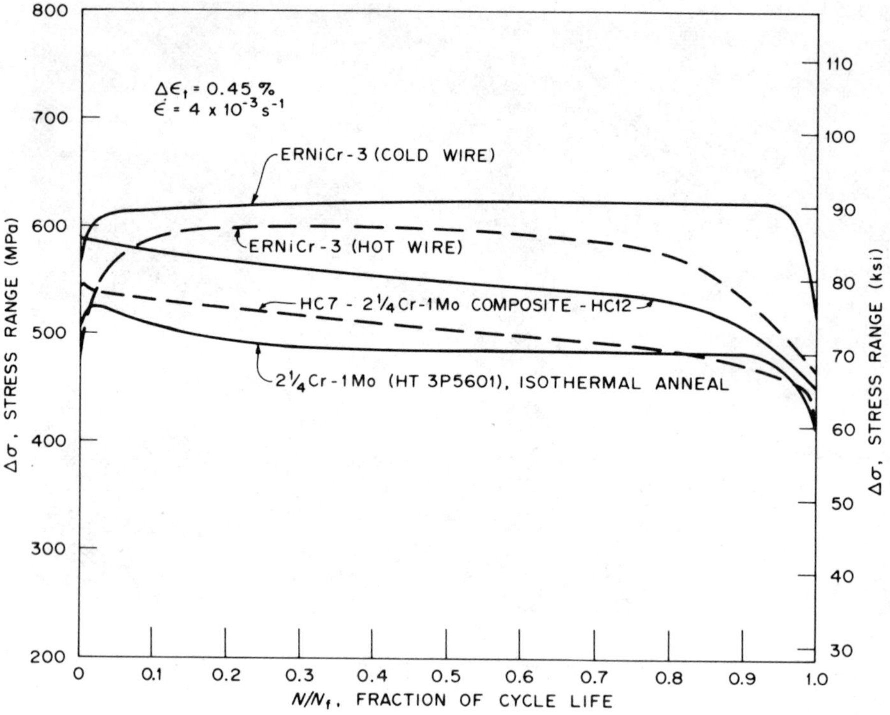

FIG. 11—*Stress range as a function of fraction of cyclic life for several transition joint materials at 811 K.*

metal were established. Although the continuous-cycle fatigue properties of as-deposited hot-wire weld metal were slightly superior to those of cold-wire-deposited weld metal, the cyclic stress-strain response differed little at 811 K. Further, neither the fatigue life nor the cyclic stress-strain response of the ERNiCr-3 weld metal depended upon the orientation of the specimen with respect to the fusion line (that is, transverse versus longitudinal).

The results of tests conducted on the hourglass-shaped composite specimens, while limited, demonstrated that specimens could be fabricated and tested to provide information concerning the strain-controlled fatigue behavior of HAZ material. An exact knowledge of the axial cyclic strain within the HAZ is not possible because of the slight stress concentration associated with the hourglass specimen configuration, across the HAZ, the strength gradient as well as the stress discontinuity associated with material differences across the fusion line. Nonetheless, reasonable reproducibility in cyclic life and stress-strain response was found indicating that specimen fabrication methods and extensometer placement with respect to the fusion line could be duplicated. Accordingly, composite specimens of this design and others will be subject to prolonged periods of thermal aging and subse-

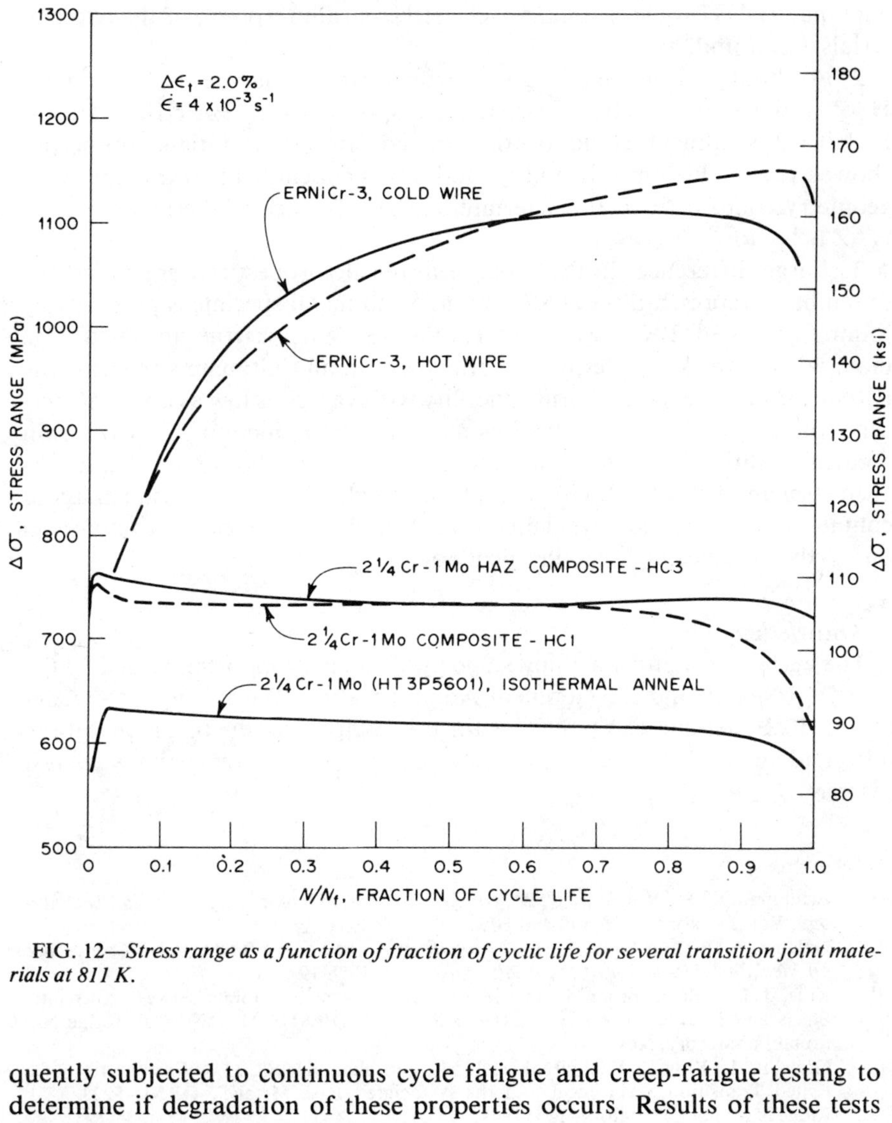

FIG. 12—*Stress range as a function of fraction of cyclic life for several transition joint materials at 811 K.*

quently subjected to continuous cycle fatigue and creep-fatigue testing to determine if degradation of these properties occurs. Results of these tests will be reported in subsequent papers.

Conclusions

Several conclusions were drawn from this work.

1. The cyclic stress-strain and continuous cycle strain-controlled fatigue properties of as-deposited ERNiCr-3 were defined over the temperature range of 295 to 866 K. A comparison made between ERNiCr-3 deposited by the automatic hot and cold wire gas tungsten-arc processes indicated that the fatigue life of hot wire material was slightly superior to that of the cold

wire material. However, similar cyclic stress-strain response of the two materials was found.

2. An hourglass-shaped fatigue specimen was employed to determine if HAZ fatigue and cyclic stress-strain response of a 2¼Cr-1Mo steel—ERNiCr-3 weldment could be determined. Resultant fatigue properties showed reasonable reproducibility, and it was concluded that this specimen geometry could be used for determining the influence of thermal aging on HAZ behavior.

3. Large differences in the short term tensile properties were found between base, representative HAZ, and all weld metal specimens prepared by joining annealed 2¼Cr-1Mo with ERNiCr-3. Whereas the annealed base material was weaker in terms of both the yield and ultimate strengths, the ERNiCr-3 had the lowest tensile ductility with values as low as about 50 percent for as-deposited weld metal as measured by reduction of area. Strain controlled fatigue behavior was similar at 811 K for both annealed 2¼Cr-1Mo steel and ERNiCr-3. However, fatigue behavior of composite material containing both base and weld metal was found to be inferior in comparison due to the presence of a metallurgical notch.

Acknowledgments

The authors gratefully acknowledge the financial support provided by the Energy Research and Development Administration. We also wish to recognize L. K. Egner and C. O. Stevens for their help in conducting many of the tests and to V. T. Houchin for his assistance in preparing the weld test plates.

References

[*1*] Brinkman, C. R., Williams, R. K., Klueh, R. L., and Hebble, T. L., *Nuclear Technology*, Vol. 28, No. 3, 1976, pp. 490-505.

[*2*] Dalcher, A. W. and Yang, T. M., *Proceedings of the Second International Conference on Mechanical Behavior of Materials*, Aug. 1976, 1978-1983.

[*3*] King, J. F., "Behavior and Properties of Welded Transition Joints Between Austenitic Steels and Ferritic Steels—A Literature Review," ORNL–TM–5193, Oak Ridge National Laboratory, Nov. 1975.

[*4*] Gray, R. J., King, J. F., Leitnaker, J. M., and Slaughter, G. M., "Examination of a Failed Transition Weld Joint and the Associated Base Metals," ORNL–5223, Oak Ridge National Laboratory, Jan. 1977.

[*5*] Brinkman, C. R., et al, "Interim Report on the Continuous Cycling Elevated-Temperature Fatigue and Subcritical Crack Growth Behavior of 2¼Cr-1Mo Steel," ORNL–TM–4993, Oak Ridge National Laboratory, Dec. 1975.

[*6*] Booker, M. K., et al, "Mechanical Property Correlations for 2¼Cr-1Mo Steel in Support of Nuclear Reactor Systems Design," ORNL/TM–5329, Oak Ridge National Laboratory, June 1976.

[*7*] Klueh, R. L. and Oakes, R. L., Jr., "Strain Rate Effects on the Tensile Properties of a Bainitic (Normalized-and-Tempered) 2¼Cr-1Mo Steel," ORNL–5314, Oak Ridge National Laboratory, April 1976.

L. Albertin[1] *and E. E. Eiffler*[1]

Effect of Residual Stress from Welding on the Fatigue Strength of Notched 347 Austenitic Stainless Steel

REFERENCE: Albertin, L. and Eiffler, E. E., "**Effect of Residual Stress from Welding on the Fatigue Strength of Notched 347 Austenitic Stainless Steel,**" *Fatigue Testing of Weldments, ASTM STP 648*, D. W. Hoeppner, Ed., American Society for Testing and Materials, 1978, pp. 235–243.

ABSTRACT: The fatigue strength of Type 347 stainless steel forging prepared according to ASTM Specification for Forged or Rolled Alloy-Steel Pipe Flanges, Forged Fittings, and Valves and Parts for High Temperature Service (A 182) was investigated in the notch-free and notched condition including a residual stress from an arc strike simulating a local weld. The notched specimens were 22.2 and 50.8 mm (0.875 and 2 in.) in diameter with a stress concentration factor, K_t, of 5.3. The notch-free specimens were standard R. R. Moore types.

In the notch-free condition, the fatigue strength at 10^6 cycles was found to be 234.4 MPa (34 ksi). For the 22.2 and 50.8-mm (0.875 and 2-in.) diameter notched specimens, the fatigue strength was 124 MPa (18 ksi). In the case of 50.8-mm (2-in.) diameter notch specimens subjected to the heat of an arc strike, the endurance strength was 83 MPa (12 ksi). Thus, the endurance strength of heated specimens was reduced to two thirds that of unheated notched specimens. While the notch sensitivity index was 0.20 for the unheated specimens, it was 0.43 for the notched specimens containing a residual stress from the heat of an arc strike.

KEY WORDS: fatigue strength at ∼ cycles, austenitic stainless steels, welding, residual stress, notch sensitivity, fatigue tests, weldments.

Structures fabricated by welding often contain discontinuities or stress concentrations in the form of grooves or fillets. Nearby welds accentuate the stress riser effect of such notches by superimposing a residual stress on the working stress. Under such conditions, components can fracture at nominal stresses far below the endurance limit of the material. This study was conducted to determine the effect of such local residual stresses from

[1] Senior engineer, Westinghouse Research and Development Center, and fellow engineer, Electro-Mechanical Division, Westinghouse Electric Corporation, Pittsburgh, Pa. 15235.

welding on the fatigue strength of notched and stabilized Type 347 stainless steel prepared according to ASTM Specification for Forged or Rolled Alloy-Steel Pipe Flanges, Forged Fittings, and Valves and Parts for High Temperature Service (A 182).

A review of published literature on the fatigue strength of the material indicated that the values varied widely.[2-4] Data on the effects of residual stresses from local welding operations in the vicinity of notches on fatigue strength could not be found.

To evaluate the magnitude of a local residual stress in a notch from a welding operation, we have chosen a cylindrical billet containing a groove and a shrunk-on sleeve. Thermal gradients from welding and residual stresses as a result of sleeve fabrication were measured. The tests indicated sufficiently high stresses in the notch which could have a detrimental effect on structural reliability. To verify the effect of these stresses on fatigue life, both notch-free and notched rotating-bending fatigue tests were conducted on various size specimens using standard and specially designed equipment. The results of these tests are presented in this paper. The data show that residual stresses from relatively low temperature effects of a nearby weld are significant and can affect design life.

Experimental Procedure

Welding and Residual Stress Measurements

To simulate the residual stress from a nearby local weld in a typical notched (K_t = 5.3) Type 347 steel cylindrical forging, a "pin-weld" mock-up shown in Fig. 1 was manufactured. The mock-up incorporated a shrunk-on sleeve which was locked by two pins 180-deg apart. The pins were secured by plug-type welds using a manual gas tungsten arc process. Tests were conducted to determine the magnitude of the thermal cycle during welding. Thermocouples were attached to the groove and on the shrunk-on sleeve as illustrated in Fig. 1. Two heat input levels were used to cover a range of manufacturing procedures. The thermocouple outputs were recorded dynamically during welding.

Residual stress measurements were made by removing the sleeve and attaching rosette strain gages in the groove at various distances away from the pin weld. The sectioning method was used to measure the residual stress. Strains relieved were measured after cutting sections into small pieces. The residual stress measurement results and the magnitude of the thermal cycle during welding were used in the subsequent laboratory specimen study to determine the effect of a residual welding stress as a result of such a local weld on the fatigue strength of notched Type 347 stainless steel forgings.

[2] Osgood, C. C. in *Fatigue Design,* Wiley, New York, 1970, p. 422.
[3] Lessells, J. M. in *Strength and Resistance of Metals,* Wiley, New York, 1954, p. 173.
[4] *Aerospace Structural Metals Handbook,* AFML-TR-68-115, Code 1309, Air Force Materials Laboratory, March 1963, p. 3.

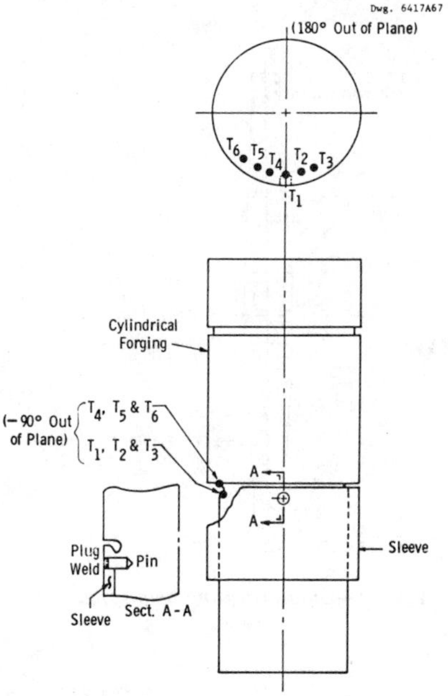

FIG. 1—*Pin-weld mock-up and location of thermocouples.*

Fatigue Tests

The unnotched fatigue strength of Type 347 steel in rotating bending was determined with conventional R. R. Moore-type specimens and fatigue testing machines. The notched rotating beam specimens had a K_t of 5.3 (a stress concentration factor used in the pin-weld mock-up) and were of the configurations shown in Figs. 2 and 3. Residual stresses from the heat of welding were introduced in the larger of the two specimens by striking an arc for about 1 min, and 12 mm (0.5 in.) away from the notch, until a ΔT of 250°C (500°F) was reached in the notch (a thermal gradient close to the maximum reached in the groove of the pin-weld mock-up).

A special rotating cantilever-beam test machine was designed and built to test the large diameter specimens. The various components of the machine are shown in Fig. 4. The machine consists of two bearing pillow blocks mounted on a heavy base plate, a motor, a bearing, a strain-gage load cell, a compression spring, and two specially machined shafts to accommodate the specimens. The larger of the shafts contains an axially drilled clearance hole to accommodate a 19.1-mm (0.75-in.) diameter draw rod. The specimen is pulled tight and held against the tapered seat with the draw rod. The tapered fit ensures that the specimen will run concentrically without play in the grip. The other end of the specimen is held in a second "floating" shaft which also has a taper and draw rod. A bearing block is mounted on the free end of the second shaft where a force can be applied with a compression spring.

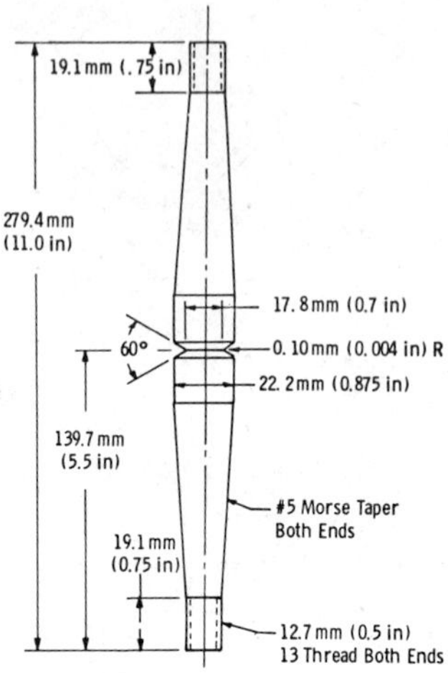

FIG. 2—*Small rotating beam specimen.*

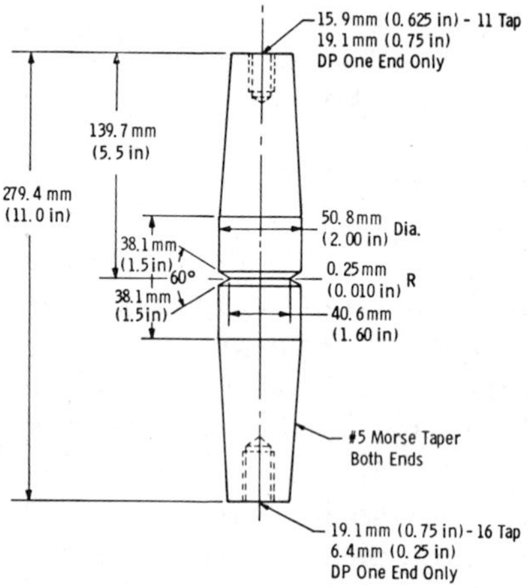

FIG. 3—*Large rotating beam specimen.*

The force is measured with a strain gaged load cell. When the specimen cracks during the test, the deflection of the free end of the second shaft increases enough to trip a limit switch which shuts off the machine. The test apparatus with a specimen in place is shown in Fig. 5. Testing is done under constant load and rotating bending.

Results

Typical plug weld thermal cycle results from the pin-weld mock-up using 135-A welding current and 12-V with a weld time of 112 s are shown in Fig.

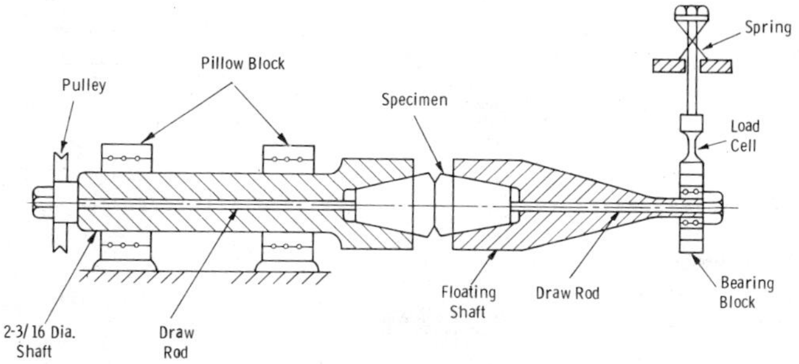

FIG. 4—*Rotating cantilever-beam test machine.*

FIG. 5—*Specially built fatigue test machine with specimen in place.*

6. The temperature histories in the groove indicate that the peak temperature increase of 292°C (557°F) occurs near the axial line through the pin. The residual stresses measured in the notch at various locations in the pin-weld mock-up are shown in Fig. 7. The maximum residual stress of 475 MPa (65 ksi) occurs at the axial line through the pin corresponding to the maximum registered temperature at this location.

The data for all rotating beam fatigue tests are given in Table 1. The data are plotted in Fig. 8 in the form of conventional *S–N* curves. For the unnotched standard R. R. Moore specimens, the fatigue strength at 10^6 cycles

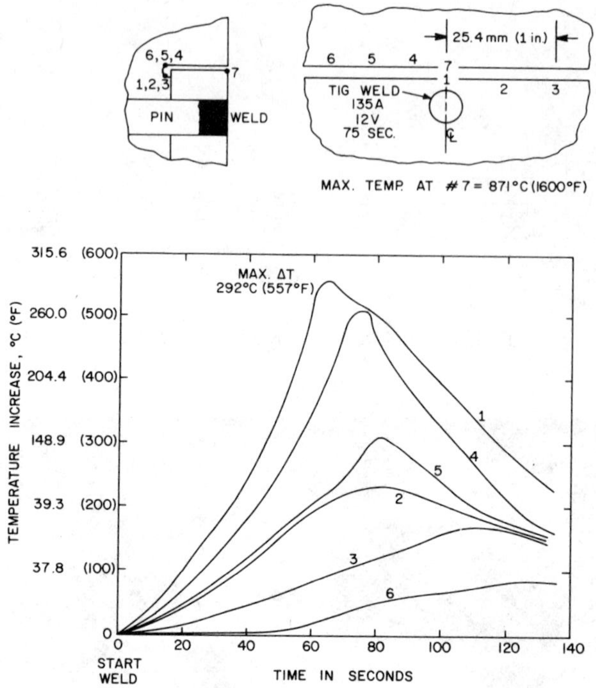

FIG. 6—*Thermal history of groove in pin-weld test.*

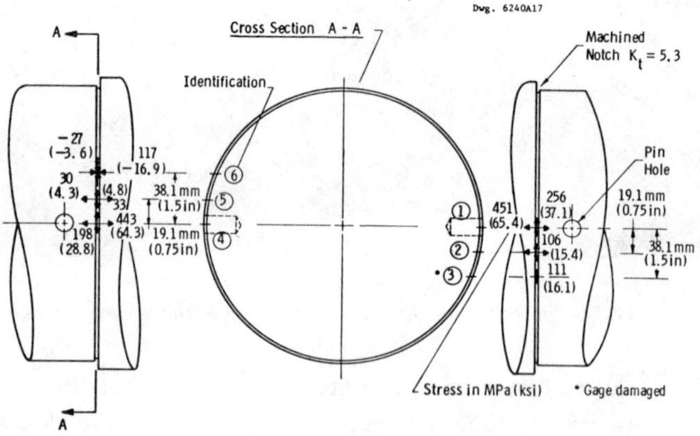

FIG. 7—*Residual stress measurement results in pin-weld mock-up.*

was 234 MPa (34 ksi) and for the 22.2-mm (0.875-in.) diameter and 50.8-mm (2.0-in.) diameter notched specimens (K_t = 5.3), it was 124 MPa (18 ksi). The thermally treated 50.8-mm (2.0-in.) diameter notched specimens produced an estimated fatigue strength of 83 MPa (12 ksi).

TABLE 1—*Fatigue data for various sizes of notched and unnotched specimens of Type 347 stainless steel at ambient temperature (rotating beam).*

Specimen	Specimen Type	Dia, mm (in.)	σ_a/σ_m[a] (= A)	K_t[b]	Notch Dia, mm (in.)	Alternating Stress, MPa (ksi)	Cycles (= N)	Frequency, Hz
E5	smooth	7.6 (0.3)	∞	1	. . .	179 (26.0)	10^{8}[c]	167
E2	smooth	7.6 (0.3)	∞	1	. . .	189 (27.5)	1.9×10^{7}[c]	167
E8	smooth	7.6 (0.3)	∞	1	. . .	193 (28.0)	9.1×10^{7}[c]	127
E4	smooth	7.6 (0.3)	∞	1	. . .	207 (30.0)	2.8×10^{5}[d]	167
E7	smooth	7.6 (0.3)	∞	1	. . .	320 (32.0)	10^{8}[c]	127
E12	smooth	7.6 (0.3	∞	1	. . .	231 (33.5)	10^{8}[c]	67
E10	smooth	7.6 (0.3)	∞	1	. . .	231 (33.5)	1.1×10^{8}[c]	67
E11	smooth	7.6 (0.3)	∞	1	. . .	241 (35.0)	3.8×10^{5}	67
E9	smooth	7.6 (0.3)	∞	1	. . .	241 (35.0)	4.2×10^{5}	67
2B1	notch	22.2 (0.875)	∞	5.3	17.8 (0.7)	138 (20.0)	1.1×10^{6}	17
1B1	notch	22.2 (0.875)	∞	5.3	17.8 (0.7)	127 (18.5)	2.3×10^{7}[c]	17
9B2	notch	22.2 (0.875)	∞	5.3	17.8 (0.7)	121 (17.5)	2.4×10^{6}	17
13B2	notch	22.2 (0.875)	∞	5.3	17.8 (0.7)	117 (17.0)	9.8×10^{6}	17
5B1	notch	22.2 (0.875)	∞	5.3	17.8 (0.7)	114 (16.5)	5.9×10^{6}	17
15L1	notch	22.2 (0.875)	∞	5.3	17.8 (0.7)	110 (16.0)	3.3×10^{7}	17
6B1	notch	22.2 (0.875)	∞	5.3	17.8 (0.7)	107 (15.6)	3.4×10^{7}	17
17L1	notch	22.2 (0.875)	∞	5.3	17.8 (0.7)	102 (14.8)	10^{8}[c]	22
1B1	notch	50.8 (2.0)	∞	5.3	0.6 (1.6)	121 (17.5)	1.4×10^{5}[e]	9
10B2	notch	50.8 (2.0)	∞	5.3	40.6 (1.6)	103 (15.0)	4.0×10^{6}	17
13L1	notch	50.8 (2.0)	∞	5.3	40.6 (1.6)	97 (14.0)	5.6×10^{6}	17
16L1	notch	50.8 (2.0)	∞	5.3	40.6 (1.6)	93 (13.5)	8.0×10^{7}[c]	17
22L2	notch	50.8 (2.0)	∞	5.3	40.6 (1.6)	90 (13.0)	2.4×10^{7}[c]	17
10B2	notch	50.8 (2.0)	∞	5.3	40.6 (1.6)	83 (12.1)	2.3×10^{6}[c]	17
7B2	notch[g]	50.8 (2.0)	∞	5.3	40.6 (1.6)	93 (13.5)	1.1×10^{6}	17
19L2	notch[g]	50.8 (2.0)	∞	5.3	40.6 (1.6)	83 (12.0)	1.1×10^{6}	17
4B1	notch[g]	50.8 (2.0)	∞	5.3	40.6 (1.6)	69 (10.0)	2.1×10^{6}	17

[a] Alternating stress divided by mean stress.
[b] Stress concentration factor.
[c] Run-out.
[d] Terminated due to bending.
[e] Terminated deep crack.
[f] Stress changed.
[g] Thermally treated notch.

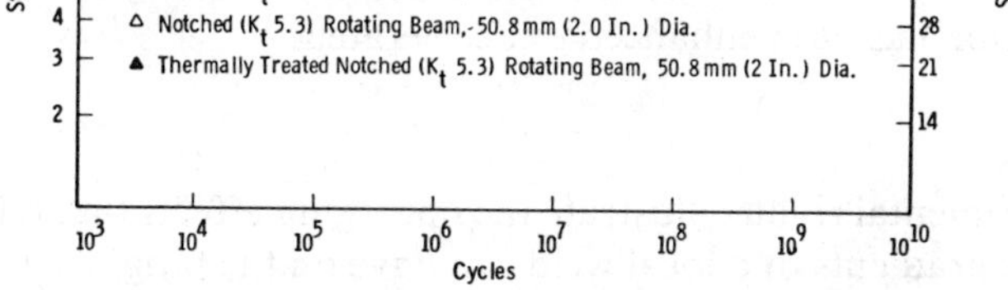

FIG. 8—*Rotating beam* S–N *curves for Type 347 stainless steel,* A = ∞.

Discussion

The relieved strains measured in the groove of a cylindrical forging containing a nearby pin-weld suggest that welding operations carried out in the vicinity of stress concentrations can leave behind high residual stresses. Although the measured residual stress at the notch was 441 MPa (64 ksi), this stress will relax during load cycling. From relaxation and strain cycling data, it has been estimated that the maximum stress that can be retained is about 303 MPa (44 ksi).[5]

The fatigue strength for the unnotched condition was determined to be 234 MPa (34 ksi) for the rotating beam tests under load cycling. Under strain cycling, this value could be much higher. The notched bars, although under load cycling, actually are strain limited at the notch.

In quenched and tempered steels, the reduction in fatigue strength is approximately in accordance with the stress concentration factor. But in austenitic stainless steel, the full reduction is not generally obtained, and the amount of reduction depends on the geometry, chiefly the notch radius. For the geometries and conditions tested, the fatigue strength reduction factor was 1.89 for the notched case and 2.83 for the notched case containing a residual stress. In terms of the notch sensitivity index, q, the aforementioned values correspond to 0.20 and 0.43, respectively.

The fatigue life results of the weld-affected 50.8-mm (2-in.) diameter notched specimens indicate the damaging effects of residual stresses (from large prestrains) in the long life region for Type 347 stainless steel. The endurance strength for welded specimens was two thirds that of the unwelded notched specimens at 10^6 cycles and could be lower at longer cyclic lives since there is no pronounced "knee" in the fatigue curve for austenitic stainless steel. Observation of the fracture surfaces of thermally treated rotating beam specimens showed that the fatigue failure origin always was in the notch surface below the weld spot. The residual stresses from welding are biaxial tension resulting from plastic compression. In the vicinity of the notch the plastic compression is magnified by the stress concentration. Therefore, a higher axial residual tension is generated at the notch on cooling. When the notch is externally load cycled, a high local stress is present, causing overstrains at the notch. These high initial overstrains result in a sufficient residual stress to reduce the fatigue strength. An explanation for this reduction, particularly in view of the effect at long lives, is that the crack initiation has been enhanced or accelerated.

Conclusions

The experimental results illustrate the damaging effects that relatively low temperature gradients of a local weld can have on the fatigue life of notched ASTM Specification A 182 Type 347 stainless steel. The lack of literature on this subject warrants further experimentation.

[5] Manjoine, M. J., "Stress Relaxation Characteristics of Type 304 Stainless Steel," International Conference on Creep and Fatigue in Elevated Temperature Applications, Institution of Mechanical Engineers, London, Sept. 1973.

Acknowledgments

We wish to thank our colleagues, A. J. Bush for the residual stress measurements, M. J. Manjoine for his advise and assistance in this project, and L. J. Ceschini for building the special fatigue machine and carrying out the fatigue tests.

B. M. Kapadia[1]

Influence of Residual Stresses on Fatigue Crack Propagation in Electroslag Welds*

REFERENCE: Kapadia, B. M., "**Influence of Residual Stresses on Fatigue Crack Propagation in Electroslag Welds,**" *Fatigue Testing of Weldments, ASTM STP 648*, D. W. Hoeppner, Ed., American Society for Testing and Materials, 1978, pp. 244–260.

ABSTRACT: An investigation of fatigue crack propagation behavior in air of as-deposited electroslag welds in two hot-rolled structural steels, prepared according to ASTM Specification for Structural Steel (A 36) and ASTM Specification for High-Strength Low-Alloy Structural Steel with 50 000 psi Minimum Yield Point to 4 in. Thick (Grade A) (A 588), has shown that the crack-growth rates (da/dN) in the welds were similar to or up to five times slower than the rate in the base steels. The retardation in crack-growth rate was observed to be substantially greater for crack propagation in the coarse-grained heat-affected-zone (HAZ) and bond-line regions than in the weld metal, but this was not consistently observed. Moreover, marked variations in the microstructure of the weld metal and HAZ did not significantly influence crack propagation. These crack-growth data for the weldments, described in an earlier paper, have been analyzed in the present paper in terms of residual stresses causing the observed retardation in crack-growth rate.

In the present work, quantitative comparisons of the crack-growth behavior of the welds were based on a cyclic-life parameter (N_1), which was defined as the number of cycles required for precracking the 1T wedge-opening-loading (WOL) specimens to a total crack length of 2.54 cm (1 in.). This parameter gave results that were consistent with those based on incremental crack-growth (da/dN) data in the earlier paper.

The observed retardation in crack-growth rate in the welds is attributed to compressive residual stresses introduced by welding. This effect was analyzed in the present paper in terms of a stress-intensity-range suppression concept, whereby the applied stress-intensity-factor range is decreased to some lower effective value. The results showed that the beneficial effect of compressive residual stresses on crack propagation appears to be of a variable nature and would diminish following a stress-relief heat treatment.

KEY WORDS: fatigue tests, weldments, fatigue (materials), crack propagation, crack growth, electroslag weldments, structural steels, residual stress, stress intensity factor, retardation

* Original experimental data were measured in U.S. customary units.

[1] Associate research consultant, United States Steel Corporation, Research Laboratory, Monroeville, Pa. 15146.

An understanding of fatigue-crack propagation in engineering structures is essential for the prediction of service lives of structures subjected to fatigue loading. This is particularly true for welded structures wherein subcritical flaws may be introduced during fabrication.

Studies [*1-5*][2] of fatigue crack propagation in weldments in steels and in aluminum alloys produced by various welding processes have generally shown that the crack-growth rates in the heat-affected zone (HAZ), as well as in the weld metal, are similar to or slower than the rates in the base metal. All but one of these studies in which crack-growth retardation was observed in weldments considered this effect to be independent of microstructure. The one exception [*4*] concerned a Type 304 stainless-steel weld with Type 308 filler metal in which crack-growth retardation was attributed to the fine duplex delta ferrite-austenite structure of the weld deposit. The remaining studies [*1–3,5*] attributed the observed crack-growth retardation to compressive residual stresses introduced by welding.

A recent investigation by U.S. Steel Research of fatigue-crack propagation in as-deposited electroslag weldments in two hot-rolled structural steels, prepared according to ASTM Specification for Structural Steel (A 36) and ASTM Specification for High-Strength Low-Alloy Structural Steel with 50 000 psi Minimum Yield Point to 4 in. Thick (Grade A) (A 588), also confirmed this general behavior. This study was conducted under National Cooperative Highway Research Program (NCHRP) Project 10-10, "Acceptance Criteria for Electroslag Weldments in Bridges."[3] The crack-growth data obtained on the welds were analyzed by using linear-elastic fracture mechanics, and the results, in the form of crack-growth rate (da/dN) versus the stress-intensity-factor range (ΔK) relationship, have been included in the final report on Phase I of the NCHRP project [*6*] and published in an earlier paper [*7*]. On the basis of these results, the following conclusions were reached.

1. Fatigue-crack-growth rates (da/dN) in the welds were similar to or up to five times slower than the rate in the base steels.
2. The retardation in crack-growth rate was most pronounced at low ΔK values, generally below about 27.5 MPa$\sqrt{m}$ (25 ksi$\sqrt{in.}$).
3. The retardation effect was substantially greater for crack propagation in the coarse-grained-HAZ and bond-line regions than in the weld metal, but this was not consistently observed.
4. The variations in the microstructure of the weld metal and the HAZ caused marked differences in the topography of the fatigue-crack surfaces,

[2] The italic numbers in brackets refer to the list of references appended to this paper.

[3] This work was sponsored by the American Association of State Highway and Transportation Officials, in cooperation with the Federal Highway Administration. The opinions and conclusions expressed or implied in this paper are those of the author, and not necessarily those of the Transportation Research Board, the National Academy of Sciences, the Federal Highway Administration, the American Association of State Highway and Transportation Officials, nor of the individual states participating in the National Cooperative Highway Research Program.

but had no significant influence on the crack-propagation rate.

Inasmuch as crack-growth retardation effects in weldments have been generally attributed in previous studies to the presence of compressive residual stresses introduced by welding, the crack-growth data obtained in the aforementioned study [7] have been analyzed further in the present paper in terms of residual stresses causing the observed retardation in crack-growth rate.

Materials and Experimental Procedure

Materials

The fatigue-crack-growth tests were conducted on 10 production-type, as-deposited electroslag butt welds in two hot-rolled structural steels, prepared according to ASTM Specifications A 36 and A 588, Grade A. Six of the weldments (1, 2, 6, 13, 17, and 19) were 2.54 to 2.54-cm (1 to 1-in.) butt joints, and the remaining four (3, 8, 18, and 20) were 10.16 to 10.16-cm (4 to 4-in.) butt joints. Further processing details of the weldments including the welding variables investigated are described elsewhere [7].

Specimen Preparation

The fatigue-crack-propagation tests were conducted on 2.54-cm (1-in.) thick 1T wedge-opening-loading (WOL) specimens. For this specimen, the stress-intensity factor, K_I, at the crack tip is given by the equation [8]

$$K_I = \frac{C_3 P}{B\sqrt{a}} \tag{1}$$

where

P = applied load,
B = specimen thickness,
a = crack length measured from the loading plane, and
C_3 = function of crack length as given in Ref 7.

The tests were conducted (*a*) with fatigue cracks oriented perpendicular to the plate surface and running parallel to the weld in both the 2.54 and 10.16-cm (1 and 4-in.) thick weldments (TL orientation), and (*b*) with fatigue cracks oriented perpendicular to the plate surface and parallel to the weld and running in the through-thickness direction in only the 2.54-cm (4-in.) thick weldments (TS orientation), Fig. 1. Fatigue cracks having TL orientation were located (*a*) at the midwidth of the weld ("weld metal"), (*b*) at the weld—base-metal interface ("bond line"), and (*c*) in the coarse- grained region of the HAZ. In the 10.16-cm (4-in.) thick weldments, these cracks were centered on the midthickness. Cracks having TS orientation (only in the 10.16-cm (4-in.) thick weldments) were located (*a*) at the midwidth of the weld, (*b*) at the bond line, and (*c*) adjacent to the weld bond line such that the crack would be expected to run through the weld, across the bond line, and then into the HAZ, or to follow a reverse course depending on the configuration of the weld cross section, Fig. 1. One specimen was tested for each of the aforementioned conditions, resulting in a total of 42 specimens. Further details of specimen preparation are discussed elsewhere [7].

Test Procedure

Testing was done in air at room temperature at cyclic-stress frequencies of 180 and 300 cpm. In all tests the fatigue crack was initiated and propagated under tension-to-sinusoidal loading at a constant maximum load of 16 kN (3600 lb) and a constant minimum load of 1.33 kN (300 lb), both of which were controlled within ± 1.0 percent. Thus, the stress ratio, R, was slightly less than 0.10. Prior to making crack-length measurements, the fatigue cracks were extended about 5.8 mm (0.23 in.) from the machined notch root, which corresponded to the total crack length, a, being approximately equal to the specimen thickness of 2.54 cm (1 in.). Details of crack-extension measurements are discussed elsewhere [7].

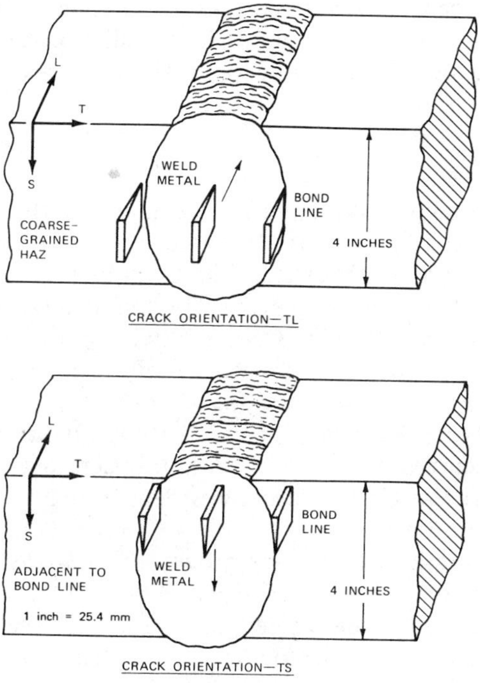

FIG. 1—*Orientations and locations of the fatigue crack in a 4-in.-thick weldment.*

Results and Discussion

Crack-Growth-Rate Results

The fatigue-crack-growth rate, da/dN, is generally related to the stress-intensity-factor range, ΔK, by the following equation [*9,10*]

$$\frac{da}{dN} = A\,(\Delta K)^n \tag{2}$$

where A and n are constants. According to this equation, a plot of da/dN versus ΔK on logarithmic coordinates is a linear relationship having a slope equal to the exponent n. The crack-growth data obtained on the weldments in the form of da/dN versus ΔK relationships were reported in an earlier paper [*7*]. The scatter band representing the crack-growth behavior of the base steels under very similar test conditions [*11,12*], which had a slope of about 3.3 (equal to the exponent, n, in Eq 2), is also included therein for reference. As noted earlier, the crack-growth rates in the weldments were similar to or up to five times slower than the rate in the base steels, and this retardation was most pronounced at low ΔK values. Thus, the da/dN versus ΔK curves for some weldments showed gradual deviation from the linear scatter band for the base steels with decreasing ΔK values. However, this effect was quite distinct from the nonlinearity ("tail") occasionally observed in such plots at very low ΔK values due to insufficient crack length, overload retardation, or other effects.

Cyclic-Life Parameter

A typical set of crack-growth data for a weldment plotted as crack length (a) versus the total number of elapsed load cycles is shown in Fig. 2. For several of the specimens (for example, the bond-line specimen in Fig. 2), valid crack-length measurements beyond the initial total crack length of 2.54 cm (1 in.) could be made only over a relatively short distance (typically about 5.1 mm or 0.2 in.) because of deviation of the crack path, and only a limited amount of crack-growth-rate (da/dN) data was obtained. Moreover, in some of these specimens, the crack-growth data obtained was invalidated by the occurrence of an accidental transient overload[4] (probably during an interruption of the test) usually evident from a "beach mark" indication on the fracture surface.

In view of this limitation of the crack-growth-rate (da/dN) data obtained, and also because the crack-growth retardation effect observed appeared to be most pronounced at low ΔK values, the number of cycles required for precracking the fatigue specimens to a total crack length of 2.54 cm (1 in.) was used in the present study for quantitative comparisons of the crack-growth behavior of the weldments. This will be referred to as the cyclic-life parameter and denoted as N_1.

[4] It has been reported [*13*] that, at low ΔK values, even a relatively small overload (order of 10 percent) can cause very marked crack-growth delays.

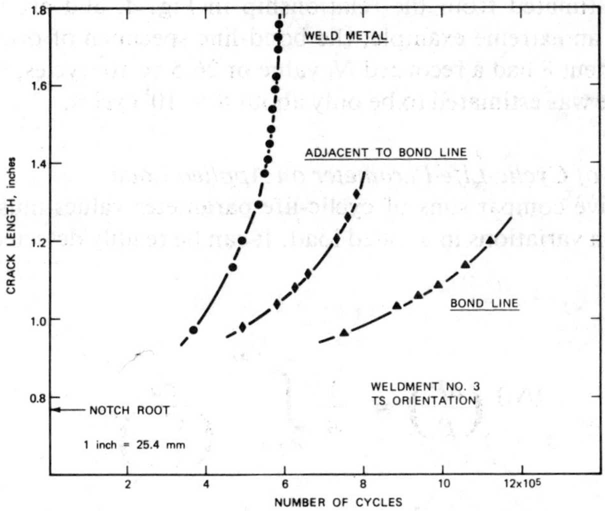

FIG. 2—*Fatigue-crack-growth behavior of Weldment 3 (TS orientation) for various crack locations.*

The use of this parameter as a measure of crack-propagation behavior was based on the assumption that the number of cycles required for crack initiation in the specimens was negligibly small. Crack-initiation data from an earlier study [*14*] suggested that for the steels investigated and the test conditions employed, including notch-root radius and stress amplitude, the crack-initiation life of the specimens would be less than 20 000 cycles, which would indeed be negligible in comparison with their propagation life. This was experimentally confirmed by crack-length measurements taken close to the tip of the notch for one of the specimens, Fig. 3.

The number of cycles required by each specimen for crack extension to total crack lengths of 2.54, 2.79, 3.05, 3.56, 4.06, and 4.57 cm (1.0, 1.1, 1.2, 1.4, 1.6, and 1.8 in.), where applicable, are shown in Table 1, along with the cyclic-load amplitude for a specimen thickness of 2.54 cm (1 in.). All values, including N_1 in a few cases, that were apparently influenced by crack-growth delay following an accidental overload, are indicated accordingly.

All valid N_1 values were found to be directly proportional to the number of cycles required for crack extension from total length of 2.54 to 3.05 cm (1 to 1.20 in.), Fig. 4, and were inversely proportional to the crack-growth rate, da/dN, at total crack length of 1 in., Fig. 5, as expected. Thus, both these relationships establish the validity of the parameter N_1 as representing crack-growth behavior of the specimens tested for crack lengths of both less than and greater than 2.54 cm (1 in.). In the few instances where N_1 values either were not determined or were not valid (because of overload effects), they were estimated from the relationship in Fig. 4, and are included in Table 1. As an extreme example, the bond-line specimen of orientation TS from Weldment 8 had a recorded N_1 value of 26.5×10^5 cycles, whereas the true N_1 value was estimated to be only about 8×10^5 cycles.

Dependence of Cyclic-Life Parameter on Applied Load

Quantitative comparisons of cyclic-life-parameter values must take into consideration variations in applied load. It can be readily detected from Eqs 1 and 2 that

$$(N_1)\left(\frac{\Delta P}{B}\right)^n = \frac{1}{A}\int_{a=0.77}^{a=1.00} \frac{da}{\left(\frac{C_3}{\sqrt{a}}\right)^n} \tag{3}$$

where

N_1 = cyclic-life parameter,
ΔP = applied-load amplitude,
B = specimen thickness,
A and n = constants for a given material,
a = crack length, and
C_3 = function of crack lengths given in Ref 7.

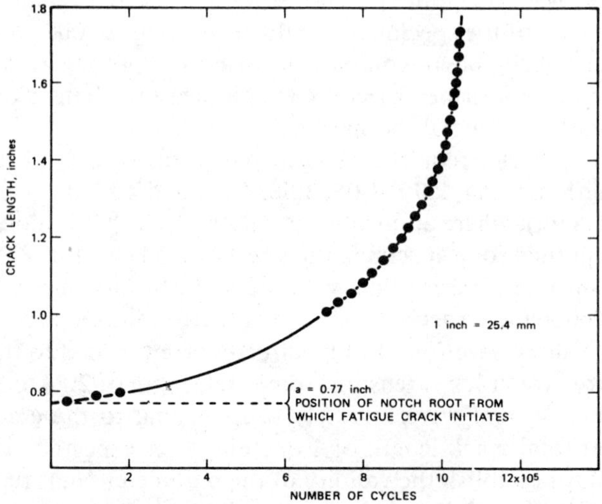

FIG. 3—*Fatigue-crack-growth curve of weld-metal specimen of Weldment 17 showing a negligible crack-initiation life under test conditions employed.*

The value of the intergral in Eq 3 is constant for a given specimen geometry and fixed crack-length limits and, therefore, Eq 3 simplifies to

$$(N_1)\left(\frac{\Delta P}{B}\right)^n = \text{constant} \quad (4)$$

According to Eq 4, a log-log plot of N_1 versus $(\Delta P/B)$ should have a negative slope equal to the exponent, n, in Eq 2.

Log-log plots of N_1 versus $(\Delta P/B)$ are presented in Figs. 6 and 7 for the 2.54 and 10.16-cm (1 and 4-in.) thick weldments, respectively. Corresponding data on the base steels, ASTM Specifications A 36 and A 588, Grade A [*11,12*] and on an additional steel from a previous study [*11*], ASTM Specification A 588, Grade B, are also included for reference. The slope of the scatter band representing these data obtained over a range of $(\Delta P/B)$ values was about 3.3, which is exactly equal to the slope of the scatter band of da/dN versus ΔK relationship for the base steels [*7*] (and to the exponent n in Eq 2).

With a few exceptions, all the N_1 values for the weldments investigated were higher than the scatter band for the base steels, Figs. 6 and 7, in agreement with the earlier crack-growth data, which indicated that almost all the weldments showed crack-growth retardation to at least some degree in relation to the base steels. Also, in agreement with the crack-growth data, the maximum crack-growth-retardation effect in terms of the N_1 parameter amounted to a factor of about 5, Fig. 6. Note that these results also show that the retardation effect was generally greater in the bond-line and coarse-grained-HAZ specimens than in the weld-metal specimens, but again, without any consistent pattern. For some reason, the retardation effect was less marked in specimens of TL orientation from the 10.16-cm (4-in.) thick weldments than in those of TS orientation from the same weldments, or in specimens of TL orientation from the 2.54 cm (1-in.) thick weldments.

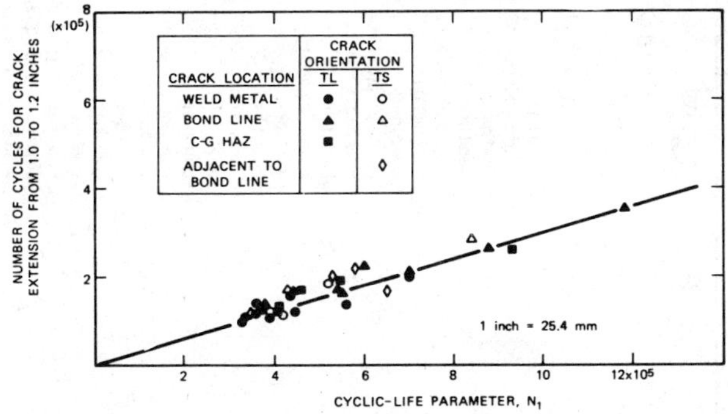

FIG. 4—*Correlation between the cyclic-life parameter,* N_1*, and the number of cycles required for fatigue-crack extension from a total length of 1.0 to 1.2 in.*

TABLE 1—*Summary of fatigue-crack-growth behavior of weldments investigated.*

Weldment	Crack Orientation	Crack Location	Cyclic-Load Amplitude[a] ($\Delta P/B$), lb/in.	Cyclic Life at Indicated Crack Lengths (in inches),[b] 10^5 cycles: (N_i)1.0[c]	1.1	1.2	1.4	1.6	1.8	Cyclic Life From Crack Length of 1.0 to 1.2 in., 10^5 cycles	Estimated N_i,[c] 10^5 cycles	Crack-Growth Rate at Crack Length of 1 in. (da/dN), 10^{-6} in./cycle
1	TL	weld metal	3466	5.60	6.40	6.95	7.45	7.60	7.70	1.35	..[f]	0.90
		bond line	3470	18.20[d]	19.80[d]	20.70[d]	. . .[e]	. . .[e]	. . .[e]	2.50	8.40	0.30
		coarse- grained HAZ	3485	18.50[d]	21.05[d]	22.40[d]	. . .[e]	. . .[e]	. . .[e]	3.90	13.00	0.25
2	TL	weld metal	3459	3.90	4.50	4.95	8.40[d]	8.70[d]	8.75[d]	1.05	..[f]	1.10
		bond line	3463	6.00	7.35	8.20	9.05	9.35	9.40	2.20	..[f]	0.50
		coarse-grained HAZ	3463	4.10	4.85	5.40	5.95	6.20	6.30	1.30	..[f]	1.00
6	TL	weld metal	3455	4.45	5.15	5.60	6.05	6.25	6.30	1.15	..[f]	1.00
		bond line	3463	8.75	10.30	11.35	12.50	13.10	13.25	2.60	..[f]	0.50
		coarse-grained HAZ	3455	5.45	6.60	7.35	8.30	8.80	9.00	1.90	..[f]	0.65
13	TL	weld metal	3638	3.35	4.10	4.40	4.80	4.95	5.00	1.05	..[f]	1.50
		bond line	3638	3.80	4.70	5.20	5.70	. . .[e]	. . .[e]	1.40	..[f]	1.00
		coarse-grained HAZ	3638	3.50	4.25	4.70	5.20	5.40	. . .[e]	1.20	..[f]	1.20
17	TL	weld metal	3051	7.00	8.20	8.95	9.95	10.35	10.45	1.95	..[f]	0.70
		bond line	3364	11.80	14.00	15.30	. . .[e]	. . .[e]	. . .[e]	3.50	..[f]	0.30
		coarse-grained HAZ	3367	0	1.50	2.40	3.55	. . .[e]	. . .[e]	2.40	8.00	0.50
19	TL	weld metal	3463	3.60	4.30	4.75	5.35	5.60	5.65	1.15	..[f]	1.20
		bond line	3466	5.50	6.45	7.10	7.70	8.00	8.10	1.60	..[f]	0.70
		coarse-grained HAZ	3481	0	0.70	1.20	1.80	2.00	2.10	1.20	3.90	1.10
3	TL	weld metal	3307	3.30	3.85	4.25	5.10[d]	5.60[d]	5.65[d]	0.95	...[f]	1.50
		bond line	3303	4.30	6.50[d]	8.00[d]	. . .[e]	. . .[e]	. . .[e]	3.70	..[f]	1.30
		coarse-grained HAZ	3300	9.30	11.00	11.90	12.85	13.20	13.30	2.60	..[f]	0.35
	TS	weld metal	3303	3.90	4.50	4.95	5.55	5.80	5.90	1.05	..[f]	0.80
		bond line	3310	8.40	10.10	11.20	. . .[e]	. . .[e]	. . .[e]	2.80	..[f]	0.40
		adjacent to bond line	3307	5.30	6.50	7.30	8.00	. . .[e]	. . .[e]	2.00	..[f]	0.60
8	TL	weld metal[g]										
		bond line	3310	7.00	8.30	9.10	10.00	10.20	10.40	2.10	..[f]	0.45
		coarse-grained HAZ	3316	6.00	7.75[d]	10.00[d]	11.20[d]	11.90[d]	. . .[e]	4.00	..[f]	0.45
	TS	weld metal	3320	7.80	8.90[d]	9.40[d]	10.10[d]	10.30[d]	10.45[d]	1.60	..[f]	0.60
		bond line	3321	26.50[d]	28.00[d]	28.85[d]	29.85[d]	30.20[d]	. . .[e]	2.35	8.00	0.45
		adjacent to bond line	3310	3.50	4.10	4.65	5.30	. . .[e]	. . .[e]	1.15	..[f]	1.20
18	TL	weld metal	3300	3.60	4.40	5.00	5.70	6.00	6.15	1.40	..[f]	0.90
		bond line	3287	4.40	5.50	6.10	6.80	7.00	7.10	1.70	..[f]	0.60
		coarse-grained HAZ	3290	4.60	5.60	6.30	7.00	. . .[e]	. . .[e]	1.70	..[f]	0.70
	TS	weld metal	3290	5.20	6.30	7.00	7.90	8.25	8.35	1.80	..[f]	0.70
		bond line	3290	0	1.70	2.80	. . .[e]	. . .[e]	. . .[e]	2.80	9.00	0.30
		adjacent to bond line	3290	5.80	7.10	7.95	. . .[e]	. . .[e]	. . .[e]	2.15	..[f]	0.55

20	TL	weld metal	3307	4.35	5.30	5.90	6.50	6.70	6.75	1.55	..[f]	0.90
		bond line	3307	5.40	6.40	7.10	7.90	8.20	8.30	1.70	..[f]	0.60
		coarse-grained HAZ	3300	4.10	4.85	5.40	6.00	6.30	6.40	1.30	..[f]	1.00
	TS	weld metal	3307	4.20	4.85	5.30	5.90	6.20	. . .[e]	1.10	..[f]	1.10
		bond line	3300	4.30	5.30	6.00	6.80	. . .[e]	. . .[e]	1.70	..[f]	0.80
		adjacent to bond line	3300	6.50	7.45	8.15	8.95	9.30	9.40	1.65	..[f]	0.80

[a] For a specimen thickness of 1 in.

[b] Life measured starting from an initial crack length of 0.77 in. (machined notch tip), except for three of the specimens in which life was measured starting from an initial crack length of 1 in.

[c] N_i is the cyclic-life parameter defined in the text. The estimated value is derived from cyclic-life values in preceding column by using the relationship in Fig. 4.

[d] Values not valid because of crack-growth delay following accidental transient overloads.

[e] Test terminated.

[f] Not estimated.

[g] Specimen was accidentally overloaded and no crack-growth data were obtained.

Conversion Factors—

1 in. = 2.54 cm, and
1 lb/in. = 1.75N/cm.

Effect of Residual Stresses on Crack-Growth Rate

As noted earlier, crack-growth-retardation effects in weldments have been generally attributed to the presence of compressive residual stresses introduced by welding [*1–3,5*].

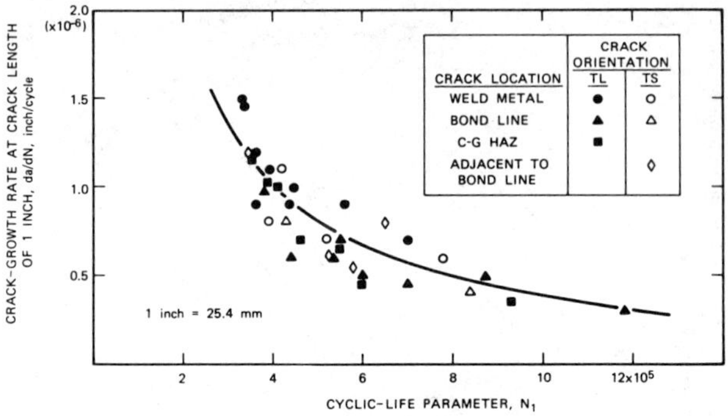

FIG. 5—Correlation between the cyclic-life parameter, N_1, *and fatigue-crack-growth rate,* da/dN, *at a total crack length of 1.0 in.*

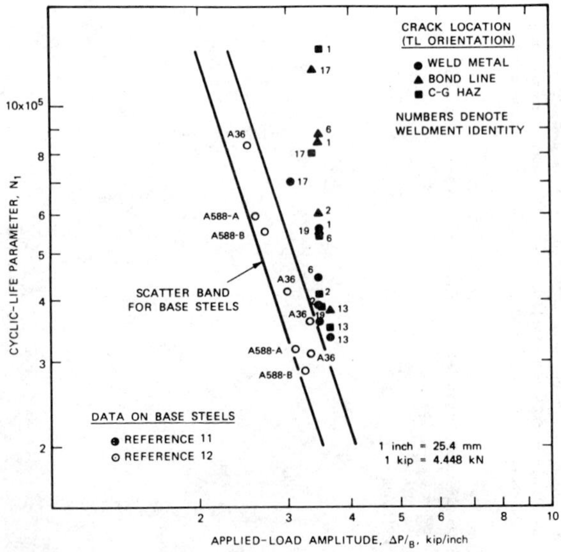

FIG. 6—*Relationship of cyclic-life parameter,* N_1, *to the applied-load amplitude per unit specimen thickness,* $\Delta P/B$, *for 1-in.-thick weldments investigated.*

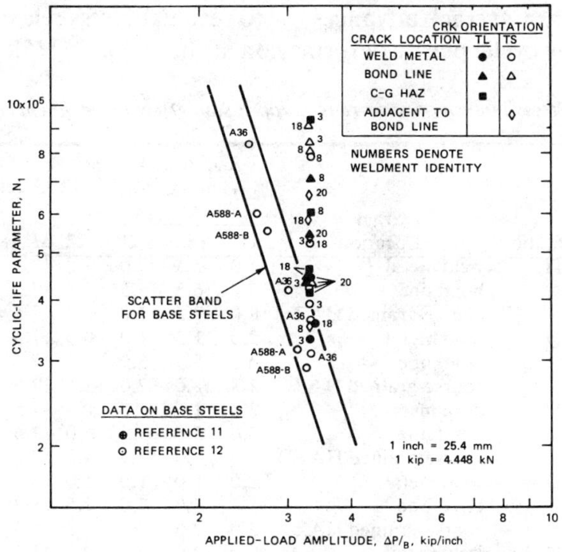

FIG. 7—*Relationship of cyclic-life parameter,* N_1, *to the applied-load amplitude per unit specimen thickness,* $\Delta P/B$, *for 4-in.-thick weldments investigated.*

Compressive residual stresses normal to the crack surface counteract the applied crack-opening loads, and thereby decrease the ΔK value calculated from knowledge of the applied loads and crack length to some lesser effective value in the vicinity of the crack tip. Thus, neglecting any secondary stresses resulting from crack-closure effects, the effective stress-intensity-factor range, ΔK_{eff}, may be simply expressed as

$$\Delta K_{eff} = \Delta K_{app} - K_{res} \tag{5}$$

where

ΔK_{app} = applied stress-intensity-factor range, and

K_{res} = stress-intensity-range suppression caused by compressive residual stresses.

Accordingly, the value of K_{res} is determined by the offset of the experimentally determined crack-growth data from the plot of the actual crack-growth behavior of the material, as shown in Fig. 8.

Values of K_{res} were thus determined at ΔK_{app} values of 18.7, 20.3, 22.0, and 24.2 MPa$\sqrt{m}$ (17, 18.5, 20, and 22 ksi$\sqrt{in.}$) from the da/dN versus ΔK plots for each weldment [7] and are reported in Table 2. The apparent variation in K_{res} with ΔK_{app} is shown in Fig. 9 for two of the 2.54-cm (1-in.) thick weldments (showing extreme behaviors), 1 and 19, and for a 10.16 cm (4-in.) thick weldment, 3. As a rule, the value of K_{res} decreased gradually with increasing ΔK_{app} values, which is consistent with the earlier observation that crack-growth retardation is most pronounced at lower ΔK values. This behavior is probably attributable to a redistribution of residual stresses in the specimen as the crack advances, or to residual-stress relaxation or "fading" caused by cyclic plastic deformation at the crack tip [*15*].

TABLE 2—*Stress-intensity-factor-range suppression effect in weldments investigated.*

Weldment	Orientation	Crack Location	K_{res}[a] at Indicated Values of ΔK_{app},[b] ksi $\sqrt{in.}$				Parameters for Crack Length of 1 in., ksi $\sqrt{in.}$		
			17	18.5	20	22	ΔK_{app}[b]	K_{res}[a]	ΔK_{eff}[c]
1	TL	weld metal	3.0	2.5	2.0	1.0	17.5	2.5	15.0
		bond line	7.0	6.5	5.5	3.5	17.5	7.0	10.5
		course-grained HAZ	8.0	7.5	7.0	6.5	17.5	7.5	10.0
2	TL	weld metal	2.0	1.5	1.0	0	17.5	2.0	15.5
		bond line	5.5	5.5	5.0	4.5	17.5	5.5	12.0
		course-grained HAZ	2.5	2.0	2.0	1.5	17.5	2.5	15.0
6	TL	weld metal	2.0	1.5	1.5	0	17.5	2.0	15.5
		bond line	5.0	5.5	5.5	6.0	17.5	5.0	12.5
		coarse-grained HAZ	4.0	4.0	3.5	4.0	17.5	4.0	13.5
13	TL	weld metal	1.0	1.0	1.0	1.0	18.3	1.0	17.3
		bond line	3.5	3.0	3.0	2.5	18.3	3.0	15.3
		coarse-grained HAZ	3.0	2.5	2.5	2.5	18.3	2.5	15.8
17	TL	weld metal	2.0	2.0	2.0	2.0	15.5	2.0	13.5
		bond line	6.5	6.0	6.0	5.5	17.0	6.5	10.5
		coarse-grained HAZ	4.5	4.5	4.0	4.5	17.0	4.5	12.5
19	TL	weld metal	1.5	1.5	1.5	1.0	17.5	1.5	16.0
		bond line	4.0	4.0	3.5	2.5	17.5	4.0	13.5
		coarse-grained HAZ	2.0	2.0	2.0	2.0	17.5	2.0	15.5
3	TL	weld metal	0	0	0	0	16.5	0	16.5
		bond line	0	0	0	0	16.5	0	16.5
		coarse-grained HAZ	5.5	5.5	5.0	4.5	16.5	5.5	11.0
	TS	weld metal	2.0	1.5	1.0	0	16.5	2.0	14.5
		bond line	5.5	5.5	5.0	3.5	16.5	5.5	11.0
		adjacent to bond line	3.5	3.5	3.5	3.0	16.5	3.5	13.0
8	TL	weld metal[d]							
		bond line	4.5	4.0	3.5	2.0	16.5	4.5	12.0
		coarse-grained HAZ	5.0	5.0	5.0	5.0	16.5	5.0	11.5
	TS	weld metal	3.5	3.0	2.5	1.0	16.5	3.5	13.0
		bond line	5.0	4.5	4.5	4.0	16.5	5.0	11.5
		adjacent to bond line	0	0	0	0	16.5	0	16.5
18	TL	weld metal	2.0	2.0	1.5	1.0	16.5	2.0	14.5
		bond line	3.0	2.0	1.5	1.5	16.5	3.0	13.5
		coarse-grained HAZ	3.0	2.0	1.5	1.5	16.5	3.0	13.5
	TS	weld metal	3.0	3.0	3.0	3.0	16.5	3.0	13.5
		bond line	5.5	5.0	4.0	3.5	16.5	5.5	11.0
		adjacent to bond line	4.0	3.5	3.5	2.5	16.5	4.0	12.5
20	TL	weld metal	2.5	2.0	1.5	1.0	16.5	2.5	14.0
		bond line	3.5	3.0	2.5	1.5	16.5	3.5	13.0
		coarse-grained HAZ	1.5	1.0	1.0	1.0	16.5	1.5	15.0
	TS	weld metal	1.0	0.5	0.5	0	16.5	1.0	15.5
		bond line	2.5	2.5	2.5	2.5	16.5	2.5	14.0
		adjacent to bond line	2.5	2.5	2.5	2.5	16.5	2.5	14.0

[a] K_{res} = stress-intensity-factor-range suppression caused by compressive residual stresses.
[b] ΔK_{app} = applied cyclic-stress amplitude.
[c] ΔK_{eff} = effective stress-intensity-factor range = $\Delta K_{app} - K_{res}$.
[d] Specimen accidentally overloaded and no crack-growth data were obtained.

Conversion Factors—
1 in. = 2.54 cm, and
1 ksi $\sqrt{in.}$ = 1.10 MPa $\sqrt{m}$.

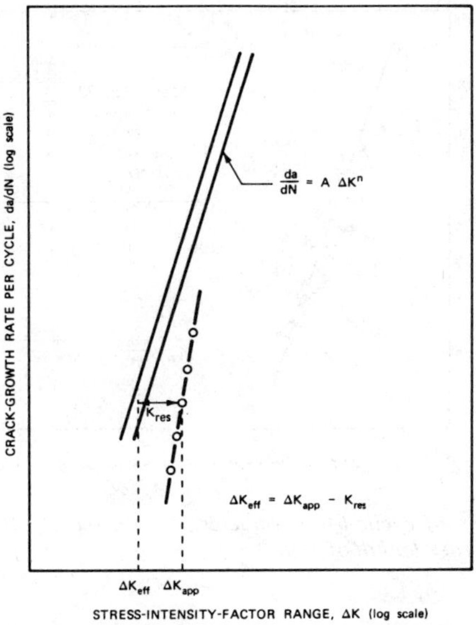

FIG. 8—*Schematic representation of stress-intensity-factor-range suppression of fatigue-crack growth under compressive-residual-stress condition.*

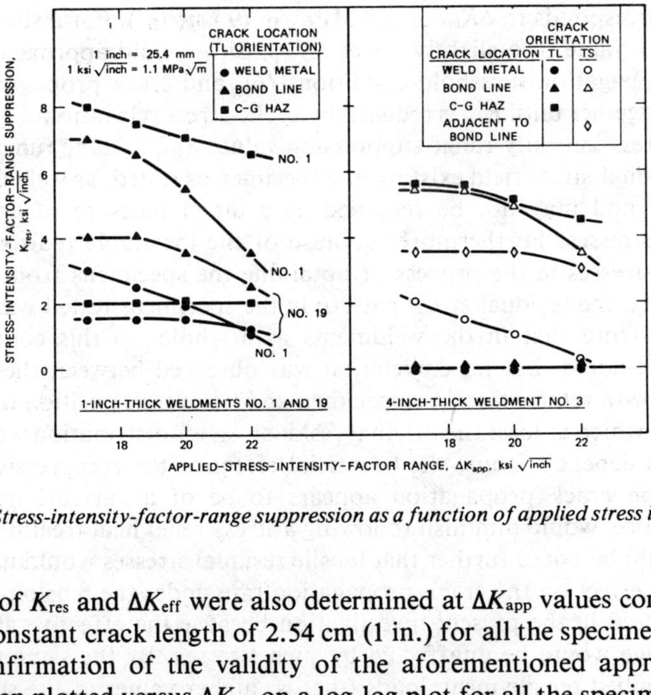

FIG. 9—*Stress-intensity-factor-range suppression as a function of applied stress intensity.*

Values of K_{res} and ΔK_{eff} were also determined at ΔK_{app} values corresponding to a constant crack length of 2.54 cm (1 in.) for all the specimens, Table 2. As confirmation of the validity of the aforementioned approach, N_1 values when plotted versus ΔK_{eff} on a log-log plot for all the specimens, Fig.

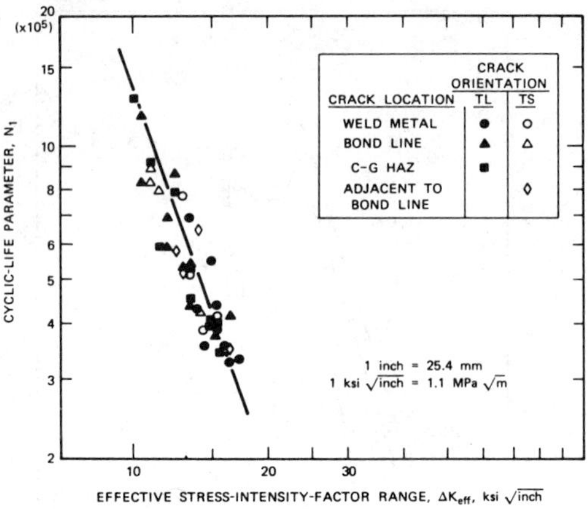

FIG. 10—*Dependence of cyclic-life parameter,* N_1, *on the effective stress-intensity-factor range,* ΔK_{eff}, *at a total crack length of 1 in.*

10, showed a relationship with approximately the same slope as that shown by the base steels in Figs. 6 and 7.

It may be noted that the maximum value of K_{res} observed was about 8.8 MPa$\sqrt{m}$ (8 ksi$\sqrt{in.}$) at a ΔK_{app} value of 18.7 MPa$\sqrt{m}$ (17 ksi$\sqrt{in.}$), Fig. 9, which corresponds to ΔK_{eff} of 9.9 MPa$\sqrt{m}$ (9 ksi$\sqrt{in.}$). For a slightly higher initial K_{res} value or a slightly lower ΔK_{app}, ΔK_{eff} would approach a fatigue-crack-propagation threshold condition [*16*], and crack propagation would cease altogether until K_{res} is reduced by cyclic-stress relaxation.

The stress-intensity-range suppression value, K_{res}, is some function of the local residual-stress field existing in a specimen as tested, as well as the stress ratio, R, and may not be regarded as a direct measure of compressive residual stresses. Furthermore, because of the inevitable redistribution of residual stresses in the process of obtaining the specimens from the whole weldments, the residual-stress pattern in the specimens tested was probably different from that in the weldments as a whole. In this connection, it should be noted that no correlation was observed between the degree of crack-growth retardation in a specimen and its overall position in the weldment (on which at least the original residual- stress distribution would be expected to depend). Thus, the beneficial effect of the compressive residual stresses on crack propagation appears to be of a variable nature and, furthermore, would diminish following a stress-relief heat treatment.

It should be noted further that tensile residual stresses would have no appreciable effect on the crack-propagation rate under the tension-to-tension loading used in the present investigation because the effective alternating-stress range would be unaffected by such stresses. By the same reasoning, higher applied tensile mean loads (that is, higher values of the stress ratio,

R) than that used in the present investigation would diminish the beneficial influence of compressive residual stresses (by decreasing K_{res} for a given residual-stress state). Finally, because crack initiation is governed primarily by the total cyclic-stress amplitude, neither tensile nor compressive residual stresses would be expected to influence significantly fatigue-crack-initiation behavior.

Summary

The results of the present investigation may be summarized as follows.

1. Evaluation of the crack-growth behavior of the weldments in terms of a cyclic-life parameter (N_1) gave results consistent with those based on incremental crack-growth (da/dN) data.

2. The observed retardation in crack-growth rate in the weldments has been attributed to compressive residual stresses introduced by welding.

3. This effect was analyzed in terms of a stress-intensity-range suppression concept, whereby the applied stress-intensity-factor range is decreased to some lower effective value.

4. The beneficial effect of the compressive residual stresses on crack propagation appears to be of a variable nature and, furthermore, would diminish following a stress-relief heat treatment.

Author's Note

It is understood that the material in this paper is intended for general information only and should not be used in relation to any specific application without independent examination and verification of its applicability and suitability by professionally qualified personnel. Those making use thereof or relying thereon assume all risk and liability arising from such use or reliance.

References

[1] Parry, M., Nordberg, H., and Hertzberg, R. W., *Welding Journal*, Oct. 1972, pp. 485–490s.

[2] Sarno, D. A., Bruner, J. P., and Kampschaefer, G. E., *Welding Journal,* Nov. 1974, pp. 486–494s.

[3] James, L. A., *Journal of Testing and Evaluation,* Vol. 1, No. 1, 1973, pp. 52–57.

[4] James, L. A., *Welding Journal,* April 1973, pp. 173–179s.

[5] El-Soudani, S. M., and Pelloux, R. M., *Welding Journal*, May 1975, pp. 144–152s.

[6] Benter, W. P., Jr., Konkol, P. J., Kapadia, B. M., Shoemaker, A. K., and Sovak, J. F., "Acceptance Criteria for Electroslag Weldments in Bridges," Phase I Final Report NCHRP 10–10, National Cooperative Highway Research Program.

[7] Kapadia, B. M. and Imhof, E. J., Jr. in *Flaw Growth and Fracture, ASTM STP 631*, American Society for Testing and Materials, 1977, p. 159.

[8] Wilson, W. K., "Review of Analysis and Development of WOL Specimen," 67–7D7–BTLPV–R1, Westinghouse Research Laboratory, 8 March 1967.

[9] Paris, P. C., "The Fracture Mechanics Approach to Fatigue," *Proceedings,* Tenth Sagamore Army Materials Research Conference, Syracuse University Press, 1964.

[10] Paris, P. C. and Erdogan, F., *Transactions, Journal of Basic Engineering,* Series D, Vol. 85, No. 3, 1963, p. 528.

[*11*] Barsom, J. M. and Novak, S. R., "Subcritical Crack Growth in Steel Bridge Members," Final Report NCHRP 12–14, National Cooperative Highway Research Program, Sept. 1974.

[*12*] Unpublished data, U.S. Steel Research.

[*13*] Bucci, R. J., written discussion in *Fracture Toughness and Slow Stable Cracking, ASTM STP 559,* American Society for Testing and Materials, 1974, p. 261.

[*14*] Barsom, J. M. and McNicol, R. C., "Effect of Stress Concentration on the Fatigue-Crack-Initiation Threshold in Various Steels," to be published.

[*15*] Forsyth, P. J. E. in *The Physical Basis of Metal Fatigue,* American Elsevier Publishing Company, New York, 1969, p. 175.

[*16*] Barsom, J. M., "Fatigue Behavior of Pressure-Vessel Steels," Welding Research Council Bulletin No. 194, May 1974.

R. R. Seeley,[1] *L. Katz,*[2] *and J. R. M. Smith*[2]

Fatigue Crack Growth in Low Alloy Steel Submerged Arc Weld Metals*

REFERENCE: Seeley, R. R., Katz, L., and Smith, J. R. M., "**Fatigue Crack Growth in Low Alloy Steel Submerged Arc Weld Metals,**" *Fatigue Testing of Weldments, ASTM STP 648*, D.W. Hoeppner, Ed., American Society for Testing and Materials, 1978, pp. 261–284.

ABSTRACT: Fatigue crack growth experiments were conducted on low alloy steel submerged arc weld metals representative of those used in the fabrication of pressure vessels. Variations of wire composition, flux type, and postweld heat treatments were used to produce weld deposits of different microstructures and tensile and toughness properties.

Fatigue crack growth experiments were conducted on 25.4-mm (1-in.) thick modified wedge opening loaded (WOL) specimens at room temperature and 288°C (550°F). A compliance calibration curve was developed at room temperature and adjusted for use with elevated temperature compliance measurements.

Results of these experiments indicate: (*a*) a slight effect of flux type on crack growth rates, (*b*) incomplete tempering or relief of residual welding stress, or both, has an appreciable effect on crack growth behavior, (*c*) comparable or lower crack growth rates at 288°C (550°F) compared to room temperature, (*d*) comparable or lower crack growth rates than those predicted in Section XI of the American Society of Mechanical Engineers *(ASME) Boiler and Pressure Vessel Code*, and (*e*) variations in crack growth rates due to wire, flux, and postweld heat treatment were less pronounced than the variations in toughness and tensile properties.

KEY WORDS: fatigue tests, fatigue (materials), weldments, pressure vessels, weld metal, welding fluxes, heat treatment, tests, tensile properties, toughness, microstructure, residual stress, boiler codes

There is a growing interest in the mechanical properties of weld metals. In the pressure vessel industry, much of this interest has been directed towards nuclear components. Some representative examples are: (*a*) American Society of Mechanical Engineers (ASME) Code Case 1592, governing the design of elevated temperature nuclear components, specifies that the allowable limit for accumulated strain in welds shall be half of that permitted

* Original experimental data were measured in U.S. customary units.

[1] Research and Development Division, Babcock & Wilcox Co., Alliance, Ohio 44601.

[2] Fossil Power Generation Division and Nuclear Equipment Division, respectively, Babcock & Wilcox Co., Barberton, Ohio.

in base metals; (*b*) extensive programs at Oak Ridge National Laboratory on the creep and rupture properties of austenitic weld metals [*1,2*];[3] and (*c*) Electric Power Research Institute sponsored work performed by Babcock & Wilcox [*3,4*], Combustion Engineering [*5*], and Effects Technology Incorporated [*6*] on the fracture toughness of low alloy steel weld metals. This increased interest, associated with nuclear components, can be attributed to a number of factors. Among these are the philosophy of design by analysis, other new design procedures such as fracture mechanics analysis, and the need to optimize plant reliability for both economic and safety considerations.

Fatigue crack growth rate properties of weld metals are attracting greater attention than before [*7–13*] since there are relatively less data for weld metals than for base metals, and many analysis procedures require crack growth properties. Few code or standards requirements for crack growth rate properties or analysis procedures exist. The *ASME Boiler and Pressure Vessel Code,* Section XI, Rules for In-service Inspection of Nuclear Power Plant Components, contains fatigue crack growth rate curves, which may be used for the analysis of flaws that might be revealed during routine nondestructive examination. Separate curves are provided for flaws exposed to the water reactor environment, and those that are not so exposed (buried flaws). These curves are based upon data for the commonly used base materials for construction, for example, SA-533-B and SA-508-2 low alloy steels. It is pointed out in Section XI that these curves are intended to be conservative estimates of crack growth rates, which can be used when data are not available from the actual material and product forms in question.

The acquisition of crack growth data frequently requires specific test methods which are tailored to the type of information desired. Often these methods must be developed by the investigator. For example, the major experimental difficulty in severe temperatures or environments is measuring or otherwise determining the crack length in the specimen. Visual observation of specimens is difficult, if not impossible, because of heating furnaces or environmental chambers. In such cases, alternate measurements are made, such as compliance or resistivity, which can be directly related to crack length. American Society for Testing and Materials (ASTM) Subcommittee E24.04 on Subcritical Crack Growth is currently developing tentative test methods for determination of high and low (threshold) fatigue crack growth rates [*14*].

Our objective was to develop constant amplitude fatigue crack growth rate properties of low alloy steel submerged arc weld metals, typical of those used in the fabrication of pressure vessels. The crack growth tests were conducted in air at 24 and 288°C (75 and 550°F). Particular attention was given to weld metal composition, flux type, and postweld heat treatments. The crack growth properties of these materials are compared to the *ASME*

[3] The italic numbers in brackets refer to the list of references appended to this paper.

Boiler and Pressure Vessel Code, Section XI, design curve for buried flaws.

Experimental Procedures

Materials and Weldment Preparation

Submerged arc groove welds were prepared with the parameters shown in Table 1. Three combinations of two weld wires and two fluxes were investigated. The welding parameters for weldments of the same wire and flux types were closely controlled so as to minimize the influence of these variables. Different parameters were, however, used for each unique wire/flux combination, because of their inherent differences in weldability.

TABLE 1—*Welding parameters.*

Variable	Wire A/ Flux A	Wire A/ Flux B	Wire B/ Flux A
Amperage, A	625	550	625
Voltage, V	34	32	34
Travel Speed, in./min (cm/min)	14 (35.6)	13 (33)	14 (35.6)
Heat input, KJ/in. (KJ/cm)	91 (35.8)	81 (31.9)	91 (35.8)
Minimum preheat temperature, °F (°C)	200 (94)	200 (94)	200 (94)
Maximum interpass temperature, °F (°C)	450 (231)	400 (205)	450 (231)
Wire Diameter, in. (cm)	0.125 (0.32)	0.125 (0.32)	0.125 (0.32)
Flux size	48 × D	35 × 200	48 × D

A common carbon steel plate material, A 515 Grade 70, 50.8 to 76.3 mm (2 to 3 in.) thick was used for the base metal. Each weldment was 127 mm (5 in.) wide, had a 22.2-mm (⅞-in.) root gap, and a 14-deg included angle. The thickness of the weldments varied from 25.4 to 76.3 mm (1 to 3 in.). Following completion of welding and postweld heat treatments, the weldments were radiographed and were free of weld defects which could bias the test results.

Two low alloy wire types designated A and B, were used. Wire A may be used to join 551 MPa (80 000 psi) minimum specified tensile strength materials, such as A 533 Grade B 1. Wire B is suitable for carbon steels, having a minimum specified tensile strength of 482 MPa (70 000 psi), such as A 515 Grade 70. The principal difference in composition of the two wire types is the higher nickel content of Wire A.

Two fused fluxes, also designated A and B, were used. Flux A was used with both the A and B wires, while Flux B was only used with Wire A. The nominal composition of these flux types is shown in Table 2. The magnesium oxide-calcium oxide-aluminum oxide-silicon dioxide (MgO–CaO–Al_2O_3–SiO_2) quarternary, and the CaO–SiO_2–titanium dioxide (TiO_2) ternary describe Fluxes A and B, respectively. Differences in the basicity ratio, CaO/SiO_2, are also apparent.

One heat of each wire type and one lot of each flux type were used to produce the welds. The composition of the deposited weld metal is shown in Table 3.

TABLE 2—*Flux compositions.*

	Flux A %	Flux B, %
SiO_2	36.0	36.0
CaO	21.0	46.0
CaF_2	5.0	...
MgO	10.0	1.0
MnO	7.0	0.5
FeO	1.5	0.3
Al_2O_3	16.0	5.0
TiO_2	...	5.0
CaO/SiO_2	0.58	1.28

TABLE 3—*Deposited weld metal chemistry, percent.*

Wire/Flux	C	Mn	P	S	Si	Cr	Ni	Mo	Cu
A/A	0.12	1.59	0.01	0.01	0.37	0.08	0.47	0.37	0.28
A/B	0.14	1.40	0.01	0.01	0.22	0.08	0.50	0.36	0.21
B/A	0.12	1.51	0.01	0.01	0.33	0.07	0.06	0.52	0.15

The postweld heat treatment test matrix containing five different treatments/wire/flux combinations is shown in Table 4. The essential variables of temperature, hold times, and cooling rates were selected so as to produce significant variations in toughness or tensile properties, or both. The weld numbers assigned in this table are used to identify the wire, flux, and heat treatment combinations throughout the remainder of this paper.

Figures 1 and 2 illustrate the typical etched and unetched structures of the weld deposits. The principal differences are the cleaner structures, that is, fewer and smaller inclusions, and the finer microstructure that results from Flux B. Microstructural differences due to the postweld heat treatments studies were not discernible by optical microscopy.

TABLE 4—*Postweld heat treatment.*

Weld Number	Wire/Flux	Heat Treatment[a]
1	A/A	48 h at 1100 to 1150°F cool 40°F/h to 600°F-air cool 180 h at 900 to 950°F cool 10°F/h to 600°F-air cool
2	A/A	48 h at 1100 to 1150°F cool 10°F/h to 600°F-air cool
3	A/B	same as Weld 2
4	B/A	same as Weld 2
5	B/A	90 h at 900 to 950°F cool 10°F/h to 600°F-air cool

[a] 1100 to 1150°F (593 to 621°C), 900 to 950°F (482 to 510°C), 40°F/h (22°C/h), 600°F (316°C), 10°F/h (5.6°C/h).

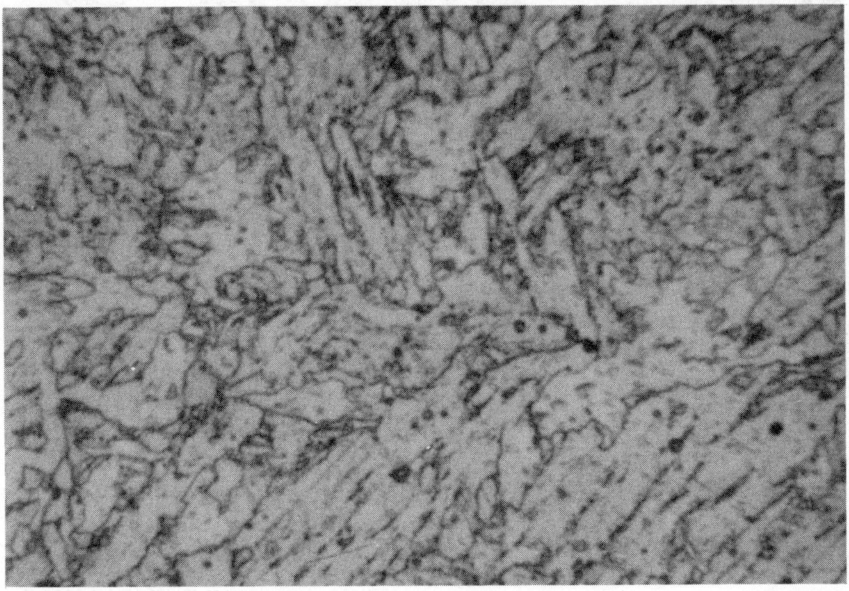

WIRE A/FLUX A
1000X

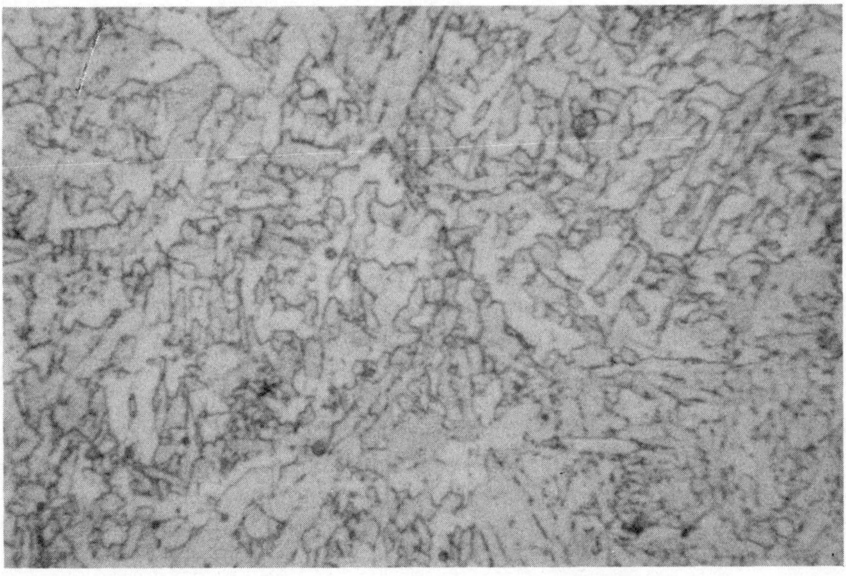

WIRE A/FLUX B
1000X

FIG. 1—*Etched weld metal microstructures.*

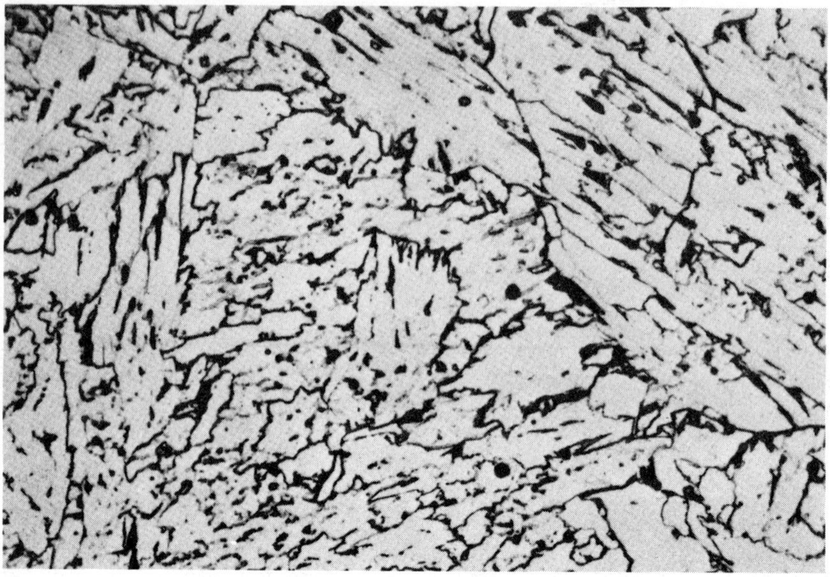

WIRE B/FLUX A
1320X

FIG. 1—*Continued.*

Specimen Preparation

Tensile 12.8-mm (0.505-in.) gage diameter, Charpy V-notch impact, drop weight and 25.4-mm (1-in.) thick (1T) smooth and side-grooved wedge opening loading (WOL) crack growth specimens were machined from the ¼ thickness (¼T) location in each weldment. Figure 3 illustrates the side-grooved WOL specimen. Figure 4 shows the orientation of these specimens in the weldment. The impact, drop weight, and fatigue specimens were oriented transverse to the welding direction. The tension specimens were oriented parallel to the welding direction since a transverse orientation would have included some base metal in the gage length and thus complicate the tension test. All specimens were centered on the weld centerline, and some specimens contained base metal in their extremities.

During the course of some of the initial high temperature fatigue crack growth tests, the cracks tended to deviate from the mid-plane of the specimens as illustrated in Fig. 5 and Table 5. To maintain the crack in the mid-plane of the specimen, the high temperature WOL fatigue specimens were subsequently side-grooved (total of 20 percent of the specimen thickness). The room temperature specimens were not side grooved since a severe problem did not occur except for Weld 5 (see Table 5).

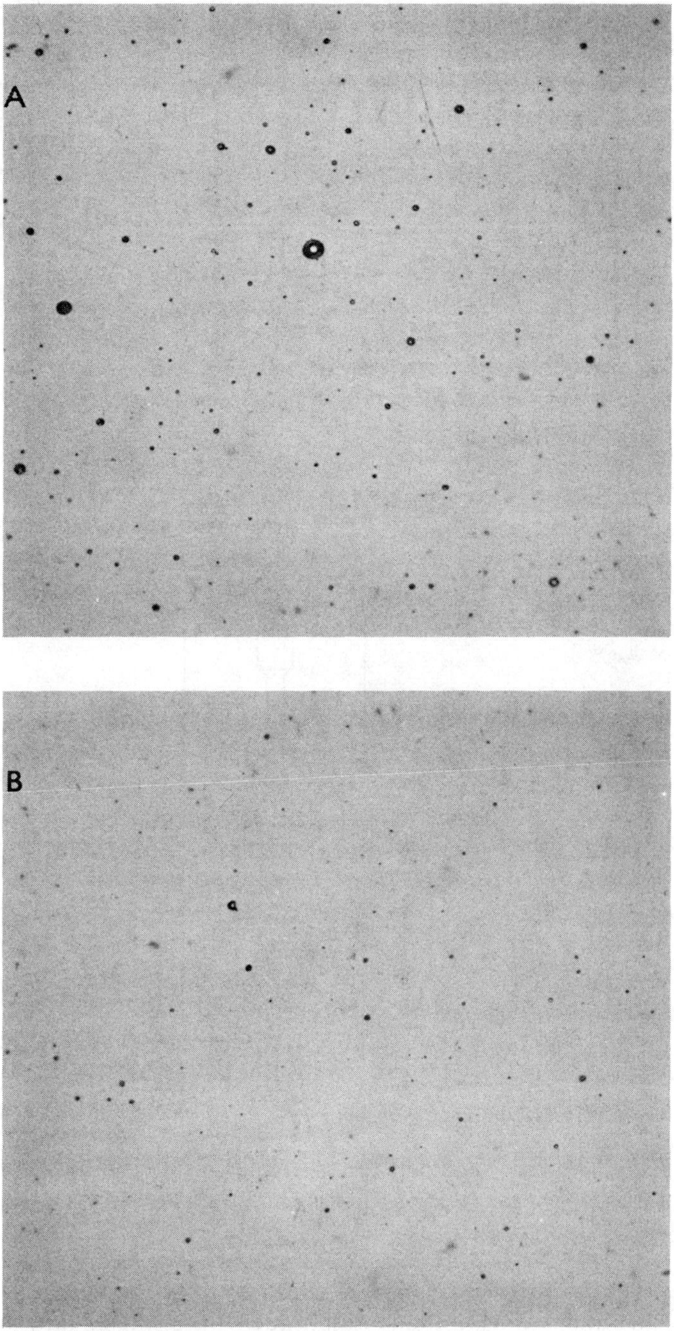

FIG. 2—*Unetched specimens of weld metal, X1000.*

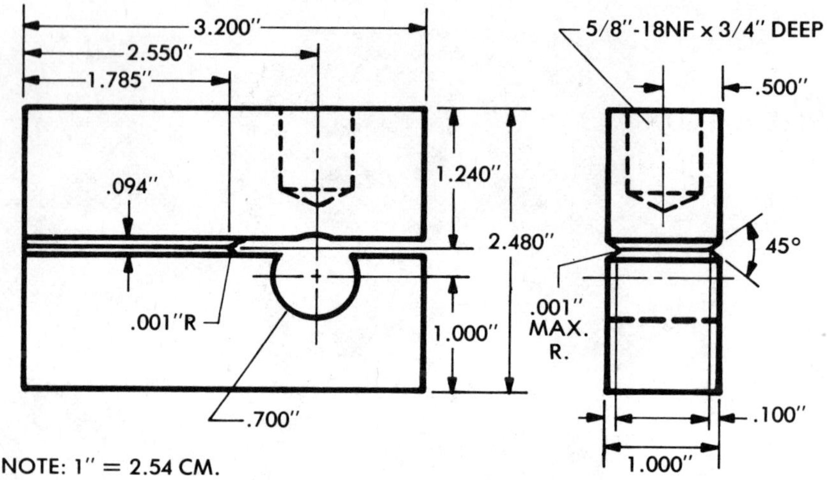

FIG. 3—*1T WOL side-grooved crack growth specimen.*

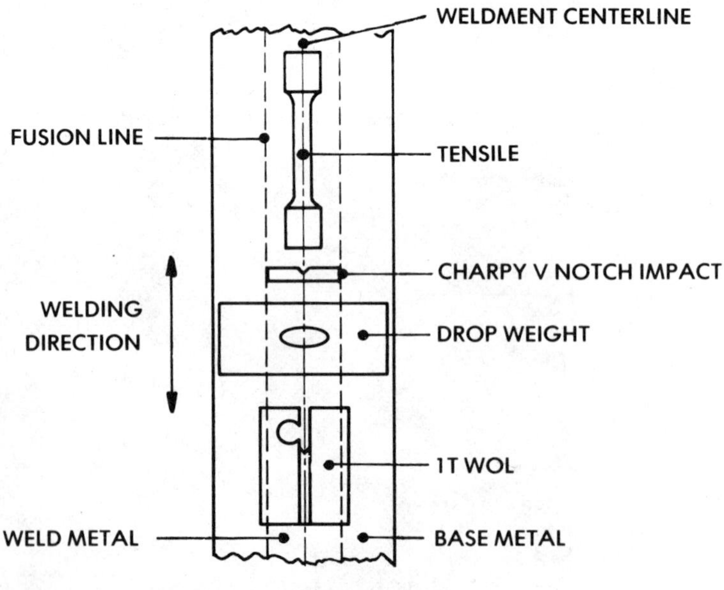

FIG. 4—*Specimen orientation.*

Mechanical Testing

Fatigue Crack Growth—All fatigue crack growth tests were conducted in air, under constant load amplitude with a sinusoidal wave form, with an R ratio (minimum load to maximum load) of 0.10 or less and at a frequency of 25 Hz. For the room temperature tests, crack lengths were visually

FIG. 5—*Fatigue crack growth specimens from a manganese-molybdenum-nickel weld.*

TABLE 5—*Fatigue crack deviation.*

Weld	Test Temperature, °F(°C)	Angle of Deviation
1	75 (29)	4
2	75 (29)	10
2	550 (288)	5
3	75 (29)	8 to 13[a]
3	550 (288)	14
4	75 (29)	10
4	550 (288)	8 to 12[a]
5	75 (29)	12 to 24[a]
5	550 (288)	18 to 24[a]

[a] Measured on a single specimen.

measured on the surface of the smooth fatigue specimens using a ×40 microscope mounted on a calibrated lead screw.

The crack lengths for the high temperature tests were determined by measuring the compliance of the specimen. Compliance of the WOL specimen is defined by the inverse slope of the load-deflection plots for each crack length (Fig. 6). A compliance calibration curve, Fig. 7, was used to relate compliance and crack length. A compliance calibration curve for the side-grooved WOL specimen was constructed using a material of similar yield strength and elastic modulus.

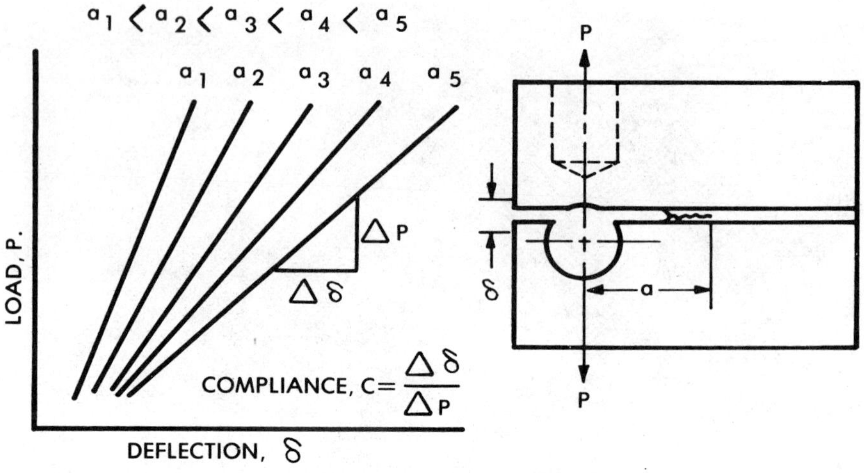

FIG. 6—*Crack length measurements by compliance.*

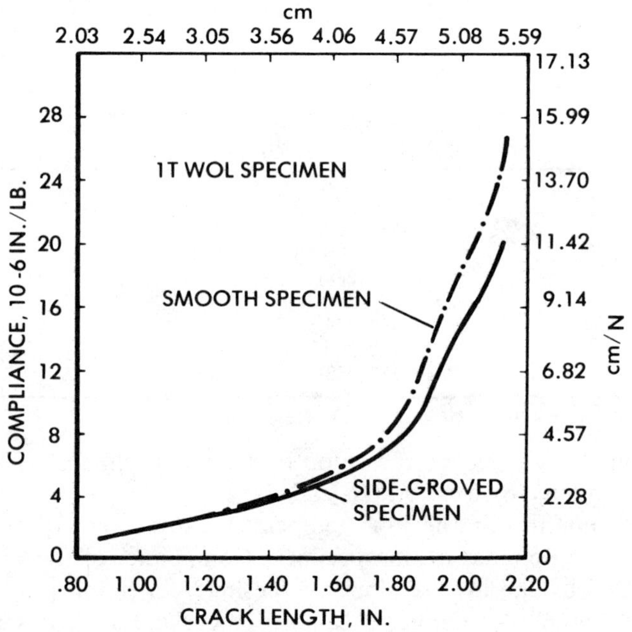

FIG. 7—*Compliance calibration curves.*

Specimen compliance measurements were made periodically during the 288°C (550°F) fatigue tests. Figure 8 illustrates the experimental setup for the high temperature crack growth tests. Specimen deflections were translated outside the furnace by a tube and rod mechanism, and measured with a displacement transducer. The compliance calibration curve, constructed at room temperature, was adjusted for the effect of temperature by the following technique. (Refer to Fig. 9.) During each high temperature fatigue test, the fatigue load amplitude was reduced resulting in a noticeable "beach" mark on the fatigue fracture surface. A compliance measurement was made immediately before the load amplitude reduction. After the test, the crack length for each mark on the fracture surface was measured directly. Thus, knowing the crack length and the compliance, a high temperature compliance data point could be plotted and the room temperature compliance curve simply adjusted upward to pass through this point. This technique is based on the concept of a normalized compliance curve described by McHenry [*15*]. A similar technique was used by Gerber et al for their crack growth experiments in a high temperature water environment [*16*].

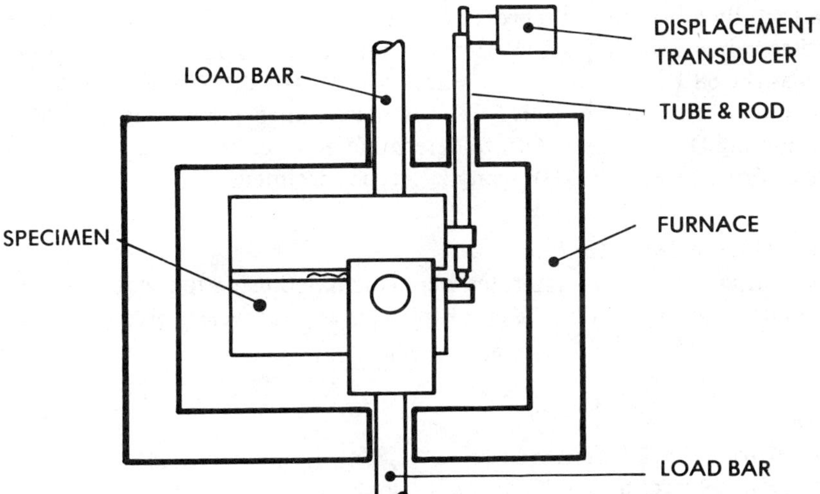

FIG. 8—*Elevated temperature test setup.*

Other Properties—The tensile properties of each weld metal were determined at 24°C (75°F). Welds 1, 4, and 5 were also tested at 288°C (550°F).

The toughness properties were characterized by both Charpy V-notch impact tests and drop weight nil-ductility transition temperature tests. The

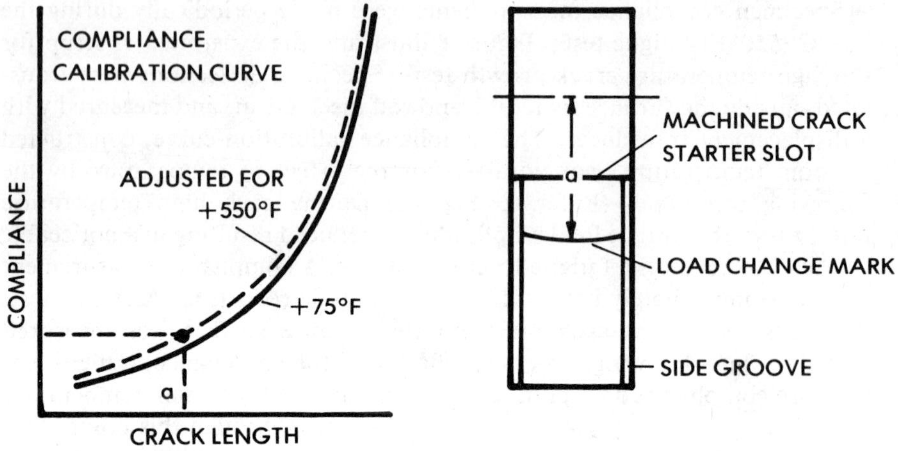

FIG. 9—*Crack length measurement at elevated temperature.*

Charpy impact properties were determined over a temperature range of − 84 to 99°C (− 120 to 210°F). (Absorbed energies in the impact tests were analyzed by a non-linear regression equation. This equation was then used to determine desired calculated mean temperatures for a given property such as the 68 J (50 ft · lb) energy temperature). The drop weight nil-ductility transition temperature (NDTT) was determined according to ASTM Conducting Drop-Weight Test to determine nil-ductility transition temperature of ferritic steels (E 208), using P–3 type specimens.

Crack Growth Data Analysis

The fatigue crack growth rates were calculated using the secant or point-by-point method in which the crack growth rate per cycle is given by

$$\frac{da}{dN} = \frac{a_{n+1} - a_n}{N_{n+1} - N_n}$$

where

a = total crack length,
N = number of applied load cycles, and
n = first, second, third, etc. crack length measurement.

The stress intensity factor range, ΔK, was calculated using an average crack length, $a_n + a_{n+1}$, between each crack length measurement, and the applied fatigue load range ΔP

$$\Delta K = \frac{\Delta P f(a/W)}{B\sqrt{a}}$$

where B is the nominal specimen thickness, and $f(a/W)$ is given by

$$30.96\,(a/W) - 195.8\,(a/W)^2 + 730.6\,(a/W)^3 - 1186.3\,(a/W)^4 + 754.6\,(a/W)^5$$

for both the smooth and side grooved specimens. (See Ref *17*.)

Results and Discussion

Crack growth rates for each of the five individual welds are shown in Figs. 10 through 14. Welds 1, 2, and 3 (Fig. 10, 11, and 12) were made with Wire A which had a different chemical composition than Wire B for Welds 4 and 5 (Figs. 13 and 14). Two high temperature and two low temperature fatigue tests were planned for each material which was considered sufficient for this work. However, not every test yielded acceptable results. Many of the test results, particularly for tests run at 288°C (500°F), exhibited appreciable data scatter. The crack growth data appears to follow the Paris power law [*18*], $da/dN = C\Delta K^n$, but because of the limited data base and observed scatter, the C and n constants were not calculated Trends in the data are, however, discernable in many instances, and these trends are used to describe the relative properties of the various material conditions.

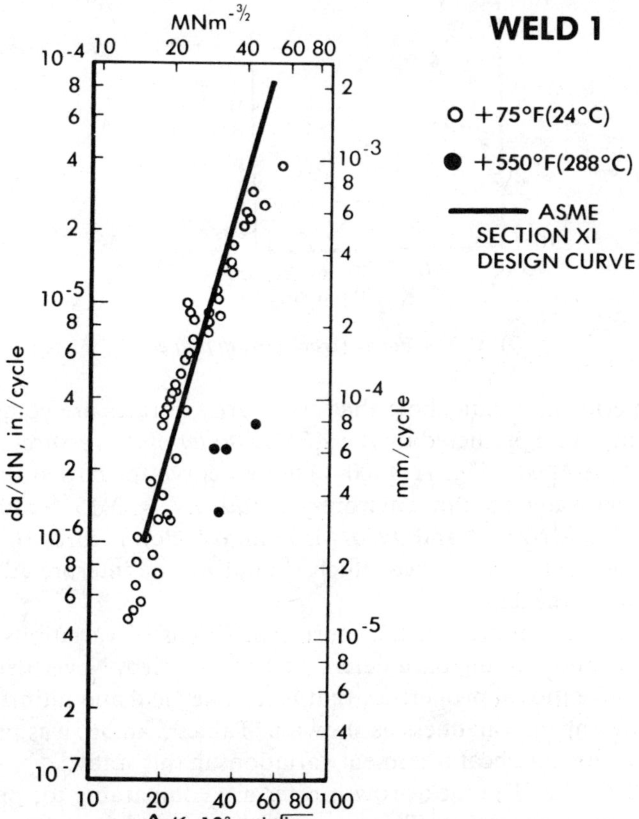

FIG. 10—*Fatigue crack growth properties.*

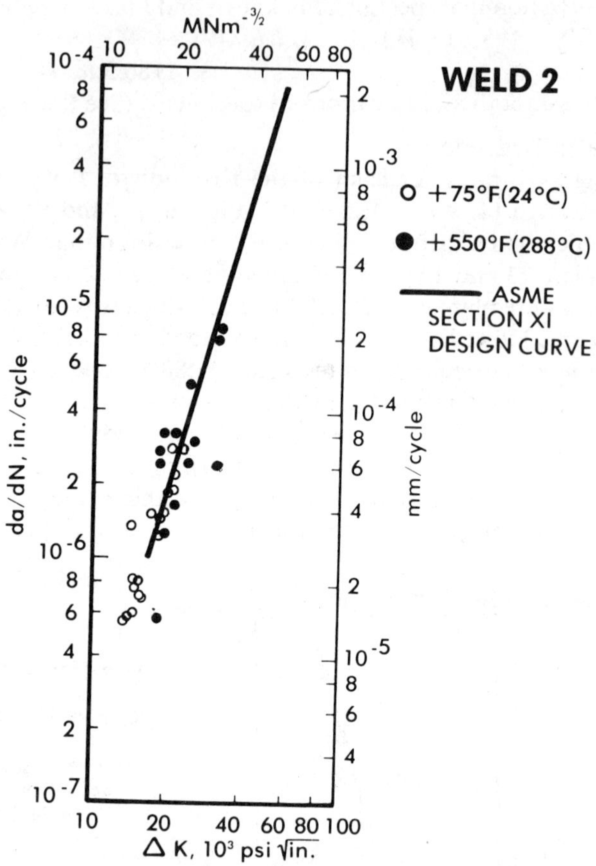

FIG. 11—*Fatigue crack growth properties.*

The experimental data show that crack growth rates are comparable to, or less than, those predicted by the *ASME Boiler and Pressure Vessel Code* Section XI (1974 ed., Fig. A-4300-1) design curve for buried flaws not exposed to the water reactor environment ($da/dN = 3.15 \times 10^{-21}\ \Delta K^{3.726}$, where ΔK is in $MNm^{-3/2}$ and da/dN is in mm/cycle). Figures 10 through 14 illustrate the Section XI curve which should not be interpreted as a curve drawn through the data.

The results are limited in the sense that extensive variations in deposit composition and welding parameters were not studied. Nevertheless, a wide range of conventional properties, that is, tensile yield and ultimate strength and Charpy impact toughness as shown in Tables 5 and 6, was produced by the composition and heat treatment variations in this study.

The 288°C (550°F) crack growth rates are comparable to, or less than, those at room temperature. This agrees with other studies of structural materials [*19*].

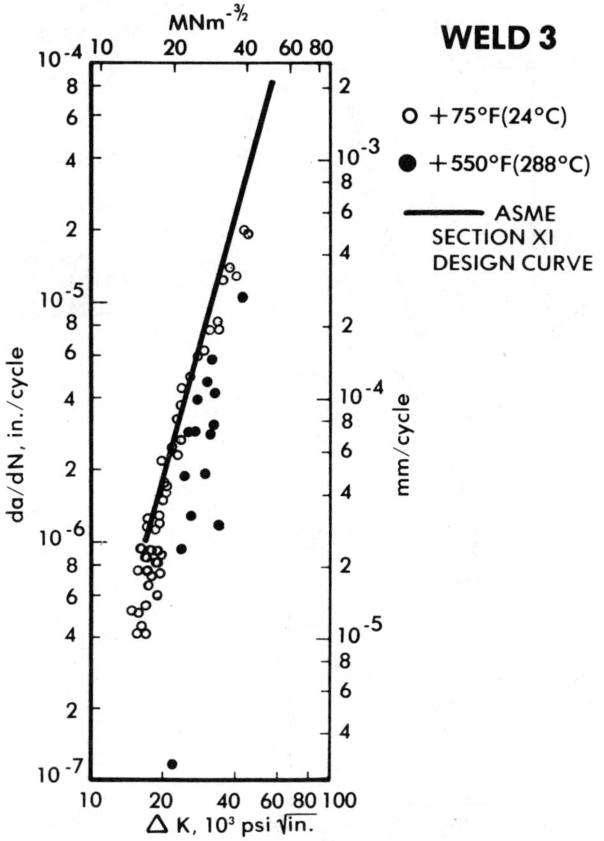

FIG. 12—*Fatigue crack growth properties.*

TABLE 6—*Tensile properties.*[a]

Weld Number	Ultimate Tensile Strength,[b] ksi	Yield Strength,[b] ksi	Reduction in Area, %	Elongation, %
1	88.8	70.5	59	24
	(79.5)	(64.5)	(57.7)	(21.5)
2	87.5	72.5	63	27
3	97.5	88.0	65	24
4	85.5	67.5	62	24
	(76.3)	(60.8)	(59.8)	(20.5)
5	102	85.5	59	24
	(97.5)	(79.4)	(46.6)	(17.0)

[a] Room temperature properties except values shown in parentheses are at 550°F (228°C).
[b] 1 ksi = 6.895 MPa.

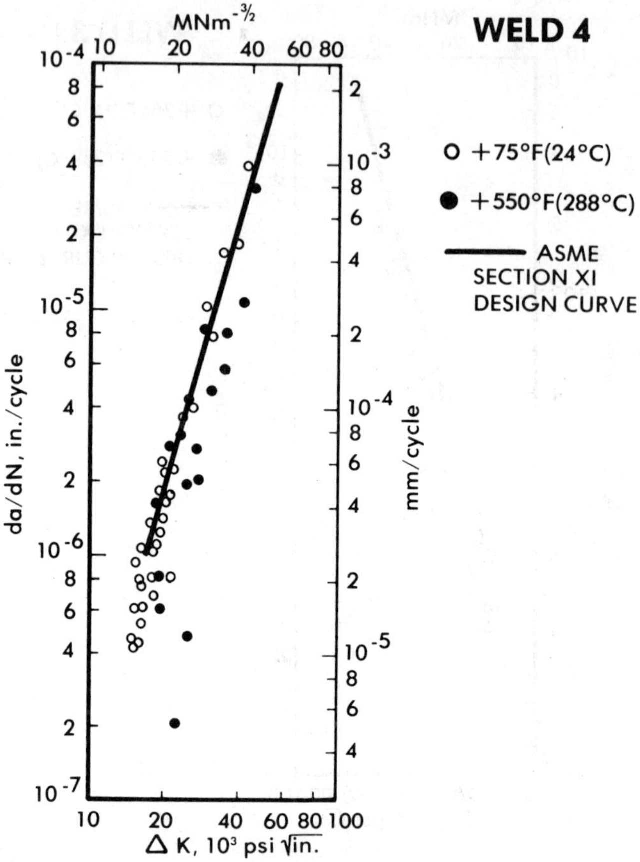

FIG. 13—*Fatigue crack growth properties.*

Several of the room temperature and high temperature tests showed a tendency for the crack to deviate from the mid-plane of the specimen, perpendicular to the axis of load application. The crack always curved toward the threaded hole side of the specimen. The gripping constraints and resultant stress distribution in the specimen would probably induce such a behavior. This effect was minimized by side-grooving the elevated temperature specimens. Table 7 lists the angle of crack deviation, ϕ, from the mid-plane of the specimen for several material conditions. Crack deviation was most prominent for Weld 5, which was postweld heat treated at 510°C (950°F). Figure 15 shows crack growth data for both smooth and side-grooved specimens. These data illustrate the severe crack deviation causes apparent higher crack growth rates.

Dawes also experienced a related problem (irregular crack fronts) in precracking weld metal fracture toughness specimens [20]. He concluded that

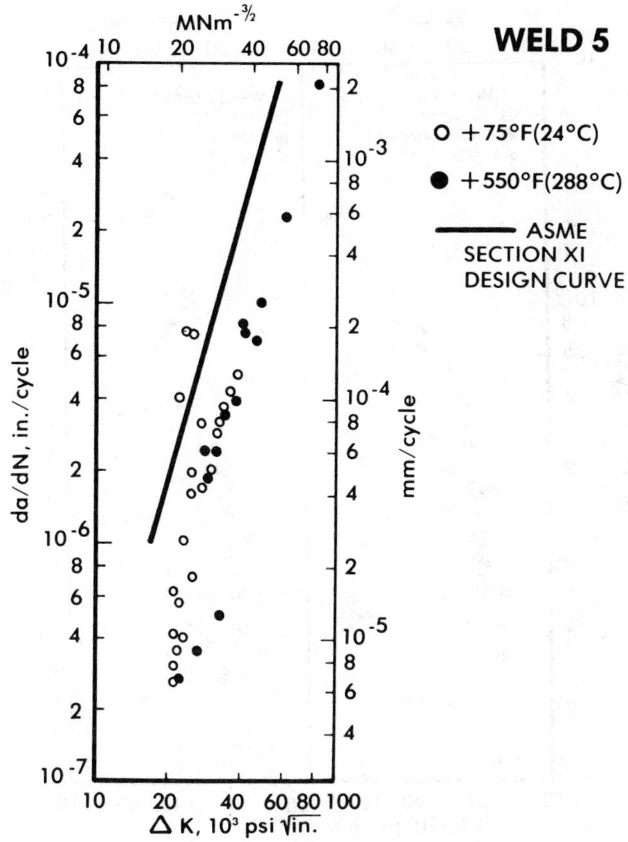

FIG. 14—*Fatigue crack growth properties.*

residual stresses were responsible for this behavior. Our results tend to confirm this since Weld 5 received the lowest temperature postweld heat treatment. Furthermore, this weld exhibits appreciably higher strength levels than its companion weld (Weld 4) which was postweld heat treated at a higher temperature indicating incomplete tempering in Weld 5.

The WOL geometry specimen was chosen for the crack growth experiments because it permitted a greater measuring capacity ($0.3 \leq a/W \leq 0.8$) at the time that the experiments were conducted than the compact tension geometry specimen in ASTM Test for Plane-Strain Fracture Toughness of Metallic Materials (E 399–74) ($0.45 \leq a/W \leq 0.55$). The compact tension specimen is, however, more symmetric about its mid-plane and thus not as sensitive to gripping constraints as is the WOL specimen. The measuring capacity of the compact tension specimen for fatigue crack growth work [*14*] has now been expanded ($a/W > 0.2$). This specimen might

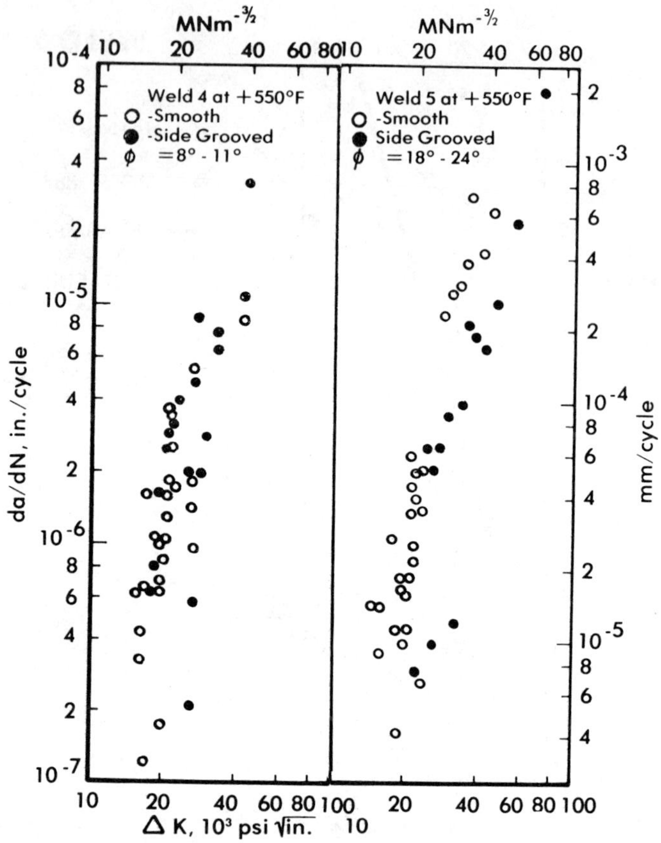

FIG. 15—*Influence of crack deviation on growth rates.*

thus be more desirable than the WOL specimen for the future crack growth work since crack deviation would be minimized.

TABLE 7—*Toughness data.*[a]

Weld	Temperature, °F 15[c]	20[c]	30[c]	50[c]	NDTT, °F	Charpy V-notch at NDTT, ft·lb	Charpy Upper Shelf ft·lb	Charpy Upper Shelf Temperature, °F[b]
1	− 47	− 23	+ 13	+ 73	− 70	12	69	+ 136
2	− 20	− 10	+ 6	+ 32	− 60	7	63	+ 102
3	− 68	− 58	− 43	− 19	− 60	19	121	+ 94
4	− 48	− 33	− 10	+ 32	− 10	30	82	+ 146
5	− 36	− 14	+ 23	+ 103	0	24	64	+ 175

[a] J = 1.36 ft·lb.
°C = 5/9 (°F-32).
[b] At onset of upper shelf behavior.
[c] Energy level, ft·lb.

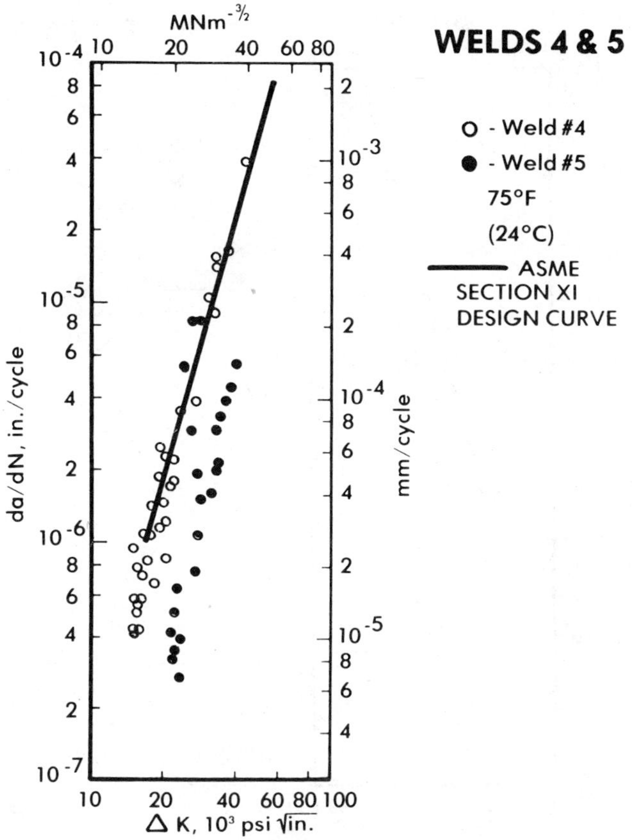

FIG. 16—*Fatigue crack growth properties.*

Figures 16 and 17 compare the crack growth rates of Welds 4 and 5 at room temperature and 288°C (550°F). The interpretation of these data is somewhat difficult. The data for Weld 5 may be misleading because of residual stresses and incomplete tempering. Higher residual stresses in Weld 5 could be responsible for a different loading condition (*R* ratio) from that of Weld 4.

The effect of postweld heat treatment on Welds 1 and 2 is shown in Figs. 18 and 19. At room temperature, there is no effect. At elevated temperature, there are insufficient data to draw any conclusion. The similarity in room temperature crack growth rates is contrasted to the differences in Charpy impact toughness of these two welds as shown in Table 6.

Flux type variations had no effect on the room temperature crack growth rates (Fig. 20) but appeared to slightly effect the rates at 288°C (550°F) (Fig. 21). The apparent lower growth rates for Weld 3 compared to Weld 2

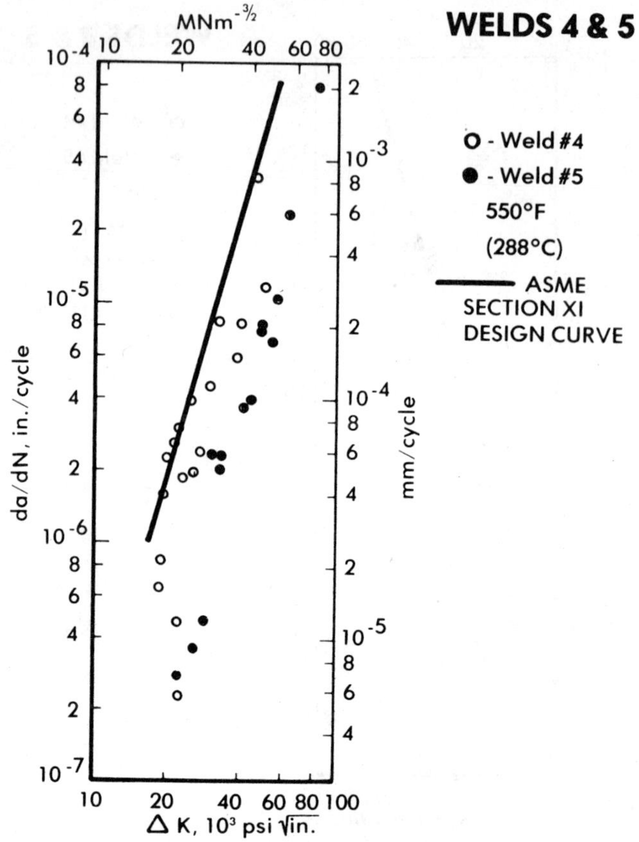

FIG. 17—*Fatigue crack growth properties.*

might be attributed to the cleaner and finer microstructure of Weld 3. These structural differences also manifested themselves in the strength and impact toughness properties shown in Tables 5 and 6 for these two weld metals. The differences in crack growth rates are, however, much less pronounced than those for impact toughness.

Conclusions

Results of these experiments indicate the following.

1. Flux type has a slight effect on crack growth rates.
2. Postweld heat treatment has no significant effect on crack growth behavior except when tempering or relief of residual welding stresses, or both, may not be complete.
3. Crack growth rates at 288°C (550°F) are comparable to, or lower than, those at room temperature.

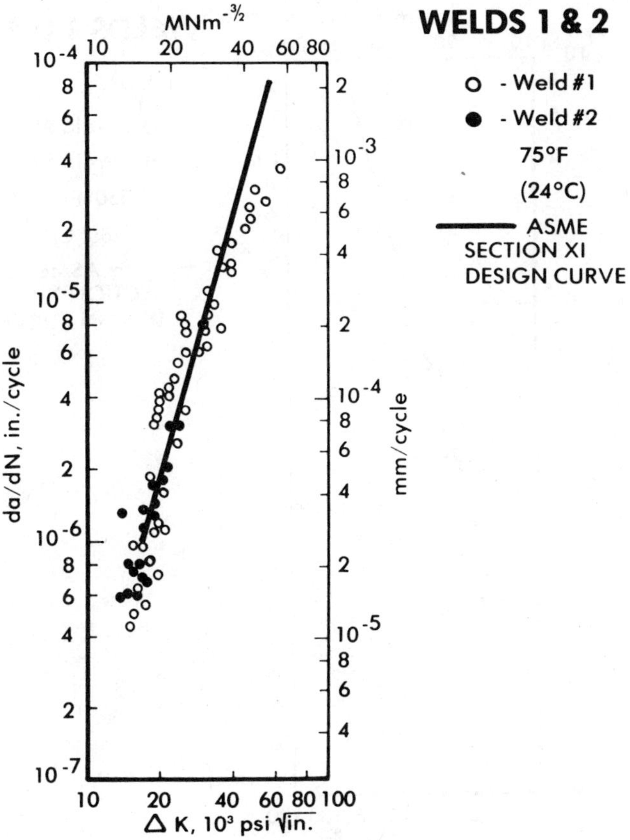

FIG. 18—*Fatigue crack growth properties.*

4. Crack growth rates for these materials are comparable to, or lower than, the design values given in Section XI of the *ASME Boiler and Pressure Vessel Code*.

5. The variations in crack growth rates due to flux type and postweld heat treatment are less pronounced than the variations in Charpy impact toughness and tensile properties.

Acknowledgments

The authors wish to acknowledge the helpful comments and suggestions of P.S. Ayers, H. A. Domian, N. L. Evanchan, J. G. Feldstein, M. Gold, V. J. Hudacko, K. E. Moore, and W. A. Van Der Sluys, and the Babcock & Wilcox Company for permission to publish this work.

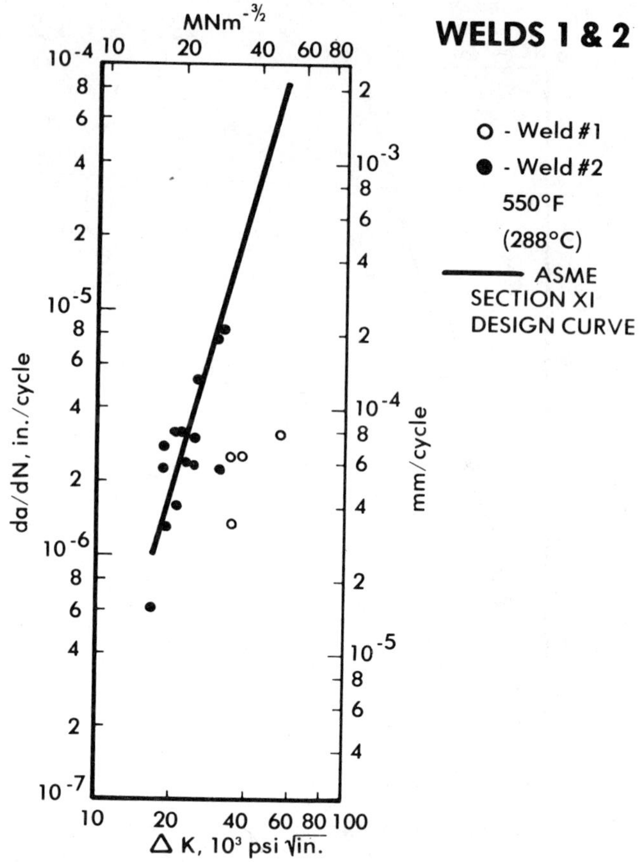

FIG. 19—*Fatigue crack growth properties.*

References

[1] King, R. T., Cole, N. C., and Goodwin, G. M., "Creep Behavior of SMA Type Weld Metal with Controlled Residual Elements," ASME Paper 75-WA/PVP-19, American Society of Mechanical Engineers.

[2] King, R. T., Stiegler, J. O., and Goodwin, G. M., *Welding Journal Research Supplement*, July 1974.

[3] Van Der Sluys, W. A., Seeley, R. R., and Schwabe, J. E., "Determining Fracture Properties of Reactor Vessel Forging Materials, Weldments and Bolting Materials," Final Report (EPRI NP-122) to Electric Power Research Institute by Babcock & Wilcox Company, July 1976.

[4] Van Der Sluys, W. A., Seeley, R. R., and Schwabe, J. E., "An Investigation of Mechanical Properties and Chemistry Within a Thick MnMoNi Submerged Arc Weldment," Topical Report (EPRI NP-373) to Electric Power Research Institute by Babcock & Wilcox Company, Feb. 1977.

[5] Marsten, T. U., Borden, M. P., Fox, J. H., and Reardon, L. D., "Fracture Toughness of Ferritic Materials in Light Water Nuclear Reactor Vessels," Final Report to Electric Power Research Institute by Combustion Engineering, Inc., Dec. 1975.

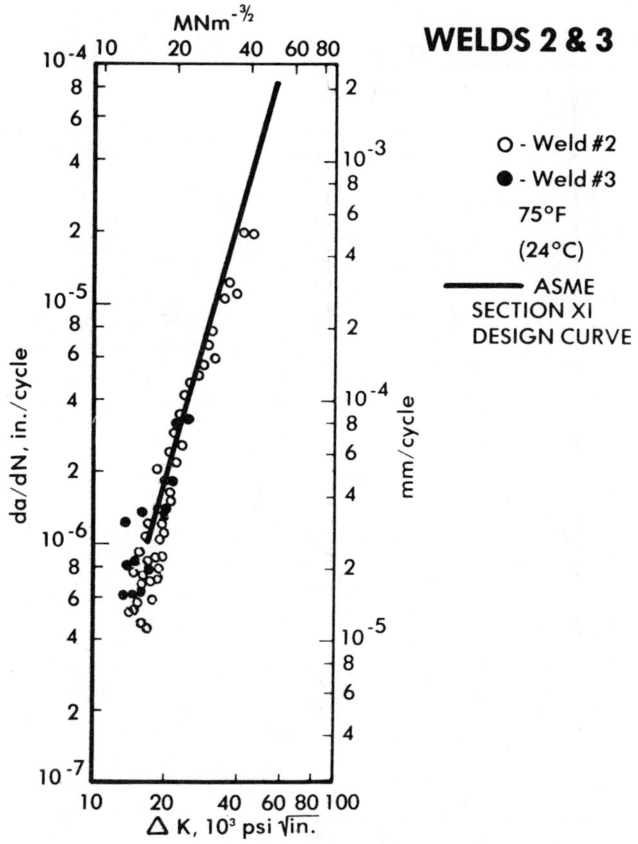

FIG. 20—*Fatigue crack growth properties.*

[6] Server, W. L., Sheckherd, J. W., and Wullaert, R. A., "Fracture Toughness Data for Ferritic Nuclear Presure Vessel Materials, Task B-Laboratory Testings," Effects Technology Final Report CR75-287 to Electric Power Research Institute, May 1975.

[7] James, L. A., *Welding Journal Research Supplement,* Sept. 1976.

[8] Solomon, H. D., *Nuclear Technology,* Vol. 29, April 1976.

[9] Shahinian, P. in *Properties of Reactor Structural Alloys After Neutron or Particle Irradiation, ASTM STP 570,* 1975, pp. 191–204.

[10] James, L. A., *Journal of Testing and Evaluation*, Vol. 1, No. 1, Jan. 1973, pp. 52–56.

[11] Socie, D. F. and Antolovich, S. D., *Welding Research Supplement*, June, 1974, pp. 267s–271s.

[12] Griffiths, J. R. and Richards, C. E., *Metal Science and Engineering*, Vol. 11, 1973, pp. 305–310.

[13] Shahinian, P., Watson, H. E., and Hawthorne, J. R., "Fatigue Crack Growth Resistance of Several Neutron Irradiated Pressure Vessel Steels and Welds," ASME Paper No. 74-Met-3, American Society of Mechanical Engineers, 1974.

[14] Hudak, S. J., Jr., Saxena, A., Bucci, R. J., and Malcolm, R. C., "Development of Standard Methods of Testing and Analyzing Fatigue Crack Growth Rate Data," Third

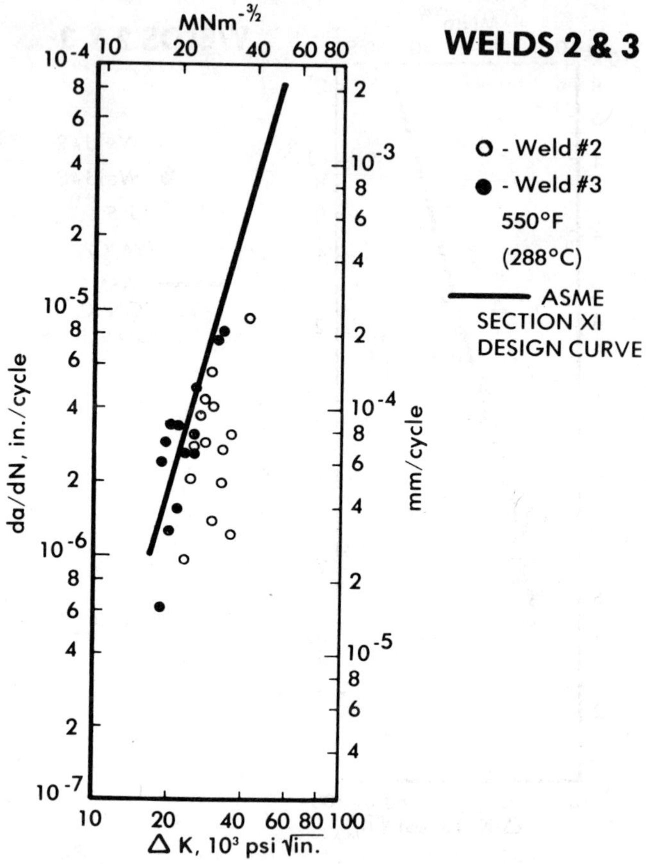

FIG. 21—*Fatigue crack growth properties.*

Semi-Annual Report by Westinghouse Research Laboratories on AFML Contract F33615-75-C-5064, 10 March, 1977.

[15] McHenry, H. I., "Fatigue Crack Propagation in Steel Alloys at Elevated Temperature," Report ERR-FW-1029, General Dynamics, Convair Aerospace Division, 15 Sept., 1970.

[16] Gerber, T. L., Heald, H. D., and Kiss, E., "Fatigue Crack Growth in SA 508 Cl 2 Steel in a High Temperature, High Purity Water Environment," ASME Paper 74-Mat-2, American Society of Mechanical Engineers.

[17] Wessel, E. T., *Engineering Fracture Mechanics*, Vol. 1, No. 1, June 1968, pp. 77.

[18] Paris, P. C., *Proceedings,* Tenth Sagamore Army Materials Research Conference, Syracuse University Press, 1964.

[19] Clark, W. G., Jr., *Journal of Materials*, Vol. 6, No. 1, March 1971, pp. 134–139.

[20] Dawes, M. G., *Metal Construction and British Welding Journal*, Feb. 1971, pp. 61–65.

Summary

This symposium provided a forum for the discussion of the many factors that play an important role in the fatigue design and evaluation of weldments. The papers by Reemsnyder and Munse, which were invited by the sponsoring committee, provided a very good assessment of the status of fatigue testing to evaluate the fatigue strength of weldments. Reemsnyder emphasized that the fatigue strength of weld joints traditionally has been evaluated by fatigue testing and assessment of the data using numerical techniques (empiricism). This theme is carried throughout the papers that were presented at the symposium. However, it became clear that attempts are being made to provide higher reliability and reduced testing costs (and time) by more thorough analysis to complement the testing.

The analysis concepts that are being developed concentrate on strain-cycling fatigue, application of fracture mechanics, and cumulative damage analysis. Most of the papers discuss various aspects of these three areas—not unlike many other areas of fatigue technology today. Initiation, propagation, and final fracture, as related to the assessment of the fatigue strength of weldments, frequently came up during the symposium. Reemsnyder and Munse highlighted this aspect and several papers dealt with a specific point of the fatigue process in weldments. The paper by Lawrence et al specifically dealt with the question of crack initiation in welds. As is the case with so many papers of this type, however, no clear-cut physical concept of fatigue-crack initiation or life prediction methodology emerges. Lawrence et al presented a great deal of data, and many concepts related to fatigue were discussed in relation to weld joints. It became clear, once again, from the Reemsnyder and Munse papers and the paper by Lawrence et al, that more effort is needed to formulate a concept of fatigue crack "initiation" that can be coupled to nondestructive testing and evaluation capability. Furthermore, a rational coupling of initiation concepts to linear elastic fracture mechanics and plastic strain-cycling analysis in relation to fatigue crack growth in the instability regime accompanied by large plastic strains must be developed.

Several other papers dealt with the traditional concepts of fatigue testing. One of the points of emphasis of several of these papers was the clear need for the development of fatigue test data in laboratory testing that could be transferred to full-scale field structure. No clear-cut transfer function (to transfer data from laboratory to service) or scaling function (from small test elements to full-scale components) emerged from the symposium. As indicated by Reemsnyder and Munse, the traditional empirical approach is still highly dominant.

However, strain-cycling fatigue concepts are now being utilized to predict fatigue life of weldments. Several of the papers, other than those already mentioned, discussed the use of strain-cycling fatigue concepts as related to fatigue life prediction for weldments. It is clear that the distribution of weld defects, type of weld, mean strain, weld type, and residual stress have an effect on the accuracy of the prediction. Throughout the symposium there was very little discussion of the characterization of weld defects (their number, size, distribution, and orientation) that could influence fatigue life. Several authors provided a discussion of the application of formalized fracture mechanics technology to fatigue crack growth in weldments. The papers by Webber and Maddox, Sandifer and Bowie, Seeley and Katz, Albrecht, and Kapadia provided valuable insight into the applicability of fracture mechanics concepts to weld-life prediction. However, the utilization of fracture mechanics in welds is complicated by the variation in either microstructure or composition (or both) throughout the weld and heat-affected zone (HAZ) areas, the presence of residual stresses in the weld region, and the complex variations in geometry in weld-joint regions. In addition, many of the fatigue crack growth concepts rely on the use of linearized plots of fatigue-crack growth data (da/dN versus ΔK). The utilization of linearized plots and fitting functions, without numerical analysis techniques, in both the strain-cycling field and fatigue-crack growth field, provides an area for future improvement of fatigue life prediction of weldments and data correlation of weldments. The symposium provided a forum for focusing on the foregoing area of fatigue behavior.

Several areas emerged as those areas that should receive emphasis in order that fatigue behavior of welds can be evaluated and predicted with greater reliability. The following items need immediate attention *in relation to the fatigue behavior of weldments:*

1. Characterization of weld defects (size, shape, distribution, and orientation)
2. Determination of residual stress magnitudes
3. Provision of methods for transferring fatigue data from simple test elements to full-scale welded structure
4. Evaluation of load sequencing effects
5. Determination of the role of weld and HAZ microstructure on fatigue behavior

Finally, in order that fatigue designers can put more confidence in their reliability estimates related to fatigue performance of welds, it is clear that numerical analysis and statistics must be introduced into the evaluation of fatigue behavior of weldments.

All of the authors did a commendable job in contributing to the symposium and stimulating activity and interest in this area. There is no doubt that continuing emphasis on this subject is needed in order to improve the safety and durability of welded structures.

D. W. Hoeppner,
editor

Index